Aman Thakur, Vineet Mehta, Priyanka Nagu and Kiran Goutam

Computer-Aided Drug Design

Also of interest

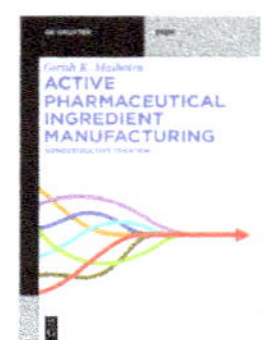

Active Pharmaceutical Ingredient Manufacturing.
Nondestructive Creation
Girish K. Malhotra, 2022
ISBN 978-3-11-070282-8, e-ISBN (PDF) 978-3-11-070284-2,
e-ISBN (EPUB) 978-3-11-070289-7

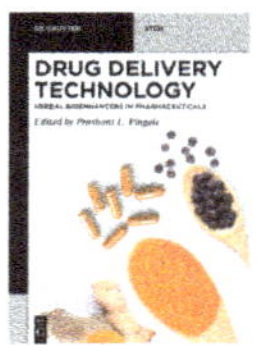

Drug Delivery Technology.
Herbal Bioenhancers in Pharmaceuticals
Prashant L. Pingale (Ed.), 2022
ISBN 978-3-11-074679-2, e-ISBN (PDF) 978-3-11-074680-8,
e-ISBN (EPUB) 978-3-11-074689-1

Medicinal and Biological Inorganic Chemistry
Ajay Kumar Goswami, Irena Kostova, 2022
ISBN 978-1-5015-2455-4, e-ISBN (PDF) 978-1-5015-1611-5,
e-ISBN (EPUB) 978-1-5015-1615-3

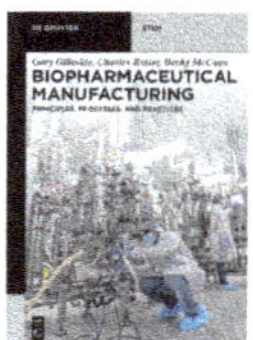

Biopharmaceutical Manufacturing.
Principles, Processes, and Practices
Gary Gilleskie, Charles Rutter, Becky McCuen, 2021
ISBN 978-3-11-061687-3, e-ISBN (PDF) 978-3-11-061688-0,
e-ISBN (EPUB) 978-3-11-061701-6

Aman Thakur, Vineet Mehta, Priyanka Nagu
and Kiran Goutam

Computer-Aided Drug Design

QSAR, Molecular Docking, Virtual Screening,
Homology and Pharmacophore Modeling

DE GRUYTER

Authors
Aman Thakur
Adarsh Vijendra Institute of Pharmaceutical Sciences
Shobhit University
Gangoh, Saharanpur 247341
Uttar Pradesh
India
amanthakur5052@gmail.com

Dr. Vineet Mehta
Department of Pharmacology
Government College of Pharmacy
Rohru 171207
Shimla, Himachal Pradesh
India
vineet.mehta20@gmail.com

Dr. Priyanka Nagu
Department of Pharmaceutics
Government College of Pharmacy
Rohru 171207
Shimla, Himachal Pradesh
India
priyanagupharma@gmail.com

Kiran Goutam
Shri Haridatt College of Pharmacy
Sewar, Bharatpur 321001, Rajasthan
India
gautam.kiran0701@gmail.com

ISBN 978-3-11-143474-2
e-ISBN (PDF) 978-3-11-143485-8
e-ISBN (EPUB) 978-3-11-143487-2

Library of Congress Control Number: 2024935223

Bibliographic information published by the Deutsche Nationalbibliothek
The Deutsche Nationalbibliothek lists this publication in the Deutsche Nationalbibliografie; detailed bibliographic data are available on the Internet at http://dnb.dnb.de.

Cover image: koto_feja/E+/Getty Images
Typesetting: Integra Software Services Pvt. Ltd.

www.degruyter.com

Acknowledgments

We wish to express our profound gratitude to the institutions and individuals whose support and guidance have been pivotal in the completion of this book, **"Computer-Aided Drug Design: QSAR, Molecular Docking, Virtual Screening, Homology, and Pharmacophore Modeling."**

We extend our deepest appreciation to the **Government College of Pharmacy, Rohru, District Shimla, Himachal Pradesh**. The institution's unwavering commitment to fostering a rigorous academic environment has been instrumental. We are particularly grateful for the access to cutting-edge facilities and the intellectual support provided by the esteemed **Director/ Principal Prof. (Dr.) Vivek Kumar Sharma and other faculty members** of the institute, which have significantly contributed to the depth and quality of this work.

Our sincere gratitude to the **Directorate of Technical Education, Vocational and Industrial Training, Sundernagar, District Mandi, Himachal Pradesh**. The Directorate's dedication to promoting excellence in technical education and vocational training has been a cornerstone of the academic and professional development of the students and faculty members of the Government College of Pharmacy, Rohru. Their support has been crucial in enabling the research and compilation of this comprehensive text.

We are profoundly grateful to **Himachal Pradesh Technical University, Hamirpur, Himachal Pradesh**. The university's emphasis on research innovation and academic excellence has provided a fertile ground for our scholarly pursuits. The collaborative environment and the university's commitment to fostering interdisciplinary research have greatly enriched our work.

We would also like to acknowledge the invaluable contributions of our colleagues, mentors, and students. Their critical insights, constructive feedback, and ongoing support have been indispensable throughout the research and writing process. Their engagement and intellectual curiosity have greatly enhanced the scientific rigor of this book.

Finally, we extend our heartfelt thanks to the family and friends for their unwavering support and understanding. Their encouragement has been a constant source of strength, allowing us to dedicate ourselves fully to this endeavor.

Thank you all for your indispensable contributions and support.

Aman Thakur
Dr. Vineet Mehta
Dr. Priyanka Nagu
Kiran Goutam

https://doi.org/10.1515/9783111434858-202

Contents

Chapter 1
Introduction to computer-aided drug design (CADD)

The power of computing has brought a new era in drug discovery, where the impossible is becoming routine through Computer-Aided Drug Design. – Eric Schmidt

1.1 Introduction

The discovery and the development of new drugs are critical endeavors in modern medicine, aiming to address a wide range of health conditions and improve the quality of life for millions of people worldwide [1–3]. Traditionally, drug discovery has been a laborious and expensive process, often taking years or even decades to identify and optimize potential therapeutic compounds [4]. However, in recent years, the field of computer-aided drug design (CADD) has emerged as a powerful and transformative approach, revolutionizing the drug discovery process. CADD, also known as computer-assisted drug design or in silico drug design, involves the use of computational tools and techniques to expedite and enhance various stages of drug development [5, 6]. By leveraging the computational power of modern computers and algorithms, researchers can analyze molecular interactions, predict biological activity, and design novel drug candidates more efficiently than ever before.

The roots of CADD can be traced back to the late 1960s and early 1970s when researchers first started using computers to model molecular structures and analyze chemical interactions [7]. Early efforts focused on simple computational methods to predict molecular properties, but with the rapid advancement of computing technology, CADD techniques became increasingly sophisticated. While CADD has made significant strides in accelerating drug discovery, challenges remain [8, 9]. Accurate modeling of complex molecular interactions, accounting for solvent effects, and predicting drug metabolism are some of the ongoing research areas. In the future, advances in artificial intelligence (AI), machine learning, and quantum computing hold great promise for further enhancing CADD capabilities. These technologies may enable more accurate predictions of binding affinities, faster virtual screening, and the exploration of larger chemical spaces [10].

CADD has revolutionized the way new drugs are discovered and optimized. By harnessing the power of computational tools and algorithms, researchers can efficiently navigate the vast chemical space, identify promising drug candidates, and bring innovative therapies to patients faster and at reduced costs. As technology continues to evolve, the impact of CADD on drug discovery is bound to grow, leading to improved treatments for a wide range of diseases and medical conditions [11].

https://doi.org/10.1515/9783111434858-001

To understand the CADD process from the basics, it is first indispensable to have a glimpse of the complete drug discovery process.

1.2 Drug discovery process

As per the considerations of the United States Food and Drug Administration (USFDA), the drug development process is broadly divided into five phases [12, 13] (Fig. 1.1).

1.2.1 Discovery and development phase

The very first step in this phase is "target identification," which includes finding the target disease against which a new drug is to be discovered. This is followed by the search for potential active compounds against the targeted disease. These active compounds, also known as hits, are obtained either by high-throughput screening (HTS) or by virtual screening techniques (applied through any CADD methodology). Starting from thousands of hits, very few are selected for further study, which have most of the desired properties considered essential for therapeutic potential. The selected hits are known as leads. In the development phase, the leads obtained from hits are subjected to further studies for evaluating their absorption, distribution, metabolism, excretion, and toxicity (ADMET) parameters. Possible drug interactions are also studied in this phase. These leads are modified further as per the requirement of potent drugs against the targeted disease. In connection with the CADD approach, this is the most crucial phase as most of the CADD approaches are applied in this phase of the drug discovery [14].

1.2.2 Preclinical studies

Preclinical studies are conducted for the evaluation of safety, toxicity, potency, efficacy, and dosing of the potential leads before the introduction of the drug to humans. Preclinical studies include various in vitro and in vivo experiments, which are known to play a pivotal role in deciding whether the drug under investigation could be therapeutically effective and safe for being administered to humans during clinical trials [15]. In vitro and in vivo studies performed in the preclinical studies can be made hybrid by accompanying the same with the CADD methodologies such as in silico ADMET predictions.

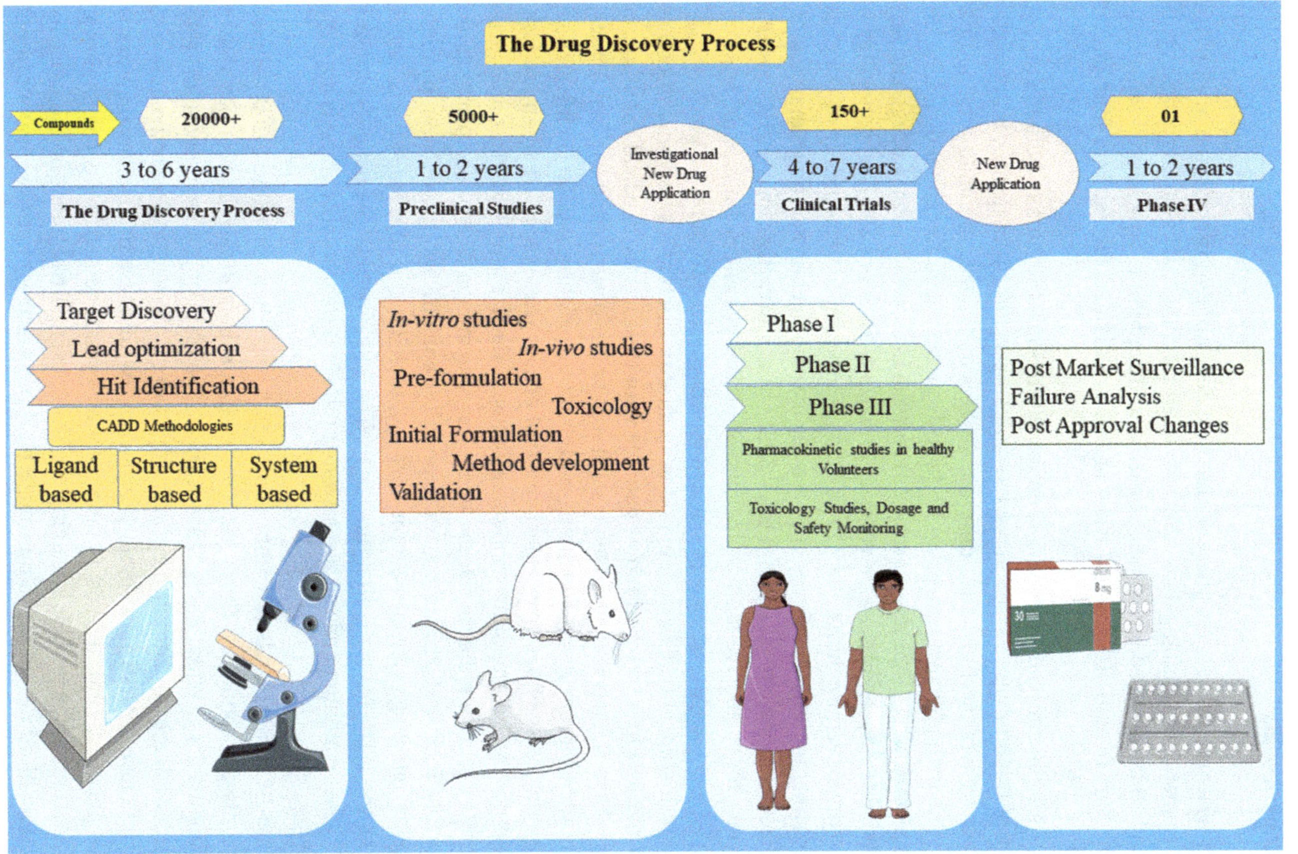

Fig. 1.1: Overview of the complete drug discovery process – from lead identification to post-market surveillance of the drug.

1.2.3 Clinical trials

Clinical trials are the longest phase of the drug development process and are carried out on human beings after taking approval from the FDA and following highly regulated and well-organized guidelines. The drug in this phase is now termed an Investigational New Drug (IND). Clinical trials of IND are divided further into four phases [16–19].

Phase I clinical trials are carried out on 20–100 healthy human volunteers. The purpose of this study is to assess the safety and efficacy of IND in healthy human subjects. Researchers carefully study different dosages of the drug to identify the optimal dose that is both effective and safe. The focus is primarily on understanding how the drug is metabolized, its pharmacokinetics (how the body processes the drug), and any potential side effects or adverse reactions. Phase 1 trials are relatively short-term, often lasting several months, but can vary, depending on the complexity of the drug and the study objectives. Success in phase I is based on achieving the primary objective of determining the safety and tolerability of the investigational drug. Positive findings at this stage pave the way for further testing in subsequent phases. Approximately, 70% of the IND passes to phase II of the clinical trial from phase I.

Phase II clinical trials are designed to further evaluate the safety and effectiveness of the IND in a larger group of patients with the targeted medical condition or disease. Phase II trials usually involve several hundred patients and can last for several months to a few years, depending on the nature of the condition being studied and the complexity of the drug's mechanism of action. These trials are often randomized and controlled, meaning that some patients receive the investigational drug while others receive either a placebo or the current standard care. The prime objective of phase II clinical trials is to ensure the safety and efficacy of the IND in patients, along with collecting relevant data for dose optimization and rare side effects associated with the IND. This is the first time that IND is introduced to the patients with the target disease condition and the success rate of the IND clearing this phase is approximately 33%.

Phase III clinical trials last from 1 to 4 years and involves 300 to 3,000 patients with the disease condition. These trials occur after a potential treatment has shown promising results in phase I and phase II trials. The main objective of phase III trials is to gather additional data on the treatment's safety, efficacy, and effectiveness in a larger and more diverse population. The sample size of phase III is larger than in the previous phases of clinical trials. In many phase III trials, neither the participants nor the researchers know who is receiving the experimental treatment or the placebo. This double-blind design helps minimize bias and ensures that the results are not influenced by expectations. The effectiveness of the new treatment is usually compared against the control group to assess its superiority, non-inferiority, or equivalence. Phase III trials are usually conducted at multiple research sites to increase the diversity of participants and improve the generalizability of the results. This phase of clinical trials has a success rate of approximately 25–30%. These trials can take several years to complete and are often the final step before a treatment can be approved and made available to the public for widespread use.

Phase IV clinical trial is the last phase of clinical trials and is carried out on a large pool of thousands of patients. The main objective of this is to evaluate the long-term efficacy and adverse effects of the drug. After the IND successfully completes the previous three phases of the clinical trials, the researcher files a New Drug Application (NDA) with the FDA. NDA provides complete data on the drug such as its safety, dosing, and adverse reactions. After receiving the NDA, the FDA reviews the complete application, and if it finds it satisfactory, it approves the NDA. Phase IV clinical trials, also known as post-marketing surveillance trials or post-approval studies, are conducted after a medication or medical device has been approved and is available on the market. These trials are designed to gather additional information about the safety, efficacy, and long-term effects of the intervention in a larger and more diverse patient population than what was studied in the previous phases. Phase IV trials are often conducted in real-world settings, reflecting the everyday use of the treatment. This trial provides valuable insights into how the treatment performs in a broader patient population, including those with coexisting health conditions and those taking other medications. Data from phase IV trials can lead to updates in the drug's labeling, including new indications, warnings, precautions, or dosing guidelines, based on the additional information gathered.

It is noteworthy that a traditional drug discovery process takes around 15–20 years for the development of a new drug moiety against any specific disease. We start from millions of molecules as our potential hits against a specific disease and we end up with one or two molecules that emerge as successful drug candidates. To reduce this time of drug discovery, wastage of resources, and to enhance the accuracy of drug discovery, CADD has become a vital part of the drug discovery process. Techniques of CADD can be applied at any stage of the drug development process, especially from target identification to the point that drug enters a clinical trial process.

In an interesting study of 2003, search for novel transforming growth factor-β1 receptor kinase inhibitors was initiated. One group of scientists from Eli-lily employed traditional drug discovery methods such as HTS, structur–activity relationship, and in vitro assay, while another group of scientists from Biogen Idec employed the CADD approaches such as virtual screening of the molecules and molecular docking. To their immense surprise, the lead developed by the Biogen scientists through the CADD approach was similar to that developed by Eli-Lilly, saving significant amount of crucial time and resources.

We now have understood the importance of the CADD in any drug discovery process – it is highly essential for reducing the time and cost of new drug development without compromising the accuracy and precision of the whole process.

1.3 Computer-aided drug design (CADD)

Computer-aided drug design is a broad term and, in the simplest context, it can be defined as any computational or in silico approach that is employed at any stage of the complete drug discovery process [4, 5]. Nowadays, there are several techniques available under the CADD, and all these techniques are broadly classified into two categories: structure-based drug designing (SBDD) and ligand-based drug designing (LBDD) (Fig. 1.2). To know and study the various techniques of CADD, a new field of medicinal chemistry has now been gaining immense popularity, which is also known as pharmacoinformatics.

1.3.1 Structure-based drug designing (SBDD)

SBDD is the approach of drug designing in which the structure of our target receptor is known and all our designing of drugs revolves around the structure of the receptor [20]. Examples of SBDD include molecular docking, molecular dynamics, and pharmacophore mapping:

- Molecular docking: Molecular docking is a widely used technique that simulates the interaction between a ligand (small molecule) and a receptor (protein) to predict the binding mode and affinity. This helps identify potential lead compounds for further optimization.
- Molecular dynamics simulation: Molecular dynamics simulations provide insights into the dynamic behavior of biomolecules, allowing researchers to study protein–ligand interactions over time and understand the stability of the complexes.
- Pharmacophore modeling: Pharmacophore models represent the essential features necessary for a ligand to interact with a target. This aids in virtual screening and the identification of molecules that fit the target's binding site.

1.3.2 Ligand-based drug designing (LBDD)

LBDD is the approach of CADD in which we have a set of known compounds against a particular disease, reported in the literature, and the designing approach of new ligands is based on the structural properties of these known compounds. Examples of LBDD include 2D-QSAR, 3D-QSAR, and ligand-based pharmacophore mapping:

- Quantitative structure–activity relationship (QSAR): QSAR models correlate the chemical structure of molecules with their biological activity [21, 22]. These models can predict the activity of new compounds based on their structural features, guiding lead optimization efforts. Based on the structural properties of the compounds, these QSAR methodologies are further classified into 2D-QSAR, 3D-QSAR, and so on.

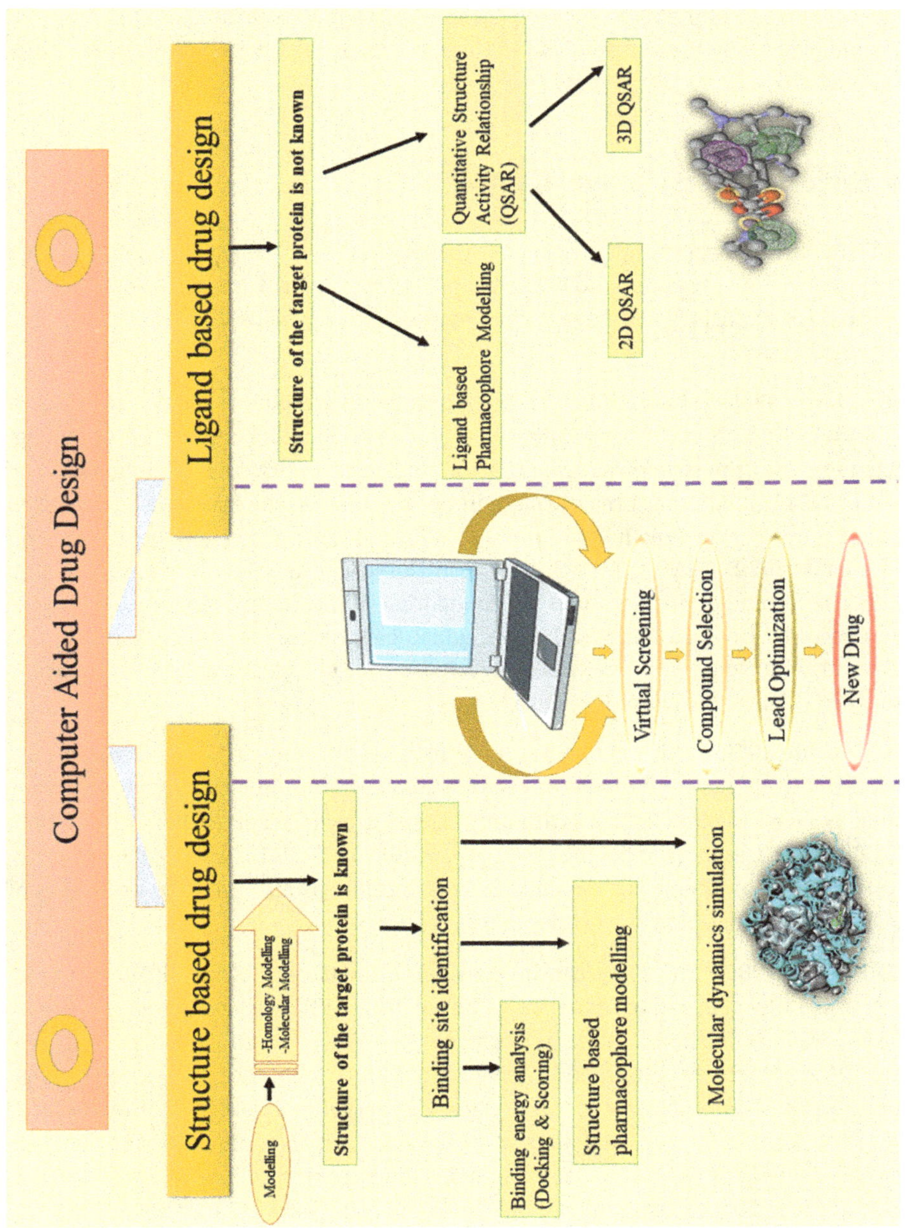

Fig. 1.2: Classification of computer-aided drug designing (CADD) into structure- and ligand-based drug designing.

To be well-versed in the knowledge of CADD and to apply its principles, it is required that one must study the subject from the basics, starting from the historical development of CADD.

1.4 Historical perspective of CADD

The CADD in the twenty-first century has come a long way and many sophisticated approaches have now come into existence for aiding the drug discovery process. But to trace back its history, we have to go back to the early 1960s when the CADD first come into existence.

CADD in the 1960s: This decade is dedicated to the development of modern QSAR. It was Corwin Hansch and Toshio Fujita who first developed the correlation of biological activity with the structural properties of molecules, which laid the foundation of QSAR in 1964 [23]. Another breakthrough in this decade was the Free-Wilson approach developed by Spencer Free Jr. and James W. Wilson in 1964, which laid the foundation of the modern 2-D QSAR approach that is still used and is relevant worldwide [24]. Another milestone of this decade is the foundation of the Cambridge Crystallographic Data Centre by Dr. Olga Kennard in 1965, which collects bibliographic, chemical, and crystal structure data for small molecules and encodes it electronically. Later, it became the Chemical Structure Database [25].

CADD in the 1970s: This decade began with the establishment of the Protein Data Bank (PDB) in 1971 at the Brookhaven National Laboratory under the dynamic leadership of Walter Hamilton. Started with only 7 structures, PDB currently has more than 2,00,000 structures of proteins that are required for many CADD approaches, like molecular docking and molecular dynamics. This decade has also seen the expansion of QSAR to environmental and toxicology studies [26].

CADD in the 1980s: The 1980s can be considered the decade of the SBDD evolution. The world's first nucleotide sequence database, the EMBL Nucleotide Sequence Data Library, was established in 1980 in Germany [27]. The first molecular docking program "UCSF Dock" was published in the year 1982 [28]. The concept of molecular mechanics came into existence and this decade has seen the emergence of many force fields such as AMBER, CHARMM, and MOPAC. In the last years of the decade, we have seen the development of other popular molecular docking software like AutoDock. This decade also witnessed immense development in the field of X-ray crystallography and NMR spectroscopy for determining protein structures. The first linear 3D-QSAR analysis, that is, comparative molecular field analysis (CoMFA) was introduced by Cramer et al. in 1988 [29]. All these advancements paved the way for the modern CADD that we are using today.

CADD in the 1990s: The first high-resolution protein structure of bacteriorhodopsin using Cryoelectron microscopy was introduced in 1990 by Henderson and his group [30]. In 1994, another variant of CoMFA, that is, comparative molecular similarity analysis indices (CoMSIA) was given by Klebe et al. [31]. This decade also marked the emergence of HIV protease inhibitors such as saquinavir, indinavir, ritonavir, etc., which were all developed by employing CADD approaches. HIV protease inhibitors, developed through CADD, are considered a major breakthrough in the treatment of HIV.

CADD in the 2000s: As the twenty-first century began, CADD has seen lots of development in all of its approaches. The distinguishing development in the first decade of the twenty-first century was the development of many databases, which proved pivotal in CADD. Major databases developed in this era are given in Tab. 1.1. These databases have laid a strong foundation for both ligand-based and structure-based virtual screening methods, which is another highlight of this decade.

Tab. 1.1: Major databases developed in the first decade of twenty-first century that played a crucial role in CADD.

S. no.	Database	Year	Purpose of the database
1.	ZINC	2004	Free database of commercially available compounds for virtual screening [32]
2.	PubChem	2004	Database of small molecules, with their biological activities [33]
3.	DrugBank	2006	Database containing the structure of drugs along with their pharmacological and pharmaceutical properties [34]
4.	ChemSpider	2007	Free chemical structure database providing fast text and structure search [35]
5.	ChEMBL	2008	A manually curated database of bioactive molecules with drug-like properties [36]
6.	SIDER	2010	Database of reported adverse drug reactions of marketed drugs [37]

CADD in the 2010s and the current trend: This decade marks its presence in the field of CADD by bagging the Nobel Prize in Chemistry in the year 2013. It was awarded to Martin Karplus, Michael Levitt, and Arieh Warshel for their work in computational chemistry [38]. Their work has brought a revolution in the field of molecular dynamics in SBDD. With the evolvement of high-performance computing systems, the focus has shifted toward AI-based drug discovery, which can be very efficient and can prove highly time- and cost-saving.

1.5 Applications of CADD in the drug discovery process

The CADD approach is applied in the drug discovery process to reduce the time and cost of the drug discovery process, besides enhancing accuracy. Numerous CADD approaches find their applications in the process of drug discovery and are discussed below:

1.5.1 Drawing molecular structures and energy minimization

This is the most basic and initial step of any CADD approach. Drawing of molecular structures and their energy minimization is a prerequisite for all the CADD techniques, whether it is molecular docking, 2D-QSAR, 3D-QSAR, molecular dynamics, Pharmacophore based virtual screening, or any other technique. The most common software available till now for serving the purpose is given in Tab. 1.2.

Tab. 1.2: Name of the popular molecular modeling software available for drawing and energy minimization of the chemical structures.

S. no.	Name of the molecular modeling software	Available at
1	ACD/ChemSketch	https://www.acdlabs.com/resources/free-chemistry-software-apps/chemsketch-freeware/
2	ChemDraw	https://revvitysignals.com/products/research/chemdraw
3	MarvinSketch	https://download.chemaxon.com/marvin
4	Avogadro	https://avogadro.cc/
5	ChemDoodle	https://www.chemdoodle.com/
6	Spartan Software	https://store.wavefun.com/Spartan_Software_s/12.htm
7	Biovia Draw	https://www.3ds.com/products/biovia/draw

1.5.2 Performing molecular docking for studying the ligand–protein interaction

This is the most widely used technique of CADD to date. Molecular docking has come a long way since starting the first docking program “DOCK” in 1982. Molecular docking is a technique that comes under SBDD approach, which works on the lock-and-key model. In molecular docking, the ligand to be investigated acts as a key and the active cavity of the receptor acts as a lock. When a ligand is placed inside the cavity of the receptor, the Gibb’s free energy of this ligand–receptor system is calculated (expressed as dock score). This dock score of the ligand is than compared with the dock score of the standard drug. Currently, there are numerous software available for docking, and are included in Tab. 1.3.

Tab. 1.3: Some common molecular docking programs available in CADD.

S. no.	Docking program	Availability	Algorithm
1	AutoDock	Free academic version	Based on Lamarkian genetic algorithm
2	Gold	Commercial	Based on genetic algorithm
3	GLIDE	Commercial	Systematic search and descriptor matching algorithm
4	UCSF DOCK	Free academic version	Based on geometric algorithm
5	FlexX	Commercial	Incremental construction algorithm
6	MDOCK	Free academic version	Based on sphere–ligand matching algorithm
7	Surflex-Dock	Commercial	Based on molecular similarity algorithm
8	HDOCK	Free online server	Based on ab initio and template-based modeling algorithm

1.5.3 2D-QSAR modeling for the development of leads

The journey of modern 2D-QSAR started in the early 1960s and has now become an indispensable part of the drug discovery process. In this technique, initially, a set of ligands whose biological activities are reported in the literature are selected. The biological activity of these ligands is then related to their 2D structural properties (Descriptors) in the form of a mathematical equation, also known as the 2D-QSAR model. This mathematical equation is then used for the calculation of the biological activity of the proposed new ligands. The list of the 2D-QSAR software available is given in Tab. 1.4.

Tab. 1.4: Available software for the development of the 2–D QSAR model.

S. no.	Program name	Availability	Available at
1	DTC-QSAR	Free open source	http://teqip.jdvu.ac.in/QSAR_Tools/
2	QSAR-Co	Free open source	https://sites.google.com/view/qsar-co
3	QSARINS	Free academic	https://dunant.dista.uninsubria.it/qsar/?page_id=37
4	QSARpro	Commercial	https://www.vlifesciences.com/products/QSARPro/Product_QSARpro.php
5	AutoQSAR	Commercial	https://www.schrodinger.com/products/autoqsar

1.5.4 3D-QSAR modeling

3D-QSAR is considered a natural extension of the traditional 2D-QSAR studies. In this approach, the 3D properties of the molecules are correlated with the biological activity for the generation of the model. CoMFA and CoMSIA are the two most exploited techniques based on the ligand-based 3D-QSAR.

1.5.5 Homology modeling

Homology modeling is the approach of obtaining a 3D structure of the target protein receptor, which is not available in the protein data bank of the Research Collaboratory for Structural Bioinformatics (RCSB). The 3D structure generated through homology modeling can be further employed, either in molecular docking or in molecular dynamics. This is based on the principle that similar proteins fold in the same manner; hence, there will be similarity in their 3D structure. Modeller, Swiss-model, and I-Tasser are among the most common software available for performing homology modeling currently.

1.5.6 Pharmacophore modeling and virtual screening

Credit for giving the concept of pharmacophore modeling is given to Paul Ehrlich; however, the term Pharmacophore was given by Lemont Kier in 1971. The International Union of Pure and Applied Chemistry (IUPAC) defines pharmacophore as "an ensemble of steric and electronic features that is necessary to ensure the optimal supramolecular interactions with a specific biological target and to trigger (or block) its biological response." Pharmacophore modeling is employed in both the SBDD and LBDD, in which a pharmacophore model is generated and then a database of compounds is screened against the model to identify the probable hits against the same. LigandScout, PharmaGist, Pharmer, Catalyst, and LiSiCA are the popular software for pharmacophore modeling and virtual screening that are currently available.

1.5.7 Molecular dynamics simulation

Molecular dynamics has gained immense popularity in the twenty-first century. All calculations in molecular dynamics are based on Newton's laws of motion. The main purpose of molecular dynamics is to understand the real-time behavior of a biological system, especially of ligand–protein, when put in biological environmental conditions. GROMACS and NAMD are among the freely available software for performing dynamics simulations, whereas Schrodinger's Desmond module is available commercially for the same purpose.

1.5.8 Prediction of absorption, distribution, metabolism, excretion, and toxicity (ADMET) parameters

Having good biological activity alone is not sufficient for any hit/lead for becoming a probable drug candidate. ADMET parameters are also of utmost importance to any drug for successfully exerting its desired pharmacological response. Therefore the in silico approaches for predicting the ADMET parameters of any designed molecule have become indispensable in the drug discovery process. SWISS-ADME, ADMET-predictor, admetSAR, cypreact, druLiTo, etc. are some of the reliable software currently available for the prediction of ADMET parameters.

The use of any approach is vague if it does not have any success story behind it. In the same context, CADD has also given us many therapeutically active drug molecules that are used clinically today. Table 1.5 provides the list of the successful drug candidates obtained for clinical applications by employing various CADD-based approaches.

Tab. 1.5: Some of the successful stories of CADD.

S. no.	Class of drugs	Major approaches used in discovery	Examples
1	Anti-HIV (protease inhibitors)	SBDD approach	Squinavir (1995, Roche), Indinavir (1996, Merck), Ritonavir (1996, Abbott), Lopinavir (2000, Abbott)
		LBDD approach	Nelfinavir (1997, Agouron Pharmaceuticals, Inc.)
2	Anti-influenza	SBDD approach	Zanamivir (1999, Glaxo Smith Kline), Oselatamivir (1999, Gilead Sciences)
3	Carbonic anhydrase inhibitors	Fragment-based and SBDD approach	Dorzolamide (1994, Merck & Co.)
4	Antihypertensive (ACE inhibitors)	Both SBDD and LBDD approaches	Captopril (1981, Squibb and Sons Pharmaceuticals)
5	Anti-HIV (integrase inhibitors)	SBDD approach (molecular docking and dynamics)	Raltegravir (2007, Merck & Co.)
6	Fibrinogen antagonist	LBDD approach	Tirofiban (1999, Merck & Co.)
7	Antihypertensive (renin inhibitors)	SBDD approach	Aliskiren (2007, Novartis and Speedel)
8	Anti-hepatitis C virus	SBDD approach	Boceprevir (2011, Schering-Plough)

Tab. 1.5 (continued)

S. no.	Class of drugs	Major approaches used in discovery	Examples
9	Anti-cancer drugs (kinase inhibitors)	SBDD approach	Imatinib (1990, Novartis), Gefitinib (2003, AstraZeneca), Erlotinib (2005, Genentech, Inc.), Acalabrutinib (2017, AstraZeneca), Lorlatinib (2018, Pfizer Inc.)
		LBDD approach	Sunitinib (2006, Pfizer Inc.), Larotrectinib (2018, Loxo Oncology, Inc.)
10	B-lactamase inhibitor	SBDD (molecular docking and dynamics) approach	Vaborbactam (2017, Rempex Pharmaceuticals)
11	Antihypertensive (AT1 receptor antagonist)	SBDD approach	Eprosartan (2001, SmithKline Beecham)
		LBDD approach	Losartan (1995, Merck & Co.), Valsartan (2002, Novartis)
12	Antibiotic (DNA gyrase inhibitor)	Both SBDD and LBDD approaches	Norfloxacin (1986, Merck & Co.)

References

[1] Bharatam PV. Computer-aided drug design. In: Poduri, R. (eds). Drug Discovery and Development. Springer, Singapore, 2021.

[2] Marshall GR. Computer-aided drug design. Annual Review of Pharmacology and Toxicology, 1987, 27(1), 193–213.

[3] Hassan BM, Ahmad K, Roy S, et al. Computer aided drug design: Success and limitations. Current Pharmaceutical Design, 2016, 22(5), 572–81.

[4] Bassani D, Moro S. Past, present, and future perspectives on computer-aided drug design methodologies. Molecules, 2023, 28(9), 3906.

[5] Taft CA, Da Silva VB. Current topics in computer-aided drug design. Journal of Pharmaceutical Sciences, 2008, 97(3), 1089–98.

[6] Baldi A. Computational approaches for drug design and discovery: An overview. Systematic Reviews in Pharmacy, 2010, 1(1), 99.

[7] Nascimento IJ, de Aquino TM, da Silva-Júnior EF. The new era of drug discovery: The power of computer-aided drug design (CADD). Letters in Drug Design & Discovery, 2022, 19(11), 951–5.

[8] Guner OF. History and evolution of the pharmacophore concept in computer-aided drug design. Current Topics in Medicinal Chemistry, 2002, 2(12), 1321–32.

[9] Atharifar H, Yildiz F, Knapp JR. Survey of the current academic and industrial trends in utilizing the CADD technology. In: 2013 ASEE Annual Conference & Exposition, 2013, 23–1121.

[10] Sharma T, Mohapatra S, Dash R, Rath B, Sahoo CR. Recent advances in CADD. In: Computer Aided Drug Design (CADD): From ligand-based methods to structure-based approaches, 2022, 231–81. Elsevier.

[11] Zhao L, Ciallella HL, Aleksunes LM, Zhu H. Advancing computer-aided drug discovery (CADD) by big data and data-driven machine learning modeling. Drug Discovery Today, 2020, 25(9), 1624–38.

[12] Van Norman GA. Drugs, devices, and the FDA: Part 1: An overview of approval processes for drugs. JACC: Basic to Translational Science, 2016, 1(3), 170–9.

[13] Deore AB, Dhumane JR, Wagh R, Sonawane R. The stages of drug discovery and development process. Asian Journal of Pharmaceutical Research and Development, 2019, 7(6), 62–7.

[14] Khanna I. Drug discovery in pharmaceutical industry: Productivity challenges and trends. Drug Discovery Today, 2012, 17(19–20), 1088–102.

[15] Aban IB, George B. Statistical considerations for preclinical studies. Experimental Neurology, 2015, 270, 82–7.

[16] Nguyen V, Sweet BV, Macek T. Defining the phases of clinical trials. American Journal of Health-System Pharmacy, 2006, 63(8), 710–1.

[17] Sedgwick P. What are the four phases of clinical research trials? BMJ, 2014, 348.

[18] Wright B. Clinical trial phases. In: A Comprehensive and Practical Guide to Clinical Trials, 2017, 11–15. Academic Press.

[19] Collins JM, Grieshaber CK, Chabner BA. Pharmacologically guided phase I clinical trials based upon preclinical drug development. JNCI: Journal of the National Cancer Institute, 1990, 82(16), 1321–6.

[20] Batool M, Ahmad B, Choi S. A structure-based drug discovery paradigm. International Journal of Molecular Sciences, 2019, 20(11), 2783.

[21] Bacilieri M, Moro S. Ligand-based drug design methodologies in drug discovery process: An overview. Current Drug Discovery Technologies, 2006, 3(3), 155–65.

[22] Sliwoski G, Kothiwale S, Meiler J, Lowe EW. Computational methods in drug discovery. Pharmacological Reviews, 2014, 66(1), 334–95.

[23] Hansch C, Fujita T. p-σ-π analysis. A method for the correlation of biological activity and chemical structure. Journal of the American Chemical Society, 1964, 86(8), 1616–26.

[24] Free SM, Wilson JW. A mathematical contribution to structure-activity studies. Journal of Medicinal Chemistry, 1964, 7(4), 395–9.

[25] Redman J, Willett P, Allen FH, Taylor R. A citation analysis of the Cambridge Crystallographic Data Centre. Journal of Applied Crystallography, 2001, 34(3), 375–80.

[26] Burley SK, Berman HM, Christie C, et al. RCSB protein data bank: Sustaining a living digital data resource that enables breakthroughs in scientific research and biomedical education. Protein Science, 2018, 27(1), 316–30.

[27] Rice CM, Fuchs R, Higgins DG, Stoehr PJ, Cameron GN. The EMBL data library. Nucleic Acids Research, 1993, 21(13), 2967.

[28] Salmaso V, Moro S. Bridging molecular docking to molecular dynamics in exploring ligand-protein recognition process: An overview. Frontiers in Pharmacology, 2018, 9, 923.

[29] Cramer RD, Patterson DE, Bunce JD. Comparative molecular field analysis (CoMFA). 1. Effect of shape on binding of steroids to carrier proteins. Journal of the American Chemical Society, 1988, 110(18), 5959–67.

[30] Henderson R, Baldwin JM, Ceska TA, Zemlin F, Beckmann E, Downing KH. Model for the structure of bacteriorhodopsin based on high-resolution electron cryo-microscopy. Journal of Molecular Biology, 1990, 213(4), 899–929.

[31] Klebe G, Abraham U, Mietzner T. Molecular similarity indices in a comparative analysis (CoMSIA) of drug molecules to correlate and predict their biological activity. Journal of Medicinal Chemistry, 1994, 37(24), 4130–46.

[32] Irwin JJ, Shoichet BK. ZINC– a free database of commercially available compounds for virtual screening. Journal of Chemical Information and Modeling, 2005, 45(1), 177–82.

[33] Kim S, Thiessen PA, Bolton EE, et al. PubChem substance and compound databases. Nucleic Acids Research, 2016, 44(D1), D1202–13.

[34] Wishart DS, Knox C, Guo AC, et al. DrugBank: A knowledgebase for drugs, drug actions and drug targets. Nucleic Acids Research, 2008, 36(suppl_1), D901–6.
[35] Ayers M. ChemSpider: The free chemical database. Reference Reviews, 2012, 26(7), 45–6.
[36] Gaulton A, Bellis LJ, Bento AP, et al. ChEMBL: A large-scale bioactivity database for drug discovery. Nucleic Acids Research, 2012, 40(D1), D1100–7.
[37] Kuhn M, Letunic I, Jensen LJ, Bork P. The SIDER database of drugs and side effects. Nucleic Acids Research, 2016, 44(D1), D1075–9.
[38] Karplus M, Levitt M, Warshel A. The Nobel Prize in Chemistry. Nobel Media AB, 2013.

Chapter 2
Quantitative structure–activity relationship

> Molecular descriptors are the alphabet of QSAR, and statistical models are the language that translates it into meaningful predictions. – Corwin Hansch

2.1 Quantitative structure–activity relationship (QSAR)

Quantitative structure–activity relationship (QSAR) is a powerful and widely used computational approach in the field of chemistry and drug design [1]. It plays a crucial role in predicting and understanding the biological activity of molecules based on their structural features [2]. The concept of QSAR has revolutionized the way researchers approach drug discovery and design, leading to the development of more efficient and targeted pharmaceutical compounds [3]. QSAR is based on the idea that the biological activity of a molecule can be correlated with its physicochemical properties and structural characteristics [4, 5]. By utilizing statistical and computational techniques, QSAR models establish quantitative relationships between the chemical structure of a molecule and its biological activity, typically against a specific target or endpoint, such as receptor binding, enzyme inhibition, or toxicity [6]. This information is invaluable for guiding the design of new molecules with improved potency and reduced adverse effects.

The roots of QSAR can be traced back to the late nineteenth and early twentieth centuries when scientists began to notice the connection between the chemical structure of compounds and their biological effects. However, it was not until the 1960s that the term “quantitative structure–activity relationship” was formally introduced by Corwin Hansch and his colleagues [7]. They were among the first to propose the application of mathematical equations to describe the relationship between chemical structure and biological activity. Their pioneering work laid the foundation for the development of QSAR as a discipline [8].

The success of QSAR heavily relies on the availability of high-quality and diverse datasets containing both chemical structures and corresponding biological activities [9]. In recent years, the accessibility of large databases and advancements in computational power has significantly bolstered the development of QSAR models [10]. Additionally, the emergence of new algorithms and machine learning techniques has further enhanced the predictive capabilities of QSAR models, making them more accurate and reliable [11]. One of the significant advantages of QSAR is its ability to prioritize molecules for experimental testing, thereby reducing the cost and time associated with drug discovery [12, 13]. QSAR models can effectively screen vast chemical libraries, highlighting compounds with the highest probability of biological activity, thus streamlining the drug development process. Furthermore, QSAR models can shed light on the underlying

https://doi.org/10.1515/9783111434858-002

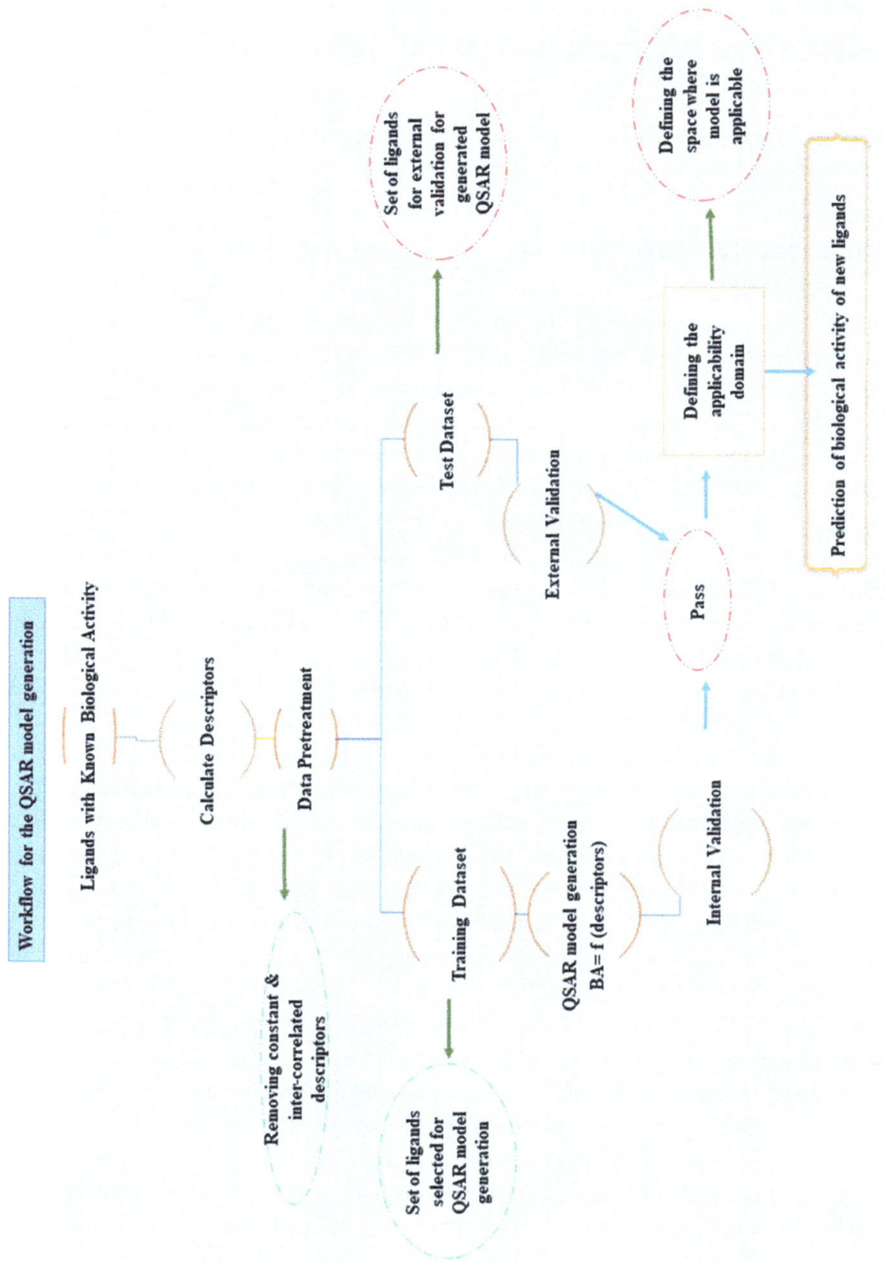

Fig. 2.1: A general workflow for the development of any QSAR model in computer-aided drug designing.

mechanisms responsible for a compound's activity, providing valuable insights for medicinal chemists to optimize molecular structures [14].

QSAR can be defined as a mathematical relationship that correlates the biological activity of a series of compounds with their physicochemical properties, steric properties, and sometimes certain structural features (combined all these are known as molecular descriptors) [15]. QSAR approach is seen as an alternative when the 3D structure of the target protein is not available for structure-based drug designing such as molecular docking [16]. The simplest representation of a QSAR model is given in the following equation:

$$\text{Biological activity} = \text{fn}(\text{descriptors}) \tag{2.1}$$

The QSAR equation generated can further be employed for predicting the biological activity of the newly proposed ligands [17]. The biological activity of the ligands can be predicted by calculating the values of the descriptors of the proposed ligands, followed by putting these values in the generated QSAR equation. The workflow of the steps involved in a QSAR study is given in Fig. 2.1.

2.2 Classification of QSAR methodologies

There are certain ways by which a QSAR methodology can be classified [18]. The most common ways of classifying the QSAR methodologies are listed in Fig. 2.2:

2.2.1 Based on the dimensions of descriptors used in the QSAR model

This is the widely accepted classification of QSAR models to date. Based on this classification, a QSAR model is further divided into:

2.2.1.1 0D-QSAR

0D-QSAR is a computational approach used in chemoinformatics and drug discovery to predict the biological activity of a molecule based on its chemical structure. The "0D" refers to the fact that it does not consider the spatial arrangement of atoms in the molecule (as opposed to higher-dimensional QSAR models like 2D- and 3D-QSAR). In a 0D-QSAR model, molecular descriptors are used to represent the chemical structure of a compound. Molecular descriptors are numerical values that capture various physicochemical properties of the molecule, such as molecular weight, hydrophobicity, hydrogen bond donor and acceptor counts, and various other structural features. It is important to note that while 0D-QSAR models can provide valuable insights into the relationship between chemical structure and biological activity, they have limitations. These models do not consider the three-dimensional spatial arrangement of atoms in the molecule, which can be critical for some biological activities. To address this limita-

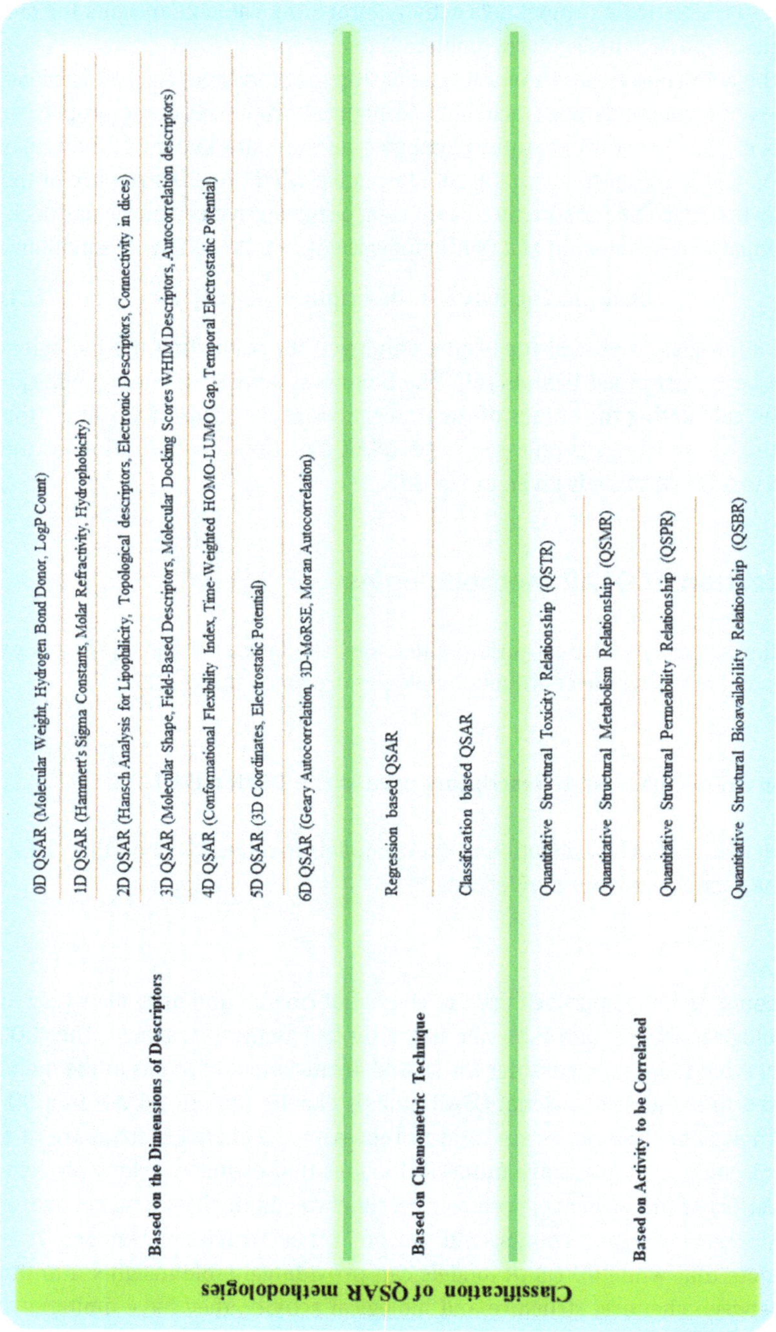

Classification of QSAR methodologies

Based on the Dimensions of Descriptors

0D QSAR (Molecular Weight, Hydrogen Bond Donor, LogP Count)

1D QSAR (Hammett's Sigma Constants, Molar Refractivity, Hydrophobicity)

2D QSAR (Hansch Analysis for Lipophilicity, Topological descriptors, Electronic Descriptors, Connectivity in dices)

3D QSAR (Molecular Shape, Field-Based Descriptors, Molecular Docking Scores WHIM Descriptors, Autocorrelation descriptors)

4D QSAR (Conformational Flexibility Index, Time-Weighted HOMO-LUMO Gap, Temporal Electrostatic Potential)

5D QSAR (3D Coordinates, Electrostatic Potential)

6D QSAR (Geary Autocorrelation, 3D-MoRSE, Moran Autocorrelation)

Based on Chemometric Technique

Regression based QSAR

Classification based QSAR

Based on Activity to be Correlated

Quantitative Structural Toxicity Relationship (QSTR)

Quantitative Structural Metabolism Relationship (QSMR)

Quantitative Structural Permeability Relationship (QSPR)

Quantitative Structural Bioavailability Relationship (QSBR)

Fig. 2.2: Different approaches for the classification of the QSAR modeling.

tion, higher-dimensional QSAR models, such as 2D- and 3D-QSAR, take into account additional structural information, leading to potentially more accurate predictions.

2.2.1.2 1D-QSAR

1D-QSAR is a computational method used in cheminformatics and drug discovery to predict the biological activity of a molecule based on its chemical structure. The term "1D" refers to the fact that it uses one-dimensional molecular descriptors to represent the molecular structure. Molecular descriptors are numerical representations of various physicochemical properties or structural characteristics of a molecule. These descriptors include simple properties such as number of rings, number of certain functional groups, pKa value, etc. This model is developed by considering the substructural properties of the ligands for correlation with the biological activity.

It is important to note that 1D-QSAR has limitations, as it only considers one-dimensional descriptors and does not take into account spatial arrangements or three-dimensional interactions of atoms in the molecule. As a result, more advanced QSAR methods, such as 2D-QSAR and 3D-QSAR, have been developed to incorporate additional structural information and improve predictive accuracy. Nonetheless, 1D-QSAR remains a valuable tool for preliminary assessments of biological activity and can be computationally less demanding than more complex QSAR methods.

2.2.1.3 2D-QSAR

2D-QSAR models are based on those descriptors that can be calculated by the graphical representation of the structures of the compounds. Topological descriptors, connectivity indices, and eigenvalue-based descriptors are among the most commonly employed descriptors for this model generation. It is important to note that QSAR models are highly dependent on the quality and diversity of the dataset and the choice of molecular descriptors and modeling techniques. Additionally, while QSAR models can be valuable tools in drug discovery and chemical design, they are not a replacement for experimental validation and should be used as supportive tools in the decision-making process [19].

2.2.1.3.1 General steps involved in developing either of 0D-, 1D-, or 2D-QSAR models

- **Data collection:** Obtain a dataset containing chemical structures and their corresponding biological activities (e.g., biological assay data or experimental results).
- **Molecular descriptors' calculation:** Calculate 2D molecular descriptors for each compound in the dataset. These descriptors represent various properties of the molecules, such as size, shape, electronic, and topological characteristics.

- **Data splitting:** Divide the dataset into two parts: a training set and a test set. The training set is used to build the QSAR model, while the test set is used to evaluate its predictive performance.
- **Model development:** Select a suitable statistical or machine learning method to build the QSAR model. Common approaches include multiple linear regression (MLR), partial least squares (PLS), support vector machines (SVMs), and random forest (RF).
- **Model validation:** Assess the predictive performance of the model using the test set. Metrics such as root mean squared error (RMSE), mean absolute error (MAE), and correlation coefficient (r) are commonly used to evaluate the model's accuracy.
- **Model interpretation:** Analyze the model to understand which molecular descriptors are most influential in predicting biological activity. This information can provide insights into the structure-activity relationship and guide further chemical design.
- **Model application:** Once validated, the QSAR model can be used to predict the activity of new, unseen molecules with similar structural features to those in the training set. This enables the identification of potential lead compounds with desired biological activity.

2.2.1.4 3D-QSAR

3D-QSAR is a computational approach used in drug discovery, environmental chemistry, and other fields of chemistry and biology to analyze the relationship between the three-dimensional (3D) structure of molecules and their biological activities. The main goal of 3D-QSAR is to predict the biological activity or potency of a compound based on its structural features and interactions with a target receptor or enzyme. These models are generated by calculating those descriptors that exist by the 3D representations of the compounds. Comparative molecular field analysis (CoMFA) and Comparative molecular similarity index analysis (CoMSIA) are the two most widely employed methods of 3D-QSAR study. WHIM descriptors, autocorrelation 3D descriptors, charged PSA, and so on are the common descriptors of 3D-QSAR [20–23].

The 3D-QSAR method builds upon the principles of traditional QSAR modeling, which involves establishing correlations between the chemical structure and the biological activity of a set of compounds. However, in 3D-QSAR, the focus shifts to considering the spatial arrangement and conformational flexibility of molecules, allowing for a more detailed understanding of the ligand-receptor interactions.

2.2.1.4.1 Overview of the key steps involved in 3D-QSAR

- **Molecular alignment:** The first step in 3D-QSAR involves aligning the 3D structures of a set of molecules. This alignment aims to superpose the molecules in a way that maximizes the overlap of their active site or pharmacophoric features, which are critical for binding to the target receptor.

- **Molecular representation:** Once the molecules are aligned, they are typically represented as 3D grids or point clouds, where various molecular properties, such as electrostatic potential, lipophilicity, and steric hindrance, are calculated for each grid point or data point.
- **Grid-based analysis:** The next step involves statistical analysis of the data points to identify correlations between the molecular properties and the biological activities of the compounds. Techniques like PLS, principal component analysis (PCA), or CoMFA are commonly used for this purpose.
- **Model validation:** It is crucial to validate the generated 3D-QSAR model to ensure its predictive power. This validation can be achieved using methods like cross-validation, external test set validation, and leave-one-out validation.
- **Interpretation and visualization:** After obtaining a reliable 3D-QSAR model, researchers can use it to interpret the structure-activity relationship of the compounds. Visualization tools can help in understanding how specific structural features contribute to biological activity or binding affinity.

2.2.1.5 4D-QSAR

4D-QSAR goes a step further and incorporates the temporal dimension, meaning it considers changes in molecular properties over time, especially in the context of molecular dynamics simulations. This approach can be particularly valuable in situations where the biological activity of a compound depends on its dynamic behavior and interactions with a target over time. A fourth dimension, ensemble averaging, is incorporated in the 3D-QSAR model to convert it into a 4D-QSAR analysis. 4D-QSAR study provides conformational and alignment freedom to a 3D-QSAR model. 4D-QSAR-based descriptors are grid cell occupancy measures of the atoms.

4D-QSAR is still an active area of research, and its application may vary depending on the specific problem and available data. It requires sophisticated computational methods and access to molecular dynamics simulations, making it more resource-intensive compared to traditional 2D- or even 3D-QSAR methods. Nonetheless, the insights gained from 4D-QSAR models can provide a deeper understanding of the molecular mechanisms behind biological activity, which can be valuable in drug design and optimization processes.

2.2.1.6 5D-QSAR

5D-QSAR is the further extension of the 4D-QSAR model by incorporating a fifth dimension, that is, an additional degree of freedom. Therefore 5D-QSAR represents different induced-fit models in 4D-QSAR.

2.2.1.7 6D-QSAR

When solvation parameters are incorporated in the 5D-QSAR descriptors it is converted further into a 6D-QSAR model [24].

2.2.2 Based on the chemometric techniques employed for QSAR model generation

This classification is based on the method used for finding the correlation between biological activity and the values of descriptors [25, 26]. They are further classified as:

2.2.2.1 Regression-based QSAR

As the name suggests, it is based on the correlation that is established based on regression. Examples include MLR, PLS, and genetic function algorithm. Out of all these, MLR models are the most widely used by scientists.

2.2.2.1.1 General overview of the working of regression-based QSAR

- **Descriptor calculation:** The first step in building a QSAR model is to calculate molecular descriptors for a dataset of molecules. Descriptors are numerical representations of various physicochemical, topological, or electronic properties of the molecules.
- **Data preparation:** The dataset should be divided into two subsets: a training set and a test set. The training set is used to build the QSAR model, while the test set is used to evaluate the model's predictive performance.
- **Model building:** Regression techniques such as MLR, PLS regression, support vector regression, or neural networks are commonly used to relate the molecular descriptors to the corresponding biological activity or property.
- **Model validation:** The performance of the QSAR model is assessed using various statistical metrics such as the coefficient of determination (R^2), RMSE, and MAE. Cross-validation techniques may also be used to assess the model's robustness and avoid overfitting.
- **Model interpretation:** After successfully building and validating the QSAR model, the model's coefficients can be analyzed to identify the key descriptors that contribute most to the biological activity or property of interest. This interpretation can provide insights into the structure–activity relationship.
- **Prediction:** Once the QSAR model is established and validated, it can be used to predict the activity or property of new molecules based on their descriptors. This prediction can aid in drug discovery and development by identifying potentially active compounds.

2.2.2.2 Classification-based QSAR

This QSAR methodology is based on the principle that the correlation between the biological activity and descriptor is established in such a way that compounds employed in the study are classified either as active or not active. Techniques such as linear discriminant analysis (LDA), logistic regression, and cluster analysis are used for the classification-based QSAR model generation [27].

2.2.2.2.1 General overview of the classification-based QSAR model generation

- **Creating the dataset:** This is the initial step of the generation of the classification-based QSAR model. In this step, a group of molecules with known biological activities is selected, and based on the biological activities, the selected molecules are categorized or classified as active or inactive against that particular disease or target.
- **Calculating the molecular descriptors:** Further, the selected molecules are subjected to the calculation of their molecular descriptors. Descriptors can either be 0D, 1D, 2D, or 3D depending upon the requirement.
- **Data preparation:** The molecules selected for the model development are divided into training and test dataset molecules. It is an unsaid rule that generally, 30% of the total molecules are put under the test dataset and the rest comes under the training dataset.
- **Development of the classification-based QSAR model:** Genetic algorithm-LDA and RF are the two most common approaches for the development of the QSAR model in this type of classification.
- **Model validation:** There are certain parameters based on which a classification-based QSAR model can be validated either internally or externally. Accuracy, sensitivity, specificity, and precision are certain parameters based on which a new classification-based QSAR model is validated.
- **Prediction by model:** Once our QSAR model is developed and validated, in the last step it is employed for the prediction of the biological activity of the unknown compounds. The most important thing in this type of QSAR model is that here prediction of the unknown compounds is only done as active or inactive against a particular disease or target.

There are certain machine learning algorithms of QSAR model development such as artificial neural network, SVM, RF, and *k*-nearest neighbors which are gaining immense popularity for the generation of both regression and classification-based QSAR models.

2.2.3 Based on the activity to be correlated in the QSAR model

This is another way of classifying the QSAR models, which is dependent on the type of biological activity and how it is correlated with the descriptors of the series of compounds. These models are further classified as:

2.2.3.1 Quantitative structure–toxicity relationship (QSTR)

These models are developed to correlate the toxicity values of the compounds with their descriptor values [28].

2.2.3.2 Quantitative structure–metabolism relationship (QSMR)

These models are developed to correlate the parameters associated with the metabolism of drugs and their structural features [29].

2.2.3.3 Quantitative structure–permeability relationship (QSPR)

These models are developed to correlate the drug permeability and the structural features of the molecules [30].

2.2.3.4 Quantitative structure–bioavailability relationship (QSBR)

For any new hit/lead to become a successful drug candidate it is indispensable to have good bioavailability. This is obtained by generating a correlation model between the bioavailability parameters and descriptors of the series of compounds [31].

2.3 History and development of QSAR

Although the modern QSAR was considered to evolve in the early 1960s, this concept was introduced way back in 1869 by the Edinburgh group's pharmacologists Alexander Crum Brown and Thomas R. Fraser in their research publication "On the connection between chemical constitution & physiological action." In this paper, they concluded that "there can be no reasonable doubt but that a relation exists between the physiologic action of a substance (Φ) and its chemical composition and constitution (C)." They suggested that physiologic action (Φ) must be a function of chemical structure (*C*) given in the following equation, which is still considered the first generalized equation of the QSAR [32]:

$$\Phi = f(C) \tag{2.2}$$

where Φ represents the physiologic action and *C* represents the chemical structure of the molecule.

In the same context, Richardson in 1869 concluded that the narcotic effect of the primary alcohols is proportional to their molecular weight [33]. In 1880, the famous British physician Lauder Brunton suggested in his book *Pharmacology and Therapeutics, or, Medicine Past and Present* that soon physicians will be able to predict the biological response of a drug through its chemical structure [34]. In an important finding, Richet in 1893 discovered an inversely proportional relationship between the toxicity and molecular weight of the organic compounds [35]. The major advancement to this came with the study of Overton [36] and Meyer [37], both of them working independently. They postulated that the toxicity of narcotic compounds in tadpoles and other aquatic species can be correlated to their oil-water partition coefficient (lipophilicity) [36, 37]. None of them could give a correlation equation for their work. The first attempt to correlate the melting point with the structure of the atom was made by Longinescu in 1903 [38]. He derived the formula of the melting point by the following equation:

$$T_{\mathrm{m}} = 8.73D\sqrt{n} \tag{2.3}$$

where D is the density and n is the number of atoms in the molecule.

In 1903 and 1904, Fuhner further established in a study on sea urchin eggs that within the homologous series of the compounds, the narcotic toxicity follows a geometric progression [39, 40]. In 1937, Hammet in his research developed a correlation between the dissociation constant and substitution on the benzene ring [41]. A simple equation was given by Hammet as follows:

$$\log K = \log K^{\circ} + \sigma\,\rho \tag{2.4}$$

where K is the rate constant, σ is a substituent constant, and ρ is a reaction constant.

The work done by various scientists from 1917 to 1935 on the correlation between chemical or physical properties and their intensity of toxicity was quantified by Ferguson in 1939. In this study, thermodynamic interpretation is given by Ferguson for the toxic effects of certain compounds when correlated with the chemical constitution [42].

The decade of 1950 had some great contributions to the development of QSAR methodologies. From a series of research, Robert W. Taft Jr. has given a linear relationship between the log of the rate of hydrolysis of substituted benzamides and electronic steric factors in 1953 [43]. The following equation was given by them:

$$\log k/k^{0} = \delta E_{\mathrm{s}} \tag{2.5}$$

where δ is a proportionality constant dependent only upon the nature of the reaction series and E_{s} is the steric substituent constant.

The first attempt to extend Hammet's equation to the biological system was done by Zahradnik in a series of research conducted from 1960 to 1962. The equation given by them is considered a "biological Hammet equation"; however, this model has little applicability within a specific homologous series [44]. The following equation was given by them:

$$\log \tau_i, - \log \tau_{\mathrm{ET}} = \alpha\beta \tag{2.6}$$

where τ_i is the biological activity of the *i*th compound, τ_{ET} is the biological activity of the ethyl substituted compound of the series, α is the substituent constant, and β represents the constant dependent on the biological activity.

The work of Hansen in 1962 provided the real correlation between the toxicity of substituents and their electronic constants (σ) and was considered a true extension of the work of Hammet [45]. The year 1964 came with the real revolution in the field of QSAR and is considered the dawn of modern QSAR. There were two major contributions this year, first by Hansch and Fujita, which provided the linear free energy relationship, also known as Hansch analysis [8], and second by Free and Wilson who gave the molecular fragment-based QSAR study popularly known as the Free Wilson approach. Hansch further extended their research and provided a breakthrough by linear addition of the physicochemical parameters in correlation with the biological activity [46]. The following equation was provided by them:

$$\log(1/C) = k1\pi + k2\sigma + k3 \tag{2.7}$$

where C is the molar concentration producing a standard response, π is the lipophilicity parameter, and σ is the steric factor constant.

In 1971, another significant contribution was made by Fujita and Ban by giving a mathematical expression to the Free Wilson approach [47] as follows:

$$\log A = \Sigma\, G_i X_i + C \tag{2.8}$$

where A is the biological activity of the substituted compound, G_i is the log activity enhancement factor of the *i*th substituent relative to that of H, and X_i is a parameter that takes a value of 1 or 0 according to the presence or absence of the *i*th substituent.

Both Hansch and Free Wilson's approaches are considered as a backbone of QSAR studies, as in the next three decades much research was conducted based on these two models. In 1977, another bilinear model for the nonlinear dependence of biological activity on the hydrophobic character was proposed by Hugo Kubiyi, also known as a mixed Hansch/Free Wilson model [48]. This equation is as follows:

$$\log 1/C = a \log P - b \log(\beta P + I) + c \tag{2.9}$$

where C is the molar concentration producing a standard response, P is the partition coefficient of the molecule, and terms a, b, c, and β are constants.

The first 3D-QSAR model was generated in 1976 by James Y. Fukunaga for the inhibition of the dihydrofolate reductase enzyme [49]. The field of 3D-QSAR became popular when the first linear 3D-QSAR model, CoMFA, was generated in 1988 by Cramer and team [50]. Another modification to this, CoMSIA, was given by Klebe in 1994 [51].

In the current scenario, QSAR has become a multidimensional field of study with the development of more sophisticated approaches such as 4D-QSAR [52], 5D-QSAR [53], and 6D-QSAR [54].

The major historical events in the development of the QSAR are given in Tab. 2.1.

Tab. 2.1: Historical development of the QSAR.

S. no.	Name of contributor (year)	Major contribution/development
1.	Fraser and Crum-Brown [32]	Postulated that there exists a correlation between the physiological action and the chemical constituent
2.	Richet [35]	Found an inverse relationship between the toxicity and molecular weight of the organic compounds
3.	Overton [36] and Meyer [37]	Postulated that the toxicity of narcotic compounds to the tadpoles and other aquatic species can be correlated to their lipophilicity
4.	Longinescu [38]	The first attempt to correlate the melting point with the structure of the compound was made
5.	Fuhner [39, 40]	Established that within the homologous series of the compounds, the narcotic toxicity follows geometric progression which suggests additivity of group contributions
6.	Hammet [41]	Developed a correlation between the dissociation constant and substitution on the benzene ring
7.	Ferguson [42]	A thermodynamic interpretation was given for the toxic effects of certain compounds when correlated with the chemical constitution
8.	Taft [43]	Gave a linear relationship between the log of the rate of hydrolysis of substituted benzamides and electronic steric factors
9.	Zahradnik [44]	Extension of Hammet's equation to the biological system
10.	Hansen [45]	Provided the real correlation between the toxicity of substituents and their steric constants (σ) and considered as a true extension of the work of Hammet
11.	Hansch and Fujita [8]	Provided the linear free energy relationship, also known as Hansch analysis
12.	Free and Wilson (1964)	Gave the molecular fragment-based QSAR study, popularly known as the Free Wilson approach
13.	Fujita and Ban [47]	Gave a mathematical expression to the Free Wilson approach
14.	Fukunaga [49]	The first 3D-QSAR model was generated for the inhibition of dihydrofolate reductase enzyme

Tab. 2.1 (continued)

S. no.	Name of contributor (year)	Major contribution/development
15.	Kubiyi [48]	A bilinear model for the nonlinear dependence of biological activity on the hydrophobic character was proposed, also known as a mixed Hansch/Free Wilson model
16.	Cramer [50]	A linear 3D-QSAR model, comparative molecular field analysis (CoMFA), was generated
17.	Klebe [51]	A variant of CoMFA, comparative molecular similarity analysis indices (CoMSIA), was generated
18.	Hopfinger [52]	Generated 4D-QSAR model
19.	Vedani [53]	Introduction of 5D-QSAR model generation was provided
20.	Vedani [54]	The first 6D-QSAR model was introduced

Over the years, QSAR has found applications in various fields, including pharmaceuticals, environmental science, agrochemicals, and material design. In drug discovery, QSAR is instrumental in identifying lead compounds with desired pharmacological properties, optimizing their structures for enhanced efficacy, and predicting potential toxicity or side effects. In environmental science and agrochemicals, QSAR aids in evaluating the ecological and toxicological risks of chemicals, enabling better decision-making for sustainable development. Additionally, QSAR is employed in designing innovative materials with specific properties, contributing to advancements in nanotechnology and materials science.

Despite its numerous successes, QSAR does face certain challenges and limitations. The accuracy of QSAR models heavily relies on the quality and representativeness of the data used for training. Moreover, extrapolating QSAR models to predict activity against new, previously unseen compounds requires cautious validation. As with any computational approach, overfitting and underfitting are potential issues that need to be carefully addressed. In conclusion, QSAR is a pivotal field that has transformed the landscape of drug discovery and various other scientific domains. By combining chemistry, biology, and computational techniques, QSAR provides a systematic and rational approach to understanding and predicting the activity of molecules. As technology continues to evolve, QSAR is expected to play an even more significant role in shaping the future of drug design and the development of innovative chemical compounds with diverse applications.

References

[1] Devinyak OT, Lesyk RB. 5-Year trends in QSAR and its machine learning methods. Current Computer-Aided Drug Design, 2016, 12(4), 265–71.

[2] Debnath AK. Quantitative structure-activity relationship (QSAR) paradigm–Hansch era to new millennium. Mini Reviews in Medicinal Chemistry, 2001, 1(2), 187–95.

[3] Devillers J, (ed). Comparative QSAR. CRC Press, 1998.

[4] Sharma S, Bhatia V. Recent trends in QSAR in modelling of drug-protein and protein-protein interactions. Combinatorial Chemistry & High Throughput Screening, 2021, 24(7), 1031–41.

[5] Kontogiorgis CA, Hadjipavlou-Litina D. Current trends in QSAR on NO donors and inhibitors of nitric oxide synthase (NOS). Medicinal Research Reviews, 2002, 22(4), 385–418.

[6] Tropsha A. Recent trends in statistical QSAR modeling of environmental chemical toxicity. Molecular, Clinical and Environmental Toxicology, 2012, 3,381–411.

[7] Hansch C, Maloney PP, Fujita T, Muir RM. Correlation of biological activity of phenoxyacetic acids with Hammett substituent constants and partition coefficients. Nature, 1962, 194(4824):178–80.

[8] Hansch C, Fujita T. p-σ-π Analysis. A Method for the correlation of biological activity and chemical structure. Journal of the American Chemical Society, 1964, 86(8), 1616–26.

[9] Verma J, Khedkar VM, Coutinho EC. 3D-QSAR in drug design-a review. Current Topics in Medicinal Chemistry, 2010, 10(1), 95–115.

[10] Du QS, Huang RB, Chou KC. Recent advances in QSAR and their applications in predicting the activities of chemical molecules, peptides and proteins for drug design. Current Protein and Peptide Science, 2008, 9(3), 248–59.

[11] Timmerman H. QSAR and Drug Design: New Developments and Applications. Elsevier, 1995.

[12] Cherkasov A, Muratov EN, Fourches D, et al. QSAR modeling: Where have you been? Where are you going to? Journal of Medicinal Chemistry, 2014, 57(12), 4977–5010.

[13] Kař S, Leszczynski J. Current trends in QSAR and machine learning models of ionic liquids: Efficient tools for designing environmentally safe solvents for the future. Handbook of Ionic Liquids: Fundamentals, Applications, and Sustainability, 2024, 369–94.

[14] Qiao LS, Cai YL, He YS, Jiang LD, Huo XQ, Zhang YL. Trend of multi-scale QSAR in drug design. Asian Journal of Chemistry, 2014, 26(18): 5917.

[15] Hansch C, Fujita T. Status of QSAR at the end of the twentieth century. In: Classical and Three-Dimensional QSAR in Agrochemistry ACS Symposium Series. American Chemical Society, Washington, DC, 1995.

[16] Neves BJ, Braga RC, Melo-Filho CC, Moreira-Filho JT, Muratov EN, Andrade CH. QSAR-based virtual screening: Advances and applications in drug discovery. Frontiers in Pharmacology, 2018, 9, 1275.

[17] Khan MT, (ed). Recent Trends on QSAR in the Pharmaceutical Perceptions. Bentham Science Publishers, 2012.

[18] Myint KZ, Xie XQ. Recent advances in fragment-based QSAR and multi-dimensional QSAR methods. International Journal of Molecular Sciences, 2010, 11(10), 3846–66.

[19] Lill MA. Multi-dimensional QSAR in drug discovery. Drug Discovery Today, 2007, 12(23–24), 1013–7.

[20] Verma J, Khedkar VM, Coutinho EC. 3D-QSAR in drug design-a review. Current Topics in Medicinal Chemistry, 2010, 10(1), 95–115.

[21] Kubinyi H. QSAR and 3D QSAR in drug design Part 1: Methodology. Drug Discovery Today, 1997, 2(11), 457–67.

[22] Akamatsu M. Current state and perspectives of 3D-QSAR. Current Topics in Medicinal Chemistry, 2002, 2(12), 1381–94.

[23] Doweyko AM. 3D-QSAR illusions. Journal of Computer-Aided Molecular Design, 2004, 18, 587–96.

[24] Damale MG, Harke SN, Kalam Khan AF, Shinde DB, Sangshetti NJ. Recent advances in multidimensional QSAR (4D–6D): A critical review. Mini Reviews in Medicinal Chemistry, 2014, 14(1), 35–55.
[25] Roy K, Kar S. How to judge predictive quality of classification and regression based QSAR models? In: Frontiers in Computational Chemistry, 2015, 71–120. Bentham Science Publishers.
[26] Guha R, Jurs PC. Determining the validity of a QSAR model– a classification approach. Journal of Chemical Information and Modeling, 2005, 45(1), 65–73.
[27] Svetnik V, Liaw A, Tong C, Culberson JC, Sheridan RP, Feuston BP. Random forest: A classification and regression tool for compound classification and QSAR modeling. Journal of Chemical Information and Computer Sciences, 2003, 43(6), 1947–58.
[28] Rai M, Paudel N, Sakhrie M, et al. Perspective on quantitative structure–toxicity relationship (QSTR) models to predict hepatic biotransformation of xenobiotics. Livers, 2023, 3(3), 448–62.
[29] Buchwald P, Bodor N. Computer-aided drug design: The role of quantitative structure–property, structure–activity and structure–metabolism relationships (QSPR, QSAR, QSMR). Drugs Future, 2002, 27(6), 577–88.
[30] Katritzky AR, Lobanov VS, Karelson M. QSPR: The correlation and quantitative prediction of chemical and physical properties from structure. Chemical Society Reviews, 1995, 24(4), 279–87.
[31] Acharya K, Werner D, Dolfing J, et al. A quantitative structure-biodegradation relationship (QSBR) approach to predict biodegradation rates of aromatic chemicals. Water Research, 2019, 157, 181–90.
[32] Brown AC, Fraser TR. On the connection between chemical constitution and physiological action; with special reference to the physiological action of the salts of the ammonium bases derived from strychnia, brucia, thebaia, codeia, morphia, and nicotia. Journal of Anatomy and Physiology, 1868, 2(2), 224.
[33] Macht DI. A toxicological study of some alcohols, with especial reference to isomers. Journal of Pharmacology and Experimental Therapeutics, 1920, 16(1), 1–10.
[34] Brunton TL. Pharmacology and Therapeutics: Or, Medicine past and Present. Macmillan, 1880.
[35] Richet C. Note sur le rapport entre la toxicit~ et les propri~t~s physiques des corps. Soc. de Biol. (Paris) Comptes Rendus, 1893, 45, 775–6.
[36] Overton CE. Osmotic properties of cells in the bearing on toxicology and pharmacology. Zeitschrift Für Physikalische Chemie, 1897, 22, 189–209.
[37] Meyer HH. Zur Theorie der Alkoholnarkose. Erste Mitteilung. Welche Eigenschaft der Anasthetica bedingt ihre narkotische Wirkung? Naunyn-Schmiedebergs Archiv Fur Experimentelle Pathologie Und Pharmakologie, 1899, 42(2–4), 109–18.
[38] Longinescu GG. Contribution à l'étude de la polymérisation des liquides organiques. Journal de Chimie Physique, 1903, 1, 289–95.
[39] Fühner H. Über die Einwirkung verschiedener Alkohole auf die Entwicklung der Seeigel. Archiv Für Experimentelle Pathologie Und Pharmakologie, 1903, 51, 1–10.
[40] Fühner H. Pharmakologische Studien an Seeigeleiern. Der Wirkungsgrad der Alkohole. Archiv Für Experimentelle Pathologie Und Pharmakologie, 1904, 52, 69–82.
[41] Hammett LP. The effect of structure upon the reactions of organic compounds. Benzene derivatives. Journal of the American Chemical Society, 1937, 59(1), 96–103.
[42] Ferguson J. The use of chemical potentials as indices of toxicity. Proceedings of the Royal Society of London. Series B-Biological Sciences, 1939, 127(848), 387–404.
[43] Taft Jr RW. Linear steric energy relationships. Journal of the American Chemical Society, 1953, 75(18), 4538–9.
[44] Zahradnik R. Correlation of the biological activity of organic compounds by means of the linear free energy relationships. Experientia, 1962, 18(11), 534–6.
[45] Hansen OR. Hammett series with biological activity. Acta Chemica Scandinavica, 1962, 16(7), 1593–1600.

[46] Hansch C. Quantitative approach to biochemical structure-activity relationships. Accounts of Chemical Research, 1969, 2(8), 232–9.

[47] Fujita T, Ban T. Structure-activity study of phenethylamines as substrates of biosynthetic enzymes of sympathetic enzymes of sympathetic transmitters. Journal of Medicinal Chemistry, 1971, 14, 148–52.

[48] Kubinyi H. Quantitative structure-activity relations. 7. The bilinear model, a new model for nonlinear dependence of biological activity on hydrophobic character. Journal of Medicinal Chemistry, 1977, 20(5), 625–9.

[49] Fukunaga JY, Hansch C, Steller EE. Inhibition of dihydrofolate reductase. Structure-activity correlations of quinazolines. Journal of Medicinal Chemistry, 1976, 19(5), 605–11.

[50] Cramer RD, Patterson DE, Bunce JD. Comparative molecular field analysis (CoMFA). 1. Effect of shape on binding of steroids to carrier proteins. Journal of the American Chemical Society, 1988, 110(18), 5959–67.

[51] Klebe G, Abraham U, Mietzner T. Molecular similarity indices in a comparative analysis (CoMSIA) of drug molecules to correlate and predict their biological activity. Journal of Medicinal Chemistry, 1994, 37(24), 4130–46.

[52] Hopfinger AJ, Wang S, Tokarski JS, et al. Construction of 3D-QSAR models using the 4D-QSAR analysis formalism. Journal of the American Chemical Society, 1997, 119(43), 10509–24.

[53] Vedani A, Dobler M. 5D-QSAR: The key for simulating induced fit? Journal of Medicinal Chemistry, 2002, 45(11), 2139–49.

[54] Vedani A, Dobler M, Lill MA. Combining protein modeling and 6D-QSAR. Simulating the binding of structurally diverse ligands to the estrogen receptor. Journal of Medicinal Chemistry, 2005, 48(11), 3700–3.

Chapter 3
Physicochemical parameters and QSAR

> Molecular descriptors are the language through which QSAR translates chemical structures into predictive models. – R. Todeschini and V. Consonni

3.1 Physicochemical parameters and QSAR

Quantitative structure–activity relationship (QSAR) is a computational modeling technique used in chemistry, pharmacology, and toxicology to predict the biological activity or properties of molecules based on their structural characteristics [1]. By analyzing the relationship between molecular structures and experimental activity data, QSAR models can estimate the potency, toxicity, or other relevant properties of new compounds without expensive and time-consuming laboratory tests [2, 3]. Utilizing mathematical algorithms, descriptors, and statistical methods, QSAR studies enable the identification of promising drug candidates, environmental pollutants, and chemical compounds, accelerating drug discovery, risk assessment, and regulatory decision-making processes [4]. Despite its advantages, the accuracy and reliability of QSAR models heavily depend on the quality and diversity of the available data [5].

Physicochemical properties play a crucial role in determining the behavior and interactions of molecules in various environments [6]. Some of the important physicochemical properties commonly considered in QSAR studies include:

- **Molecular weight:** The total atomic weight of a molecule, which is essential in understanding its mass and size.
- **Partition coefficient (log *P* or log *D*):** A measure of a compound's tendency to partition between a hydrophobic (lipid) phase and a hydrophilic (aqueous) phase. It provides insight into the compound's solubility and permeability.
- **Hydrophobicity:** Related to the partition coefficient, it describes the tendency of a compound to avoid water interactions and prefer hydrophobic environments.
- **Polarizability:** A measure of how easily the electron cloud of a molecule can be distorted by external electric fields, influencing its interactions with other molecules.
- **Electrostatic charge:** The distribution of electric charge in a molecule affects its interactions with charged entities, such as proteins or ions.
- **Hydrogen bond donor/acceptor count:** The number of hydrogen bond donors and acceptors in a molecule impacts its ability to form hydrogen bonds, affecting various biological interactions.
- **Topological polar surface area:** A measure of the accessible surface area of a molecule that is polar and, therefore, potentially involved in interactions with biological targets.

https://doi.org/10.1515/9783111434858-003

- **Refractivity/polarizability:** Another measure related to molecular polarizability, providing information about the compound's refractive index.
- **Molecular flexibility:** Describes the ability of a molecule to undergo conformational changes, influencing its binding and activity.
- **Steric parameters:** Parameters that account for spatial arrangements of atoms and groups within the molecule, impacting its interactions with other molecules.
- **Dipole moment:** The measure of the overall molecular polarity, which is important in understanding molecular interactions.
- **Ionization constant (pK_a):** The measure of acidity or basicity of a compound, affecting its behavior in different pH environments.

These physicochemical properties, among others, are used as descriptors in QSAR models to establish relationships between molecular structures and biological activities. By building predictive models based on these relationships, QSAR allows for the estimation of various properties and biological activities of compounds, which can be valuable in drug design, environmental assessment, and other fields of research.

The very first question that should come to mind is why physicochemical properties are so important in a QSAR study. The basic explanation to this question is that the origin of the physicochemical properties is due to the electronic arrangement of the molecule and its interaction with the biological system. Hence any change in the physicochemical property of a drug candidate will directly influence its biological response. For the successful development of any mathematical relationship, it is indispensable to study the changes in the physicochemical properties of the drug candidates. Types of basic physicochemical properties that influence biological activities are given in Tab. 3.1.

Tab. 3.1: Types of physicochemical parameters employed in QSAR model generation.

S. no.	Physicochemical parameters	Examples
1.	Lipophilicity parameters	Partition coefficient, distribution coefficient, lipophilic substituent constituent, chromatographic parameters
2.	Electronic parameters	Hammett constant, dipole moments, field and resonance parameters
3.	Steric parameters	Molar refractivity, parachor, molar volume, Taft's steric factor, Verloop steric factors

3.1.1 Lipophilicity parameters

The term lipophilicity is often used interchangeably with hydrophobicity, as both have the same meaning. Lipophilicity parameters are most studied and exploited for defining the QSAR studies [7]. To define the hydrophobicity/lipophilicity parameters of a drug molecule and its dependence on biological activity, the following parameters are studied and measured:

3.1.1.1 Partition coefficient (*P*)

Partition coefficient in the simplest terms is defined as the ratio of the concentration of unionized drugs in organic and aqueous solvents at equilibrium by the following equation:

$$P = \frac{\text{Concentration in an organic solvent}}{\text{Concentration in aqueous solvent}} \tag{3.1}$$

This is an important physicochemical parameter as it is an indicator of the ability of a drug to cross the cell membrane [8]. For QSAR studies, the octanol/water system is preferred due to its high resemblance with the biological system [7]. The therapeutic effect of any drug is possible only if it reaches the systemic circulation after crossing the cellular membrane, which is lipophilic in nature. This journey of the drug molecule inside the biological system can be understood by a simple kinetic model that has been described in Fig. 3.1. A lipid compartment (representing the cell membrane) lies between two aqueous compartments. Any compound that has a low *P* value (hydrophilic molecule) will remain in the first compartment and will not cross the cell membrane. Likewise, compounds possessing high *P* value (lipophilic molecule) will be retained in the lipid compartment. Hence, an optimal value of *P* is desirable for a drug to cross the cell membrane, which in turn indirectly corresponds to its biological activity.

From the earlier studies, a linear relationship has been established between the biological activity and the partition coefficient of the drug molecules (Fig. 3.2). These earlier studies were done on small homologous series having a small range of *P* values [9]. The equation which emerged from these studies is given below:

$$\log\left(\frac{1}{C}\right) = a \log P + b \tag{3.2}$$

where *C* represents the concentration required for producing the biological response, *P* is the partition coefficient, and *a* and *b* are constants. This equation resembles the equation of a straight line ($y = mx + c$).

Hansch with his coworkers suggested that with a large range of *P* values, rather than linear relationships, there exists a parabolic relationship between the biological activity and *P* value [10] (Fig. 3.3).

They enumerated the following equation:

$$\log\left(\frac{1}{C}\right) = -k_1 (\log P)^2 + k_2 (\log P)^2 + k3 \tag{3.3}$$

where *C* is the concentration for producing a biological response, *P* is the partition coefficient, and k_1, k_2, and k_3 are constants obtained through the least square method.

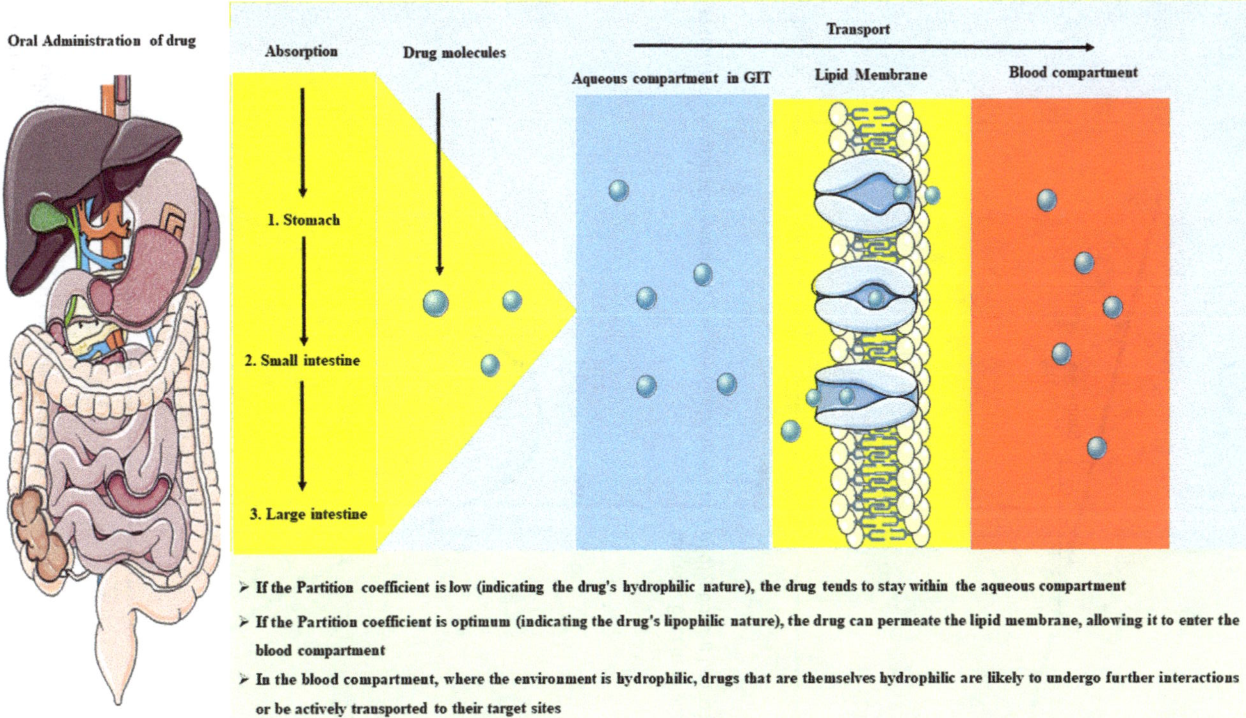

Fig. 3.1: Dependence of the partition coefficient on the absorption of the drug. There should be an optimum partition coefficient of the drug. In drugs having a high partition coefficient, the drug is hydrophobic and will remain in the lipophilic membrane whereas in drugs with a low partition coefficient, the drug is hydrophilic and it will remain in the GIT.

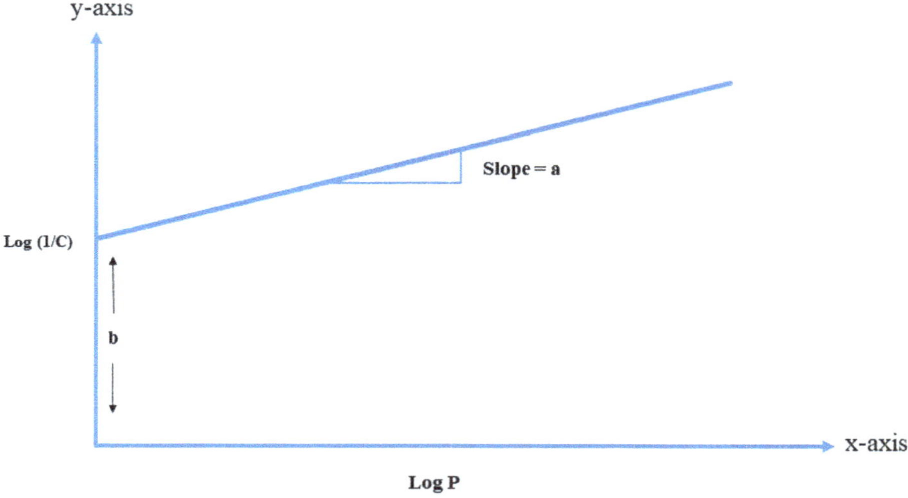

Fig. 3.2: Relationship between the log *P* (partition coefficient) value and the biological activity (log(1/*C*)) of the drug for a small range of the log *P* values.

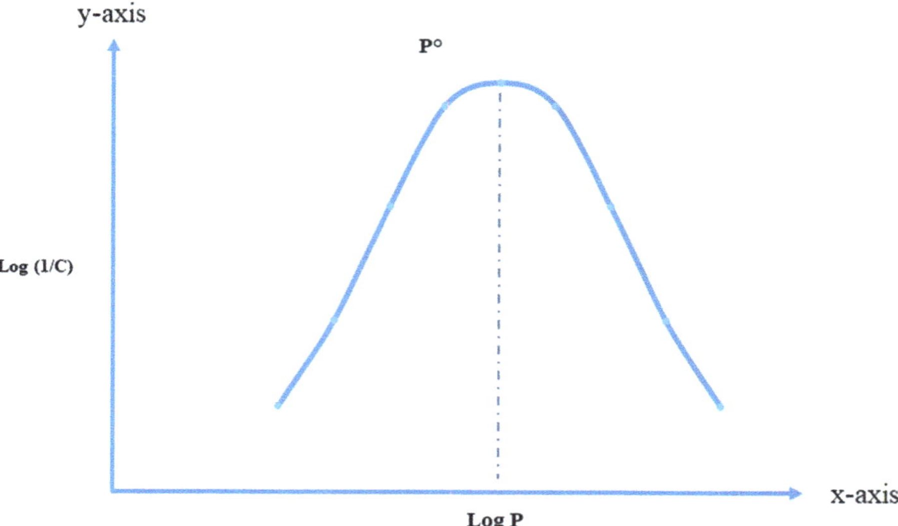

Fig. 3.3: Relationship between the log *P* (partition coefficient) value and the biological activity (log(1/*C*) of the drug for a small range of the log *P* values. P^0 is the optimum value of the partition coefficient where the maximum biological response is obtained.

The value P_0 in Fig. 3.3 represents the optimal value of the partition coefficient for a drug molecule. Any value less than P_0 will account for high hydrophilicity and the drug will face hindrance in crossing the cell membrane. Similarly, higher P_0 results in a lipophilic drug which makes it difficult for the drug to reach the bloodstream and also results in high plasma protein binding.

3.1.1.2 Lipophilic/hydrophobic substituent

In simplest words, π is the assignment of a numerical value to a substituent for defining the extent of lipophilicity changes it induces in a system at equilibrium. This was another appreciable piece of research put forward by Hansch and coworkers. It is calculated by the following equation:

$$\pi = \log P_X - \log P_H \tag{3.4}$$

where P_X is the partition coefficient value of the molecule having substitution and P_H represents the partition coefficient of an unsubstituted molecule in the two-solvent systems. As this π value is calculated at equilibrium, it is also considered a free energy-related constant. Consider an example of a benzene ring as a parent molecule and nitrobenzene as a substituted benzene. The value of π for the nitro group can be calculated by the following equation:

$$\pi = \log P_{\text{nitrobenzene}} - \log P_{\text{benzene}} \tag{3.5}$$

Table 3.2 depicts the π value of some substituents calculated by Hansch and coworkers in their research published in 1964. These values were calculated in an octanol/water solvent system using benzene as a parent molecule.

Tab. 3.2: Values of π of some of the substituents calculated by Hansch and coworkers in 1964.

S. no.	Substituent	Value of π
1	4-Fluoro	0.14
2	4-Chloro	0.71
3	4-Bromo	0.86
4	4-Methyl	0.56
5	4-CH_2OH	−1.03
6	4-CH_2COOH	−0.72
7	4-OCH_3	−0.02
8	4-NO_2	−0.28
9	4-OH	−0.67
10	4-$COCH_3$	−0.55

From Tab. 3.2, it becomes evident that any substituent having π value greater than 0 will enhance the lipophilicity and any substituent having π value less than 0 will shift the molecule toward hydrophobicity.

3.1.1.3 Distribution coefficient (*D*)

Ionization plays a crucial role in the pharmacokinetic and pharmacodynamic properties of the drug. One of the major drawbacks of considering log *P* values for lipophilicity is that it does not take into account the ionization phenomenon of the drugs. It is a proven fact that most of the drugs available undergo ionization in the biological system at various stages. To consider drug ionization, a new phenomenon of distribution coefficient (*D*) is considered crucial for studying the lipophilicity of the drugs [11]. The value of *D* can be calculated from the following equation:

$$D = \frac{\text{The concentration of all the species in partition solvent (ionized and unionized)}}{\text{The concentration of all the species in water (ionized and unionized)}} \tag{3.6}$$

As ionization is a pH-dependent phenomenon, the value of *D* and log *D* will also be dependent on the pH. A pH-dependent equation for calculating the log *D* is given by the following equation:

$$\log D = \log P - \log 10\,(\mathrm{p}K_\mathrm{a} - \mathrm{pH}) \tag{3.7}$$

where *D* is the distribution coefficient and $\mathrm{p}K_\mathrm{a}$ is the dissociation constant.

The major assumption that is made while calculating log *D* is that the entire drug molecule in the organic phase will be in unionized form. Figure 3.4 demonstrates the log *D* value of the acidic drug, ibuprofen, at different pH values.

3.1.2 Electronic parameters

QSAR models are built by correlating the electronic, structural, and physicochemical parameters of compounds with their corresponding biological activities [12]. QSAR models rely on a set of electronic parameters, which are descriptors that characterize the electronic properties of molecules. These parameters play a crucial role in building predictive models and understanding the relationships between the molecular structure and its activity. Depending on the specific research objectives and the complexity of the molecules under study, various electronic properties are considered to be important in building predictive QSAR models. Electronic parameters are an essential component of QSAR models, as they provide information about the electronic properties and behavior of molecules [13].

These parameters are of utmost importance for defining the electronic properties (distribution of electrons in a molecule) of the molecules [14]. These electronic effects

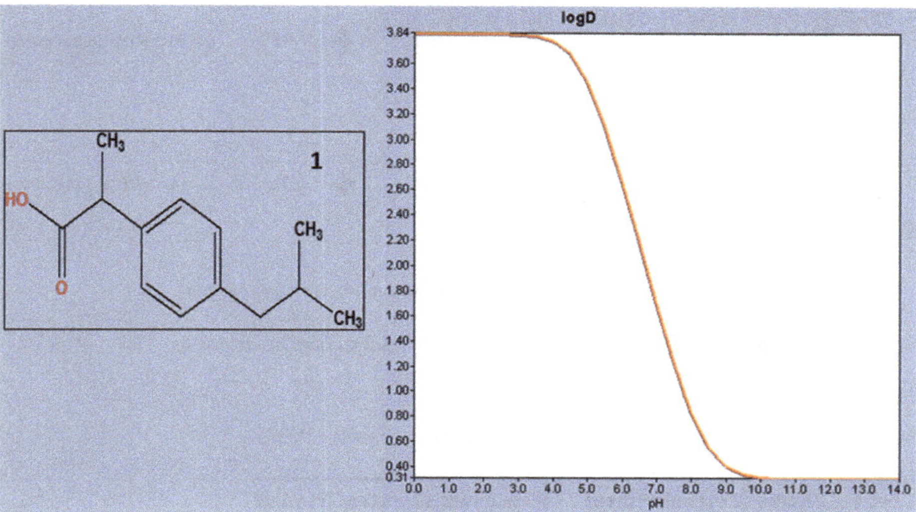

Fig. 3.4: Log *D* value versus pH of ibuprofen obtained from MarvinSketch of Chemaxon.

play their role in the ionization and polarity of a drug molecule. The transport of a drug through the cell membrane is primarily governed by its ionization and interaction with the receptor is affected by the polarity. Various electronic parameters are studied for this purpose, which are discussed below:

3.1.2.1 Hammett's constant (σ)

The effect of electron-withdrawing or -donating substituent on the rate of hydrolysis was studied by Hammett and his coworkers in 1937 [15]. In this peculiar work, they quantified this effect and assigned a numerical value (σ) to each substituent based on its ability to withdraw or donate electrons in the parent molecule. They used benzoic acid as a parent molecule for their studies. They defined the σ value for a particular substituent as the difference in the values of the log of equilibrium constants of substituted benzoic acid and unsubstituted benzoic when undergoing hydrolysis (Fig. 3.5).

Mathematically, the Hammett constant (σ) is calculated from the following equation:

$$\sigma = \log K_X - \log K \text{ or } \sigma = \frac{\log K_X}{\log K} \tag{3.8}$$

where K_X is the equilibrium constant of the substituted benzoic acid and K is the equilibrium constant for unsubstituted benzoic acid under hydrolysis.

The values of the Hammett constant (σ) calculated by them in their experiment are given in Tab. 3.3.

The negative value of σ is an indicator of the presence of an electron-donating group whereas a positive value of σ is an indicator of the presence of an electron-withdrawing

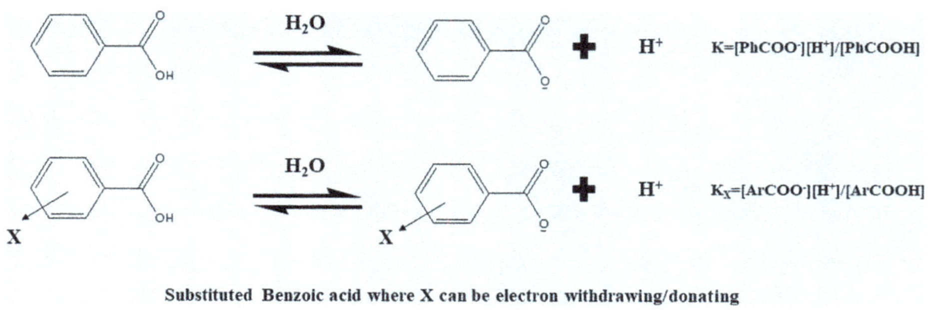

$$K=[PhCOO^-][H^+]/[PhCOOH]$$

$$K_X=[ArCOO^-][H^+]/[ArCOOH]$$

Substituted Benzoic acid where X can be electron withdrawing/donating

Fig. 3.5: Value of equilibrium constants of substituted and unsubstituted benzoic acids.

Tab. 3.3: Value of σ of different substituents calculated by Hammett in 1937.

S. no.	Name of the substituent	Hammett's constant (σ) value
1	*para*-Amino	−0.66
2	*meta*-Amino	−0.161
3	*para*-Methyl	−0.17
4	*meta*-Methyl	−0.069
5	*para*-Methoxy	−0.268
6	*meta*-Methoxy	0.115
7	*para*-Fluoro	0.062
8	*meta*-Fluoro	0.337
9	*para*-Nitro	1.27
10	*meta*-Nitro	0.71

group. Here, it is important to understand that there are two main effects shown by the substituent on a benzene ring. First is the inductive effect (I) and second is the resonance effect (R). At the *meta* position, only the I effect plays its role, whereas at *para* both I and R effects are operational with the R effect being the dominant. If we carefully observe Tab. 3.3, it becomes evident that the value of methoxy at *para* is dominated by the $+R$ effect (electron donating), making it a negative value (−0.268), and the same at *meta* is dominated by the $-I$ effect (electron withdrawing) only, thus giving it a positive value (0.115). The value of σ at the *ortho* position is not as accurate, as at *ortho*, the steric hindrance factor also plays an important role and should also be considered while quantifying the electronic behavior of *ortho*-substituent.

These differences in the value of σ at different positions of benzene were studied rationally for the first time by Swain and Lupton in 1968 [16]. They introduced two independent parameters, F and R, which are calculated based on values of σ at the *ortho* and *para* positions. The mathematical equations they derived are as follows:

$$\sigma_m = 0.6F + 0.27R \tag{3.9}$$

$$\sigma_p = 0.56\,F + R \quad (3.10)$$

where σ_m and σ_p depict the value of the Hammett constant at *meta* and *para* positions, respectively.

The values of F and R parameters were recalculated by Hansch, providing the following equations:

$$F = 1.369\,\sigma_m - 0.373\,\sigma_p - 0.009 \quad (3.11)$$

$$R = \sigma_p - 0.921F \quad (3.12)$$

The value of σ is not only limited to aromatic compounds but can also be calculated for aliphatic substituent (σ^*), as demonstrated by Taft Jr. in 1956 by using the following equations:

$$\sigma^* = \frac{\log\,(k/K_0)_{\text{base}}}{\log\,(k/K_0)_{\text{acid}}} \times 2.48 \quad (3.13)$$

$$\sigma^* = \left(\log\,(k/K_0)_{\text{base}} / \log\,(k/K_0)_{\text{acid}}\right)/2.48 \quad (3.14)$$

where k and k_0 represent the rate constants of base- and acid-catalyzed hydrolysis of RCOOR′ and CH_3COOR'.

The first true attempt for correlating the Hammett constant (σ) with biological activity was made by Hansen in 1962 [17]. From a series of compounds, he derived the relationship between the biological activity and the Hammett constant (σ) by the following equation:

$$\log\left(\frac{1}{C}\right) = 1.454\sigma + 1.787 \quad (3.15)$$

3.1.2.2 Dipole moment (μ)

The dipole moment (μ) is one of the molecular descriptors used to characterize chemical compounds [18]. It provides information about the distribution of electric charge within a molecule, indicating the presence of polar covalent bonds and the overall polarity of the molecule. μ is a measure of the separation of positive and negative charges within a molecule and indicates the molecule's overall polarity. The higher the electronegativity difference between two atoms in a bond, more will be the value of μ. It is an important property as it affects a molecule's interactions with other molecules, including proteins and receptors in biological systems [20]. The μ is a simple product of the charge and distance between the bonds and can be mathematically calculated from the following equation:

$$\mu = q \times r \quad (3.16)$$

where q is the charge and r is the distance between the atoms in a bond.

The very first use of μ as a descriptor in QSAR was done by Lein and coworkers and from their research they have given the following equation for the relation between μ and the Hammett constant (σ):

$$\mu = -5.99\,\sigma_m - 0.53 \tag{3.17}$$

$$\mu = -3.65\,\sigma_p - 1.27 \tag{3.18}$$

When developing a QSAR model, it is essential to consider molecular descriptors like the dipole moment in combination with other relevant descriptors to create robust and accurate models. The dipole moment of a molecule can have various effects on QSAR, influencing intermolecular interactions, solubility, permeability, biological activity, ADMET properties, and conformational flexibility.

3.1.2.3 Hydrogen bonding parameters

Hydrogen bonding is a crucial intermolecular interaction that plays a significant role in various biological and chemical processes [21]. In the context of QSAR, hydrogen bonding parameters are used to describe and quantify the strength and nature of hydrogen bond interactions between molecules.

Some hydrogen bonding parameters commonly used in QSAR:

- **Hydrogen bond donor count:** This parameter represents the number of hydrogen bond-donating groups (usually hydrogen atoms) present in a molecule. These are typically atoms with polar hydrogen bonded to an electronegative atom, such as oxygen or nitrogen.
- **Hydrogen bond acceptor count:** This parameter represents the number of hydrogen bond-accepting groups in a molecule. These are typically electronegative atoms, such as oxygen, nitrogen, or sulfur, with available lone pairs of electrons to form hydrogen bonds.
- **Hydrogen bond strength:** This parameter characterizes the strength of individual hydrogen bonds. It is often described in terms of bond dissociation energies or other energetic measures.
- **Hydrogen bond distance:** The distance between the hydrogen atom and the hydrogen bond acceptor atom in a hydrogen bond is an important geometric parameter that influences the strength of the interaction.
- **Hydrogen bond angle:** The angle formed by the hydrogen atom, the hydrogen bond acceptor atom, and the hydrogen bond donor atom is another crucial geometric parameter that affects the hydrogen bond's stability.
- **Hydrogen bond potential:** This parameter quantifies the potential of a molecule to participate in hydrogen bonding interactions. It considers both the presence of hydrogen bond donor and acceptor groups and their spatial arrangement in the molecule.

– **Hydrogen bond index (HBI):** HBI is a numerical index that combines various factors, such as hydrogen bond strength, distance, and angle, to evaluate the overall hydrogen bond potential of a molecule.

These parameters are often utilized in QSAR studies to establish correlations between the hydrogen bonding properties of molecules and their biological activities or other physicochemical properties. By considering these hydrogen bonding parameters along with other molecular descriptors, QSAR models can provide insights into the structure–activity relationships and aid in the design of new bioactive compounds [22].

3.1.3 Steric parameters

Steric parameters play a crucial role in QSAR studies, which are used to establish relationships between the structure of molecules and their biological or chemical activities [23]. Steric parameters focus on the spatial arrangement and interactions of atoms and groups within a molecule. These parameters are essential in understanding how the size and shape of a molecule influence its biological activity, binding affinity, and interactions with biological targets. The movement of a drug molecule within the biological system and its interaction with the receptor depends on steric parameters [24]. The movement of large bulky drug molecules across the cell membrane is restricted; besides, interaction with the binding sites of the target receptor is also hindered. Some of the important steric parameters are discussed below.

3.1.3.1 Taft's steric factor (E_s)

Taft's steric factor (E_s) is the first parameter that was defined in the QSAR study for quantifying steric properties in 1956 [25]. The steric factor, proposed by R. W. Taft, is one of the parameters used in QSAR models to account for the steric hindrance or spatial interactions between a molecule and its biological target or receptor. Steric hindrance refers to the resistance encountered by molecules when they approach or bind to a receptor due to the bulkiness of certain substituents or functional groups. E_s was reported by Taft in an acid-catalyzed hydrolysis reaction of aliphatic esters and was defined by the following equation:

$$E_S = \log\left(\frac{k}{k_0}\right) \tag{3.19}$$

where E_s is Taft's steric factor, k is the rate constant of the substituted aliphatic ester, and k_O is the rate constant of CH_3COOR. In the same study, E_s values of *ortho*-substituted benzoates were also calculated. In Tab. 3.4, the E_s values of some different *ortho*-substituted benzoates are given.

Tab. 3.4: The values of E_s of some of the substituents calculated by Taft in 1956.

S. no.	Name of substituent	Value of E_s
1	*o*-CH_3	+0.97
2	*o*-C_2H_5	+0.86
3	*o*-Cl	+0.18
4	*o*-Br	+0.01
5	*o*-I	−0.20
6	*o*-NO_2	−0.72
7	*o*-C_6H_5	−0.90

After analyzing the above table, it can be suggested that the value of E_s for different *ortho* substituents of benzoates follows the order: *o*-CH_3 > *o*-C_2H_5 > *o*-Cl > *o*-Br > *o*-I > *o*-NO_2 > *o*-C_6H_5.

The steric factor in QSAR equations helps quantify the steric effects of different substituents in a molecule and how they influence the molecule's biological activity. It takes into account the spatial arrangement and size of specific groups attached to the molecule, which can affect how well the molecule interacts with the target. Taft's steric factor is a crucial parameter in QSAR modeling that considers the steric hindrance effects of substituents on molecular activity. By incorporating this factor into QSAR models, researchers can gain insights into the steric requirements of a molecule for optimal binding to a target, aiding in the rational design of new and more potent compounds.

3.1.3.2 Molar refractivity (MR)

Molar refractivity (MR) is another steric parameter that plays an important role in QSAR study and is the specific refraction of one molar compound. MR is a measure of the polarizability of a molecule and is related to its molecular volume and electron distribution. It quantifies how a molecule interacts with electromagnetic radiation, particularly light, and quantifies the degree of polarization and size of a compound [26, 27]. Mathematically it is calculated as follows:

$$MR = \frac{M(n^2 - 1)}{\rho\,(n^2 + 2)} \tag{3.20}$$

where M is the molar mass of the compound, n is the refractive index, and ρ represents the density of the molecule. The term $(n^2 - 1)/(n^2 + 2)$ represents the correction factor which corresponds to the measure of how easily a substituent can be polarized. This correction factor plays an important role if the substituent has any lone pair or π electrons.

In QSAR, MR is used as a molecular descriptor, which is a numerical representation of a molecule's chemical structure. MR is a valuable descriptor in QSAR studies as it helps capture important molecular features related to the molecule's physical interactions and can contribute to more accurate predictions of biological and physicochemical properties. By incorporating MR along with other descriptors, such as lipophilicity, molecular weight, and topological indices, QSAR models can be built to predict various properties of interest, such as biological activity, toxicity, and physicochemical properties [28, 29].

3.1.3.3 STERIMOL parameters

STERIMOL stands for "steric parameters for a linear free energy relationship" and was developed by Christopher Hansch and his colleagues in the late 1960s. The STERIMOL parameters primarily focus on the steric aspects of molecules, which refer to the spatial arrangement and size of substituents or atoms around a particular site of a molecule [30–33].

The three key STERIMOL parameters are:

- *L*: Represents the length parameter and is a measure of the van der Waals radius of a substituent or atom. It characterizes the size of a group or atom.
- *B*: The breadth parameter signifies the spatial width of a substituent or atom. It describes how bulky or slender a group is.
- *S*: The shape parameter quantifies the geometric properties of a group or atom. It indicates whether the substituent has a linear, branched, or cyclic structure.
- *T*: The tolerance parameter represents the minimum value of *S* and is used to describe the "flatness" of the molecule.

By using STERIMOL parameters, researchers can quantify the steric characteristics of different molecular fragments and understand how these steric properties influence the biological activity of compounds.

The major disadvantages of Taft's steric parameter (E_S) were that they were available for a limited number of substituents and had to be calculated experimentally, which was a tedious task. To overcome this, Verloop and his coworkers wrote a computer program to calculate new steric parameters in 1976 and termed them STERIMOL parameters [30]. These are also known as Verloop steric parameters. These STERIMOL parameters were calculated from Van der Waals descriptors (Van der Waals radii, bond length, bond angle, and conformations) of the substituent.

The main assumption behind calculating STERIMOL parameters is that an ideal drug molecule fits completely inside the curvature of the receptor's cavity. The one very important thing one must understand is that all the Verloop parameters are calculated for the substituent of a drug only. To understand this, let us consider Fig. 3.6 in which a drug molecule with one of its substituents is made to fit inside the cavity of the receptor.

The first parameter, length "*L*," is calculated as its length along the *X*-axis. The other Verloop parameters are related to the width of the substituent (B_1 to B_4) named

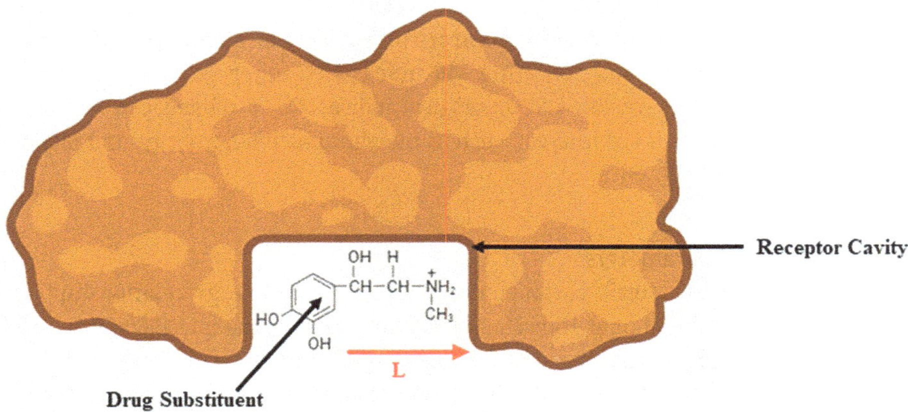

Fig. 3.6: The drug along with its substituent is present inside the cavity of a receptor. The length of the substituent is measured along one axis.

in increasing order of their distance, that is, B_1 has the smallest value and B_4 will have the largest. As per Verloop "the value B_1, B_2, B_3, and B_4 are in ascending order, taken as the distance to the X-axis of the paired parallel tangential planes perpendicular to the Z and the *Y* axis."

Consider a drug molecule having a substituent interacting in the cavity of the receptor (Fig. 3.7a). In the very first step, a boundary surface diagram of the substituent will be created where *O* is the point of attachment with the parent molecule (Fig. 3.7b). Along the *X*-axis in Fig. 3.7b the length parameter "*L*" is represented. Four more distance parameters can be seen from the point of origin *O* (1.4 Å, 1.6 Å, 2.2 Å, and 2.3 Å). From the definition, these distance parameters are the width parameters of the substituent and will be numbered in ascending order (smallest will be B_1, i.e., 1.4 Å and longest will be B_4, i.e., 2.3 Å).

Verloop parameters calculated for some of the substituents by the STERIMOL program are given in Tab. 3.5.

It can be observed clearly that the calculated value of all B_1 to B_4 is the same for hydrogen, that is, 1.00 Å. This can be explained by the fact that hydrogen is a spherical atom and hence its distance will be the same in all directions from the point of contact.

STERIMOL parameters help in assessing how the steric features of the molecule may influence the binding to the target site. Larger or bulkier molecules may experience steric hindrance, affecting their biological activity. Incorporating STERIMOL parameters into QSAR models can improve their accuracy in predicting biological activity. STERIMOL parameters are useful for aligning molecules in a QSAR study, ensuring that the relevant structural features are appropriately superimposed. This alignment is important because it helps compare and analyze the steric properties of

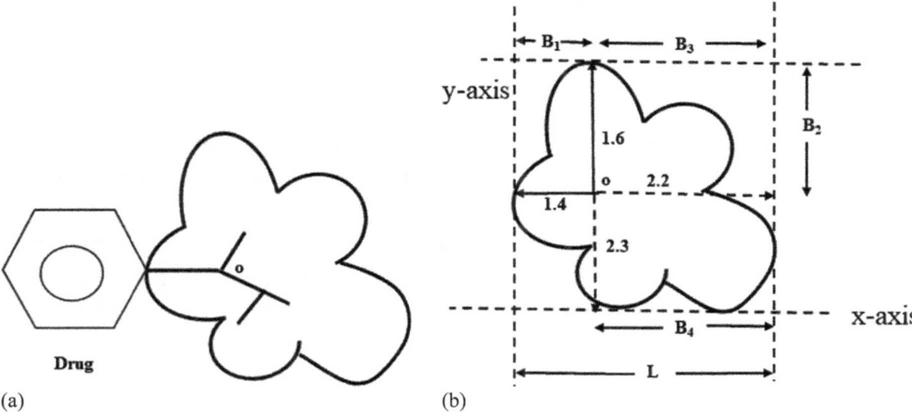

Fig. 3.7: (a) A surface diagram is drawn around the substituent of an imaginary drug molecule. (b) Length of the substituent along all the axis from the center of the surface of the substituent.

Tab. 3.5: Verlopp/STERIMOL parameters of some of the substituents.

S. no.	Substituent	*L* (Å)	B_1 (Å)	B_2 (Å)	B_3 (Å)	B_4 (Å)
1	H	2.06	1.00	1.00	1.00	1.00
2	CH_3	3.00	1.52	1.90	1.90	2.04
3	CH_2OH	3.97	1.52	1.9	1.9	2.70
4	CF_3	3.30	1.98	2.44	2.44	2.61
5	OH	2.74	1.35	1.35	1.35	1.93
6	OCH_3	3.98	1.35	1.90	1.90	2.87
7	NO_2	3.44	1.70	1.70	2.44	2.44

different molecules consistently. During the development of new compounds with desired biological activity, STERIMOL parameters can guide the structural modification process. By analyzing the effect of changes in substituents on the STERIMOL parameters, researchers can make informed decisions about which modifications might be most beneficial.

3.1.3.4 Other steric parameters

With more explorations in QSAR, certain other steric parameters evolved through the passage of research. Some important steric parameters used in QSAR studies include:

- **Steric hindrance:** Steric hindrance refers to the obstruction of molecular interactions due to bulky groups or atoms in close proximity. It affects the accessibility of a molecule's functional groups and can influence the molecule's reactivity and binding affinity [34].

- **Steric bulk:** Steric bulk measures the spatial size of a substituent or functional group attached to the molecule. It is usually quantified by the volume occupied by the substituent and helps in understanding how the size of different groups impacts the molecular properties [35].
- **Van der Waals volume**: The Van der Waals volume represents the space occupied by a molecule due to the attractive and repulsive forces between atoms. It considers the effective size of atoms or groups in a molecule [36].
- **Molecular surface area (MSA):** MSA calculates the accessible surface area of a molecule, taking into account the shape and size of the molecule. It can be useful in predicting interactions with enzymes or receptors that depend on surface complementarity [37].
- **Steric energy:** Steric energy is the energy associated with the repulsion or overlap of electron clouds between atoms or groups in a molecule. High steric energy may lead to instability or unfavorable interactions [35].
- **Shape index:** The shape index is a numerical value that characterizes the three-dimensional shape of a molecule. It can be used to compare the shapes of different molecules and assess their structural similarity [38].

Steric parameters are often combined with other physicochemical properties in QSAR models to gain a comprehensive understanding of how molecular structure relates to biological or chemical activity. By considering these steric factors, researchers can design and modify molecules to optimize their properties for specific applications, such as drug design, toxicity prediction, and environmental studies.

References

[1] Abdel-Ilah L, Veljović E, Gurbeta L, Badnjević A. Applications of QSAR study in drug design. International Journal of Engineering Research & Technology (IJERT), 2017, 6(06), 582–7.
[2] Puzyn T, Leszczynski J, Cronin MT, (ed). Recent Advances in QSAR Studies: Methods and Applications. Springer, 2010.
[3] Roy K. Advances in QSAR Modeling. Applications in Pharmaceutical, Chemical, Food, Agricultural and Environmental Sciences, Springer: Cham, Switzerland. 2017, 555, 39.
[4] Trinajstić N, Nikolić S, Carter S. QSAR: Theory and Application. Kemija U Industriji, 1989, 469–84.
[5] De P, Kar S, Ambure P, Roy K. Prediction reliability of QSAR models: An overview of various validation tools. Archives of Toxicology, 2022, 96(5), 1279–95.
[6] Wenlock MC, Barton P. In silico physicochemical parameter predictions. Molecular Pharmaceutics, 2013, 10(4), 1224–35.
[7] Dearden JC. Partitioning and lipophilicity in quantitative structure-activity relationships. Environmental Health Perspectives, 1985, 61, 203–28.
[8] Leo A, Hansch C, Elkins D. Partition coefficients and their uses. Chemical Reviews, 1971, 71(6), 525–616.

[9] Hoekman D. Exploring QSAR fundamentals and applications in chemistry and biology, volume 1. hydrophobic, electronic and steric constants, volume 2. Journal of the American Chemical Society, 1995, 117, 9782. Journal of the American Chemical Society, 1996, 118(43), 10678.

[10] Hansch C, Steward AR, Anderson SM, Bentley DL. Parabolic dependence of drug action upon lipophilic character as revealed by a study of hypnotics. Journal of Medicinal Chemistry, 1968, 11(1), 1–11.

[11] Manallack DT. The pKa distribution of drugs: Application to drug discovery. Perspectives in Medicinal Chemistry, 2007, 1, 1177391X0700100003.

[12] Hammett LP. Physical Organic Chemistry: Reaction Rates, Equilibria, and Mechanisms. Tokyo Yagumo, 1943.

[13] Martin YC. Quantitative Drug Design: A Critical Introduction. CRC Press, 2010.

[14] Shorter J. Multiparameter extensions of the Hammett equation. In: Correlation Analysis in Chemistry: Recent Advances. Springer US, Boston, MA, 1978, 119–73.

[15] Hammett LP. The effect of structure upon the reactions of organic compounds. Benzene derivatives. Journal of the American Chemical Society, 1937, 59(1), 96–103.

[16] Swain CG, Lupton EC. Field and resonance components of substituent effects. Journal of the American Chemical Society, 1968, 90(16), 4328–37.

[17] Hansen RS. Thermodynamics of interfaces between condensed phases 1. The Journal of Physical Chemistry, 1962, 66(3), 410–5.

[18] Li WY, Guo OR, Lien EJ. Examination of the interrelationship between aliphatic group dipole moment and polar substituent constants. Journal of Pharmaceutical Sciences, 1984, 73(4), 553–8.

[19] Lien EJ, Liao RC, Shinouda HG. Quantitative structure-activity relationships and dipole moments of anticonvulsants and CNS depressants. Journal of Pharmaceutical Sciences, 1979, 68(4), 463–5.

[20] Lien EJ, Guo ZR, Li RL, Su CT. Use of dipole moment as a parameter in drug–receptor interaction and quantitative structure–activity relationship studies. Journal of Pharmaceutical Sciences, 1982, 71(6), 641–55.

[21] Bohacek RS, McMartin C. Definition and display of steric, hydrophobic, and hydrogen bonding properties of ligand binding sites in proteins using Lee and Richards accessible surface: Validation of a high-resolution graphical tool for drug design. Journal of Medicinal Chemistry, 1992, 35(10), 1671–84.

[22] Abraham MH, Lieb WR, Franks NP. Role of hydrogen bonding in general anesthesia. Journal of Pharmaceutical Sciences, 1991, 80(8), 719–24.

[23] Hansch C, Leo A. Substituent constants for correlation analysis in chemistry and biology. Wiey, 1979.

[24] Balaban AT, Chiriac A, Motoc I, Simon Z. Steric Fit in Quantitative Structure-activity Relations. Springer Science & Business Media, 2012.

[25] Taft RW. Separation of polar, steric and resonance effects in reactivity. Steric Effects in Organic Chemistry, 1956, 556–675.

[26] Moriguchi I. Quantitative structure-activity studies. I. Parameters relating to hydrophobicity. Chemical and Pharmaceutical Bulletin, 1975, 23(2), 247–57.

[27] Selassie CD, Verma RP. History of quantitative structure-activity relationships. Burger's Medicinal Chemistry and Drug Discovery, 2003, 1, 1–48.

[28] Roy K, Saha A. QSPR with TAU indices: Molar refractivity of diverse functional acyclic compounds. Indian Journal of Chemistry Section B-organic Chemistry Including Medicinal Chemistry, 2005, 44, 1693–1707.

[29] Padrón JA, Carrasco R, Pellon RF. Molecular descriptor based on a molar refractivity partition using Randic-type graph-theoretical invariant. Journal of Pharmacy and Pharmaceutical Sciences, 2002, 5(3), 258–66.

[30] Verloop A, Tipker J. Use of linear free energy related and other parameters in the study of fungicidal selectivity. Pesticide Science, 1976, 7(4), 379–90.

[31] Verloop A. The STERIMOL approach: Further development of the method and new applications. In: Pesticide Chemistry: Human Welfare and Environment. Pergamon, 1983, 339–44.
[32] Draber W. STERIMOL and its role in drug research. Zeitschrift für Naturforschung C, 1996, 51(1–2), 1–7.
[33] Verloop A. Lead optimization: Quantitative structure-activity relations. Philosophical Transactions of the Royal Society of London. B, Biological Sciences, 1981, 295(1076), 45–55.
[34] Kubinyi H. What is QSAR? JACS, 1964, 86, 1616.
[35] Williford CJ, Stevens EP. Strain energies as a steric descriptor in QSAR calculations. QSAR & Combinatorial Science, 2004, 23(7), 495–505.
[36] Roy K, Narayan Das R. A review on principles, theory and practices of 2D-QSAR. Current Drug Metabolism, 2014, 15(4), 346–79.
[37] Prasanna S, Doerksen RJ. Topological polar surface area: A useful descriptor in 2D-QSAR. Current Medicinal Chemistry, 2009, 16(1), 21–41.
[38] Kier LB. A shape index from molecular graphs. Quantitative Structure-Activity Relationships, 1985, 4(3), 109–16.

Chapter 4
Quantitative models in QSAR

In the realm of Quantitative Structure-Activity Relationships, data becomes the brush, and models the canvas, painting a picture of molecular behavior – Alexander Tropsha

4.1 Introduction

Quantitative structure–activity relationship (QSAR) models are mathematical and computational models that aim to establish a quantitative relationship between the structural features of chemical compounds and their corresponding biological activities or properties [1]. These models provide a way to predict the activity of new compounds based on their structural characteristics, which is particularly valuable in drug discovery, environmental chemistry, toxicology, and related fields [2]. Quantitative models in QSAR involve the following key components:

- **Molecular descriptors:** These are numerical representations of the structural features of chemical compounds. Descriptors can include information about atom types, bond lengths, molecular weight, electronic properties, and more. These descriptors capture the molecular characteristics that may influence biological activity.
- **Biological activity data:** Experimental data that quantifies the biological activity or property of a set of compounds. This data serves as the basis for building and validating QSAR models.
- **Modeling algorithms:** Various mathematical and statistical techniques are used to establish the relationship between molecular descriptors and biological activity. These techniques can range from linear regression to more complex methods like neural networks and support vector machines.
- **Training and validation:** QSAR models are developed using a training set of compounds with known activities. The model's performance is then validated using a separate set of compounds (validation set) that the model has not seen before. This helps assess how well the model can predict the activity of new compounds.
- **Model evaluation:** The performance of QSAR models is evaluated using metrics such as correlation coefficients, root mean square error, and cross-validation scores. A good QSAR model should accurately predict the activity of compounds not included in the training data.
- **Interpretation:** QSAR models can provide insights into the relationship between molecular structure and activity. By analyzing the model's coefficients or feature importance, researchers can identify which structural features contribute most to the observed activity.

https://doi.org/10.1515/9783111434858-004

- **Applicability domain:** QSAR models often define a region in the descriptor space where they are most accurate. This region is known as the applicability domain. Predictions falling within this domain are considered reliable, while those outside might be less trustworthy.

Quantitative models in QSAR help researchers prioritize compounds for further testing, design new molecules with desired properties, and make informed decisions about the potential risks or benefits of chemical substances [3]. The development of accurate and robust QSAR models requires careful consideration of data quality, model complexity, and validation methods [4].

As already discussed, the real dawn of the QSAR is considered in the early 1960s with the work of Hansch and Fujita. It was during this period that scientists started to correlate the biological activity of the compounds with their structural properties (now known as molecular descriptors) more rationally. There have been many methods or models presented by scientists after this period for the correlation between biological activity and molecular properties. The most important methods that still hold validity in the current era are discussed below in detail.

4.2 Hansch analysis

Hansch analysis, a cornerstone of QSAR studies, is a pioneering approach that elucidates the relationship between chemical structure and biological activity by integrating molecular properties and statistical modeling [5, 6]. Named after the chemist Corwin Hansch, this method revolutionized drug design and predictive toxicology [7]. At its core, Hansch analysis involves constructing linear regression models that correlate the biological activities of a set of compounds with their physicochemical properties or molecular descriptors. These descriptors encompass a range of attributes, including electronic, steric, and hydrophobic properties, which collectively characterize a molecule's structure [8]. The resulting quantitative equations reveal the quantitative contribution of individual descriptors to the observed activity, thereby uncovering structure–activity relationships.

Hansch analysis has proven particularly valuable in rational drug design, enabling researchers to tailor molecular structures to enhance or modify desired biological activities [9]. The approach also offers insights into the underlying mechanisms governing compound interactions with biological targets. While Hansch analysis originated as a linear model, its principles have been extended to nonlinear and more advanced methodologies, further enriching the predictive capabilities of QSAR models [10]. Overall, Hansch analysis serves as a foundational framework in QSAR, empowering researchers to decipher the intricate links between chemical structure and biological response. Its historical significance and ongoing relevance in contemporary drug discovery and toxicological assessments underscore its enduring importance in the field of computational chemistry [11].

In the initial development of QSAR, scientists tried to correlate the biological activity of drug molecules with either their lipophilicity parameters or with the electronic parameters, respectively, as in the following equations [12, 13]:

$$\log\left(\frac{1}{C}\right) = a\log P + b \tag{4.1}$$

$$\log\left(\frac{1}{C}\right) = k_1\sigma + k_2 \tag{4.2}$$

where P is the partition coefficient representing the lipophilicity of the drug molecule and σ is the Hammett constant affecting the electronic changes in the molecule due to different substituents. A, b, k_1, and k_2 are the constants. Both of the above equations were able to correlate the biological activity with either of these parameters locally but none applied to the broader range of molecules and other QSAR models for different biological activities.

It was the year 1962 when Hansch and Fujito provided a breakthrough in QSAR studies [14]. The main idea behind their concept was that the site of drug administration is different than its site of action [15, 16]. A drug has to pass through many phases to reach the site of action. Adsorption and desorption take place with lipid membranes and proteins, whereas partitioning of drug molecules occurs between the different liquid systems (Fig. 4.1). For any drug after reaching the site of action, interaction with the receptor is essential to generate any biological effect. This drug-receptor interaction is governed by the electronic parameters of the drug molecule. Hence, for defining any QSAR model, it is important to correlate both lipophilic and electronic parameters with biological activity. This gave birth to the famous model of the QSAR, popularly known as Hansch analysis. The very first forms of the equation given by Hansch are given below:

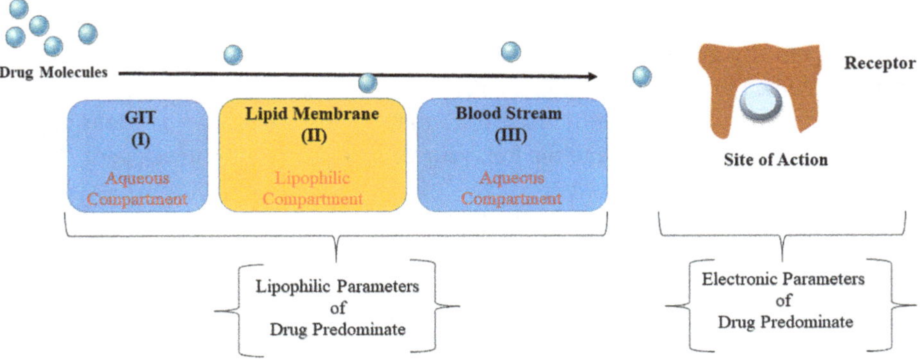

Fig. 4.1: Movement of the drug inside the body. Partition/lipophilic parameters of the drug govern its absorption whereas the electronic parameters determine its binding with the receptor.

$$\log\left(\frac{1}{C}\right) = k_1\pi + k_2\sigma + k_3 \tag{4.3}$$

$$\log\left(\frac{1}{C}\right) = -k_1\pi^2 + k_2\pi + k_3\sigma + k_4 \tag{4.4}$$

where π is the lipophilic/hydrophobic constant and σ represents the Hammett constant. Equation (4.3) is applicable only when the study is done on small range values of π; otherwise equation (4.4) is applicable when the study is done on broader range values of π.

Hansch analysis is also known as the "extra-thermodynamic approach" or "linear free energy-related approach" as the physicochemical parameters used in the Hansch analysis are derived from the rate or equilibrium constants.

Later on, another parameter, Taft's steric parameter (E_s) was included in the Hansch analysis. It quantifies the steric hindrance or spatial bulkiness of substituents attached to a molecule [12]. This parameter is named after the American chemist Robert W. Taft. The steric parameter is crucial in understanding how the three-dimensional arrangement of atoms affects a molecule's interactions with other molecules or biological receptors. In QSAR, it helps predict how changes in the size or shape of substituents on a molecule can influence its biological activity. The resulting modified equations are depicted in the following equations:

$$\log\left(\frac{1}{C}\right) = k_1\pi + k_2\sigma + k_3E_s + k_4 \tag{4.5}$$

$$\log\left(\frac{1}{C}\right) = -k_1\pi^2 + k_2\pi + k_3\sigma + k_4E_s + k_5 \tag{4.6}$$

Equation (4.5) is applicable when the study is done on small range values of π and equation (4.6) is applicable when the study is done on broader range values of π. The constants k_1 to k_5 used in equations (4.3)–(4.6) are derived from the multiple linear regressions (MLRs).

The interesting fact of the Hansch analysis is that it originated from the studies conducted on plants. The correlation between the plant growth (biological activity) with different lipophilicity and electronic parameters of substituted phenoxy acetic acid was studied to develop the initial equations of Hansch analysis [16]. The effect on the biological activity due to different substitutions at *meta* and *para* positions of phenoxy acetic acid was studied and the following QSAR equation was developed:

$$\log\left(\frac{1}{C}\right) = -2.14\ \pi^2 + 4.08\ \pi + 2.78\ \sigma + 3.36 \tag{4.7}$$

From the above equation, it is evident that plant growth is enhanced with an increase in the value of σ and also, initially, with the increase in the value of π (by keeping σ constant) plant growth increases. However, for a very high value of π the biological activity becomes 0. The values of parameters used in this QSAR model along with observed and calculated biological activity are given in Tab. 4.1.

Tab. 4.1: Calculation of biological activity (log(1/*C*) from the σ and π values of different substituents of substituted phenoxy acetic acid derivatives.

meta *para*

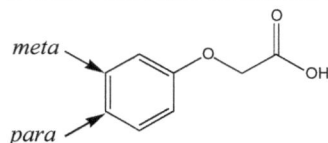

Substituted phenoxy accetic acid

S. no.	Substituent	σ-Value	π-Value	Observed log (1/*C*)	Calculated log (1/*C*)
1.	*m*-CF_3	0.55	1.09	6.8	6.5
2.	*p*-Cl	0.37	0.80	6.3	6.4
3.	*m*-I	0.28	1.08	6.1	6.3
4.	*p*-F	0.34	0.20	5.0	6.3
5.	*m*-Br	0.23	0.97	5.9	6.0
6.	*m*-SF_3	0.68	1.67	6.2	6.0
7.	*m*-Cl	0.23	0.82	5.9	5.7
8.	*m*-NO_2	0.78	0.04	5.7	5.3
9.	*m*-SCH_3	−0.05	0.59	4.9	5.3
10.	*m*-C_2H_5	−0.15	0.87	4.9	5.3
11.	*m*-*n*-C_3H_7	−0.15	1.52	4.2	4.7
12.	*m*-OCH_3	−0.27	0.13	3.1	4.7
13.	*m*-CN	0.63	−0.23	4.1	4.5
14.	*m*-CH_3	−0.17	0.44	4.3	4.3
15.	*m*-CH_3CO	0.52	−0.08	4.5	4.0
16.	*m*-F	0.06	0.18	4.2	3.5
17.	H	0.00	0.00	3.4	3.5
18.	*m*-OH	−0.36	−0.73	−1.8	3.7
19.	*m*-COOH	0.27	−0.13	3.5	3.0
20.	*m*-*n*-C_4H_9	−0.15	2.03	2.4	0.0

From the above table, a noteworthy point to be studied is that *meta*- and *para*-fluorine-substituted phenoxy acetic acid have remarkable differences in their biological activity (log(1/*C*)). Here in this case, the value of π is similar for both of the substituents and this change in the biological activity is attributed to the change in their value of σ at *para* and *meta* substitutions. The value of σ of fluorine at the *para* substitution is 0.34, whereas the same for the *meta* substitution is 0.06 producing a remarkable difference in the overall biological activity of the compound. This difference in the values of σ can be explained based on the electronic interactions as we all know fluorine exerts $a + R$ effect on the benzene ring and this effect is more predominant at the *para* position as compared to the *meta* position. The same explanation can be sought for the difference in the values of biological activity of *para*–chloro- and the *meta*-chloro-substitutions. From the above table, it is evident that the value of π increases the size of the atom as it is lowest for fluorine (0.20) and largest for iodine (1.08). The length of the alkyl group also affects the value of π as the increase in the length results in the higher values of π.

4.2.1 Importance of Hansch analysis in drug discovery

It can be concluded that Hansch analysis holds significant importance in QSAR-driven drug discovery due to its ability to elucidate the relationship between chemical structure and biological activity. This methodology provides valuable insights that empower drug researchers to design more effective and efficient compounds. Here are several reasons why Hansch analysis is crucial in QSAR for drug discovery:

- **Rational drug design:** Hansch analysis helps scientists understand how specific structural features of molecules contribute to their biological activities. This insight enables the rational design of new compounds with improved activity profiles by making targeted modifications to molecular structures.
- **Efficient compound optimization:** By identifying the key structural elements that impact activity, Hansch analysis guides researchers in optimizing molecules with minimal trial and error. This accelerates the drug discovery process by focusing efforts on compounds most likely to exhibit desired properties.
- **Scaffold modification:** The analysis aids in modifying existing molecular scaffolds to enhance desired activities. Substituent effects on biological interactions can be quantified, leading to informed choices about where to introduce modifications for optimal outcomes.
- **Lead compound selection:** Hansch analysis aids in selecting lead compounds with the highest likelihood of success. It allows researchers to prioritize compounds with favorable structural attributes, saving resources and time during the lead identification phase.
- **Mechanism of action understanding:** By quantifying the relationship between structural features and activity, Hansch's analysis can shed light on the mechanisms of action of bioactive compounds. This understanding contributes to the development of more targeted therapies.
- **Compound toxicity assessment:** The analysis not only predicts activity but also can offer insights into potential toxicity. This is vital for avoiding compounds with unfavorable safety profiles early in the drug discovery process.
- **Reduction of experimental efforts:** Hansch analysis reduces the need for extensive experimental testing of every compound by providing a computational framework for predicting activity. This is especially valuable when dealing with large compound libraries.
- **Cost efficiency:** By guiding compound design and prioritization, Hansch analysis contributes to cost-effective drug development. It minimizes the wastage of resources on compounds with poor potential.

- **Empirical model building:** While more advanced methods exist, Hansch analysis offers a straightforward approach to QSAR modeling that can serve as a foundation for more complex studies.

Overall, Hansch analysis aids in making informed decisions during drug discovery by linking molecular structure to biological activity. Its contributions to rational compound design, lead selection, and mechanism understanding are crucial for the efficient development of new pharmaceutical agents.

4.3 Craig plot

Craig plot is named after the scientist Paul N. Craig; it is a plot between the hydrophobicity substituent (π) and the Hammett constant (σ) (Fig. 4.2) [17]. In this plot σ value is plotted on the *y*-axis and the π values are plotted on the *x*-axis. Craig's plot provides us the advantage of the ease of choosing the right substituent. The Craig plot suggests that there is no linear relationship between the σ and π values of a substituent and there should be careful selection of the substituent based on these values in a rational drug design. The Craig plot is a graphical representation used in QSAR studies, particularly in the context

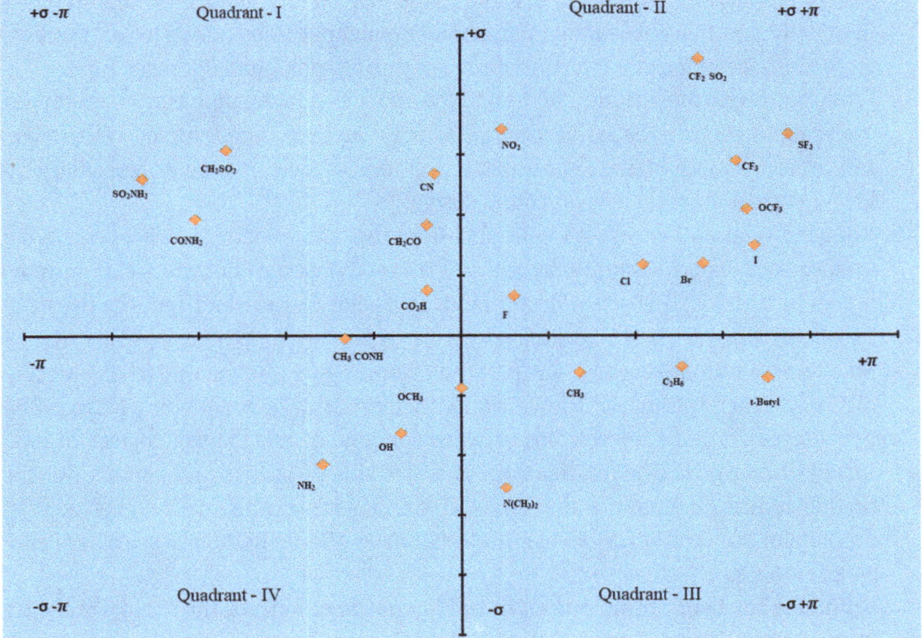

Fig. 4.2: The Craig plot for determining the values of lipophilic and electronic parameters of the drug. On the *x*-axis is given the π value and on the *y*-axis is given the σ value of the drug.

of drug design. It visually displays the relationship between the electronic and steric properties of substituents on a molecular scaffold and their corresponding effects on biological activity. The importance of the Craig plot in QSAR-based drug design lies in its ability to provide insights into the optimal structural modifications required to enhance a compound's activity.

4.3.1 Importance of Craig plot in drug discovery

- **Substituent optimization:** The Craig plot helps identify the optimal substituents or modifications that lead to enhanced biological activity. By plotting electronic and steric parameters on a graph, it becomes easier to pinpoint regions associated with higher activity. This guides researchers in selecting the most favorable substituents for further compound design. It allows the scientists to select the substituent with diverse properties based on the quadrant they are coming from (e.g., $-NO_2$ and $-CH_3$ groups lie in two different quadrants and are expected to have different properties). It also helps in replacing one similar substituent with another based on their closeness in the values on the Craig plot (e.g., $-OCF_3$ and –I groups are closely related as per the Craig plot).
- **Visualizing trade-offs:** The plot visualizes the balance between electronic and steric effects. It shows where specific regions of the plot correspond to favorable changes in activity and helps researchers understand potential trade-offs between these effects. This aids in making informed decisions during compound optimization.
- **Compound customization:** The Craig Plot provides a systematic way to customize molecular structures based on desired activity. By observing the plot, researchers can predict how changes to substituents will impact the compound's bioactivity, allowing for a more targeted compound design.
- **Reduced trial and error:** With insights from the Craig plot, researchers can avoid unnecessary experimentation by focusing on modifications likely to yield improvements in activity. This streamlines the drug design process and reduces the need for extensive synthesis and testing.
- **Mechanism understanding:** The plot can offer insights into the underlying mechanisms by which specific substituents affect activity. This understanding helps refine hypotheses about the interactions between compounds and their biological targets.
- **Efficient compound prioritization:** The plot aids in prioritizing compounds for further testing. Compounds that fall within regions associated with higher activity on the plot are more likely to exhibit the desired effects, helping researchers allocate resources effectively.
- **Guidance for lead optimization:** In lead optimization efforts, the Craig plot assists in making decisions about which modifications to pursue, leading to the development of compounds with improved activity profiles.

- **Visualization of structure–activity relationships:** The Craig plot condenses complex structure–activity relationships into a visual representation. This can aid in communicating findings to interdisciplinary teams and stakeholders.

The Craig plot serves as a valuable tool in QSAR-based drug design by providing a clear visualization of the interplay between electronic and steric effects. Its role in guiding compound customization, optimization, and decision-making makes it an important asset in the development of pharmaceutical agents with enhanced bioactivity.

4.4 Topliss scheme

The Topliss scheme, a fundamental concept in the field of drug design and medicinal chemistry, is an approach that streamlines the optimization of molecular structures to enhance their biological activity [18, 19]. Developed by Edward M. Topliss, this scheme offers a systematic and pragmatic method for modifying chemical compounds while maintaining or improving their desired pharmacological properties. The scheme is rooted in the principle of substituent similarity, where structurally related compounds are designed based on the premise that similar substituents will elicit comparable effects on the target receptor or biological system.

The essence of the Topliss scheme lies in its hierarchical framework, which categorizes substituents based on their impact on biological activity and potential structural changes. The scheme consists of three main levels: Topliss diamond, Topliss tree, and Topliss matrix. The Topliss diamond, the simplest level, groups substituents into classes based on their electronic and steric properties. The Topliss tree takes this further by branching out into specific chemical functionalities, while the Topliss matrix systematically explores combinations of substituents to design compounds with desired activities. The Topliss scheme offers several advantages in drug design. It aids in the identification of privileged scaffolds and substituents that contribute favorably to biological activity. It also guides researchers in optimizing compounds by emphasizing modifications that are most likely to lead to improvements in activity. This approach significantly reduces the trial-and-error nature of compound optimization, allowing for a more focused and efficient drug design process. Overall, the Topliss scheme serves as a valuable strategy in the development of pharmaceutical agents by providing a structured framework for compound modification and optimization. Its emphasis on systematic exploration and rational design makes it an indispensable tool in the pursuit of novel and potent therapeutic compounds.

Whenever a drug designing approach is opted for, the chief objective is to find a potent drug moiety through optimum substitutions. The Hansch analysis proposed by Hansch and Fujita in the QSAR analysis is one such rational approach. Although Hansch analysis proved to be a very effective approach in rational drug designing, the use of many mathematical and statistical calculations sometimes becomes a hindrance for the medicinal chemist in applying the Hansch analysis in the drug discovery process. To overcome this hindrance, a nonmathematical approach to the Hansch analysis was proposed by John Topliss in 1972, popularly known as Topliss scheme.

In search of an appropriate substituent that can be most potent in terms of biological activity, Topliss proposed two schemes, first for the aromatic substitutions (scheme I) and second for the side chain substitutions (scheme II). Both of the schemes are based on Hansch analysis but employ a nonmathematical approach.

4.4.1 Topliss scheme I

In Topliss scheme I, the search for a potent substituted aromatic drug molecule is initiated from an unsubstituted phenyl ring, and its biological activity is measured. The effect of different substitutions on the biological activity is observed and related to their values of π, σ, and E_S qualitatively. The very basic concept of the Topliss scheme is that in most cases the biological activity increases with an increase in the lipophilicity (π value) character. Starting from the unsubstituted phenyl ring, we move up to 4-Cl substitution. The reason for taking the chloro-substitution is the ease of synthesis and increasing order of π value ($\pi = 0.70$ and $\sigma = 0.23$). There will be three possibilities for the potency of the 4-Cl-substituted compound: it could be more potent, it could be equipotent, or it could be less potent than the unsubstituted compound (Fig. 4.3). Let us discuss each case to understand the concept of the Topliss scheme:

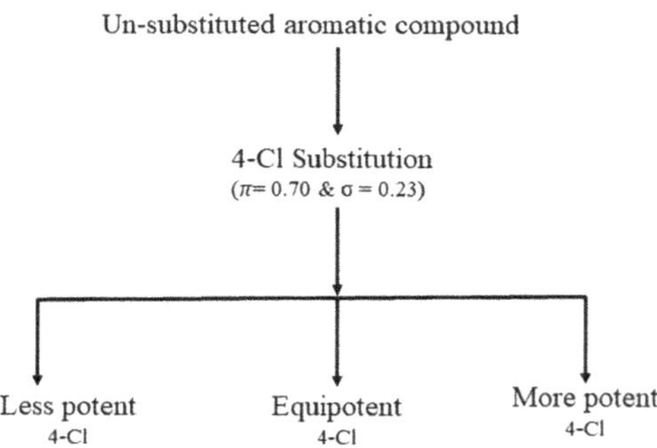

Fig. 4.3: Topliss scheme I–part I; conditions when the unsubstituted aromatic compound is substituted with the chlorine atom.

4.4.1.1 4-Chloro-substituent is more potent

This is the most favorable outcome of the substitution. This increased biological activity can be due to the increased value of either π or σ, or both. Now the choice of the next analog is the incorporation of a 3-Cl substituent ($\pi = 0.76$ and $\sigma = 0.37$). This will result in the further enhancement of the π and σ characters of the new substituted compound. Here too, the new 3,4-dichloro-substituted compound can either be more potent, less potent, or equipotent as compared to the 4-Cl analog (Fig. 4.4). The increased potency will be attributed to the increased total lipophilic (π = 0.70 + 0.76) and electronic parameters (σ = 0.23 + 0.37). The choice of next analog will be 3-CF_3 (π = 1.21 and σ = 0.43) and 4-Cl bisubstituted compound and will be expected to have increased potency as it will further enhance the values of π and σ characters. The last choice in this series will be 3-CF_3 and 4NO_2 (π = 0.24 and σ = 0.78) substituted for a more potent compound.

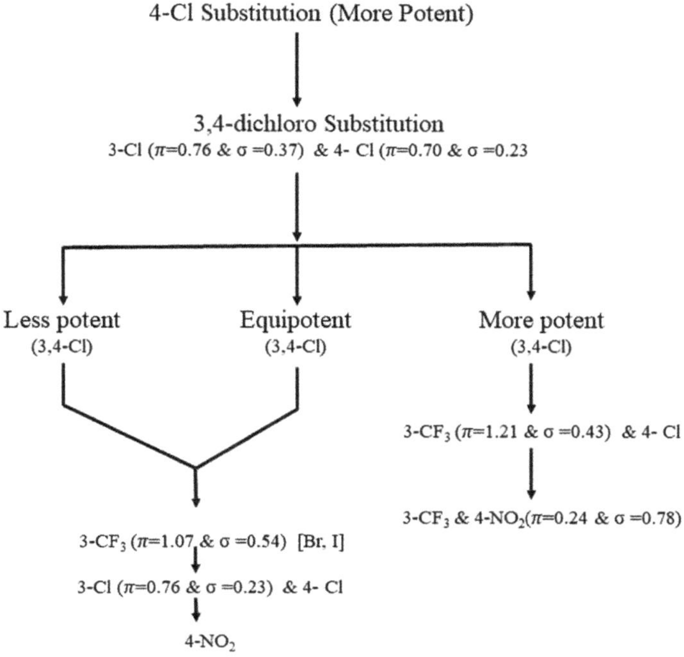

Fig. 4.4: Topliss scheme I–part II; conditions when the 4-Cl-substituted aromatic ring is substituted with another chlorine atom.

In the other scenario, if the 3,4-dichloro-substituted compound is less or equipotent, then it can be attributed to increased steric hindrance due to *meta* substitution or increased lipophilic character than the optimum value. This can be countered by removing the *meta* substitution and replacing the 4-Cl with 4-CF_3 as it will result in no steric hindrance at *meta* substitution and more lipophilicity compared to 4-Cl and less than 3,4-dichloro-substituted compound. Further potency of 2,4-dichloro-substitution can also be assessed if the decreased activity is due to the steric hindrance at the *meta* position. The last analog in the series will be the 4-NO_2 analog to check whether the potency is dependent on the value of σ as this substituent has a high σ value.

4.4.1.2 4-Chloro-substituent is equipotent

Now coming back to the initial 4-Cl substituent it came out to be equipotent to the previous unsubstituted analog. This can be explained in a way that for good potency, $+\pi$ and $-\sigma$ values are required. Therefore, to enhance the activity, such substituent is proposed that fulfills this criterion. 4-CH_3 ($\pi = 0.60$ and $\sigma = -0.17$) substitution is the best choice for this reason (Fig. 4.5). If this assumption is proved to be right then the next analog could be 4-$C(CH_3)_3$ ($\pi = 1.68$ and $\sigma = -0.20$). However, if due to the steric hindrance of such a bulky group at the *para* position results in a decrease of potency then further 3-CH_3 ($\pi = 0.54$ and $\sigma = -0.07$) and 4-CH_3 substituent can be synthesized.

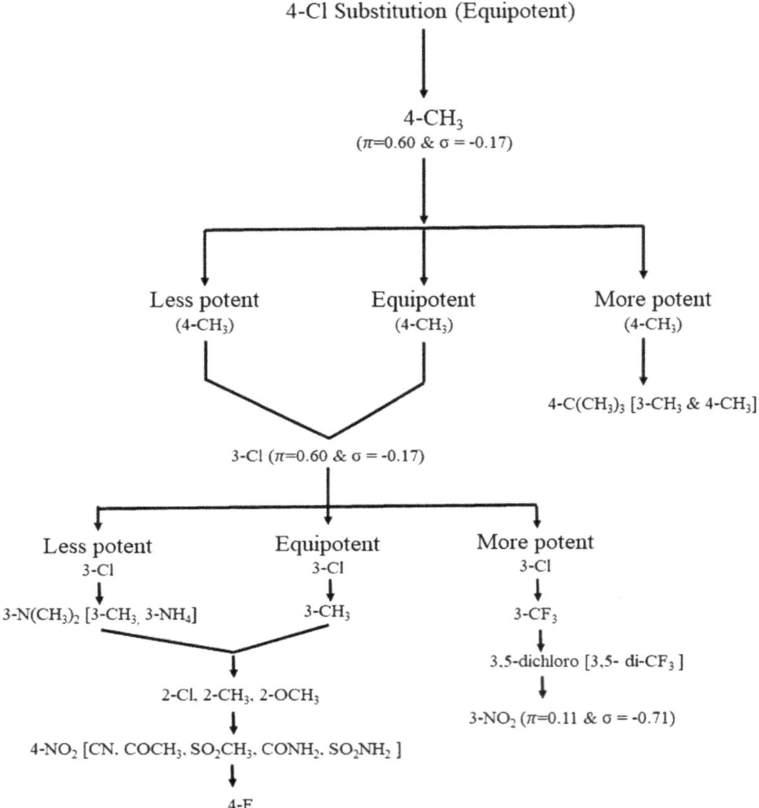

Fig. 4.5: Topliss scheme I–part III; conditions when the 4-Cl-substituted aromatic ring is substituted with a methyl group.

Taking another case, if the 4-CH_3-substituted compound is less or equipotent than the previous 4-Cl analog, then this decrease or no change in the potency can be explained due to the steric hindrance at the *para* position and decrease in the lipophilic (π) character. The steric hindrance, thus, can be countered by introducing the next analog which can be 3-Cl-substituted, and if the potency increases then further analogues can be synthesized with 3-CF_3 and further with 3,5-dichloro and 3-NO_2 substituents.

On the other hand, if potency remains the same with 3-Cl substitution then the next choice of substitution will be 3-CH_3. If no success is achieved still, then the substitution is shifted to the second position (Cl, CH_3, or OCH_3). If no success is achieved still, then it can be explained based on an increased σ is required for enhancing potency with low π value. This can be achieved by analyzing the 4-NO_2 substitution. 4-F substitution in this scheme can also be considered if there is not much change is required in the π and σ values. Coming back to the initial situation, if 3-Cl substitution results in decreasing the potency, then it can be ascribed to the requirement of $-\sigma$ value and hence 3-$N(CH_3)_2$ or 2-NH_2 along with 3-CH_3 are the choice of substitutions. The remaining choice of analogs is similar as discussed above and in Fig. 4.4.

4.4.1.3 4-Chloro-substituent is less potent

Decreased potency in this case can be explained based on steric hindrance at the *para* position. The potency can be increased by $-\pi$ and $-\sigma$ values. Out of these, the $-\sigma$ value is the most probable explanation and can be countered by changing the substitution to 4-OCH_3 ($\pi = -0.04$ and $\sigma = -0.27$). If the potency enhances from this, then further analogs (4-$N(CH_3)_2$) or 3-CH_3, or 4-$N(CH_3)_2$) can be considered which reinforces the $-\sigma$ value further (Fig. 4.6). If no change or decrease in the potency is observed by 4-$N(CH_3)_2$ then further analogs such as 4-NH_2, 4-OH, 3-CH_3, or 4-OCH_3 should be considered.

If 4-OCH_3 substitution results in a decrease or no change in the potency, there may be some unfavorable reasons for decreased or equipotent at *para* substitution. Hence, 3-Cl substitution is the next choice of analog for analyzing the potency, and from 3-Cl further analogs are synthesized as mentioned in Figs. 4.4 and 4.5. The complete Topliss scheme I is given in Fig. 4.7.

4.4.2 Topliss scheme II

Topliss scheme II is used for the side chain substituent represented by –COR, NHR, –CONHR, –NHCOR, and so on, where R is the substituent. It must be noted that these substituents are not directly attached to the aromatic ring but are in the side chain. In scheme II, we start with our methyl-substituted analog as our base compound. The potency is measured of the analog by substituting $-CH_3$ with the isopropyl (i-C_3H_7) ($\pi = 1.30$, $\sigma^* = -0.19$, and $E_s = -0.47$) group (Fig. 4.8). Here too, after substitution one of the three conditions may arise as discussed below:

4.4.2.1 Isopropyl-substituted compound is more potent

If the i-C_3H_7-substituted compound is found to be more potent, this can be attributed to its enhanced π value. To increase the potency further, cyclopentyl ($\pi = 2.14$, $\sigma^* = -0.20$,

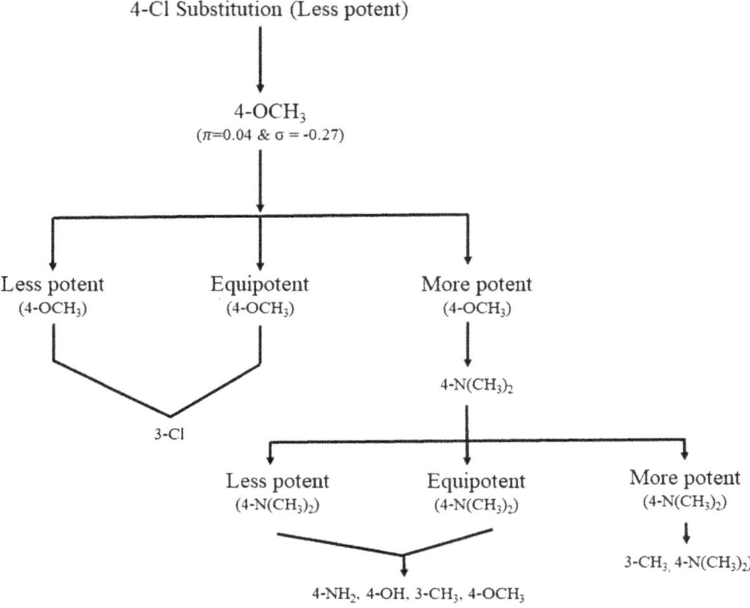

Fig. 4.6: Topliss scheme I–part IV: conditions when the 4-Cl-substituted aromatic ring is substituted with a methoxy group.

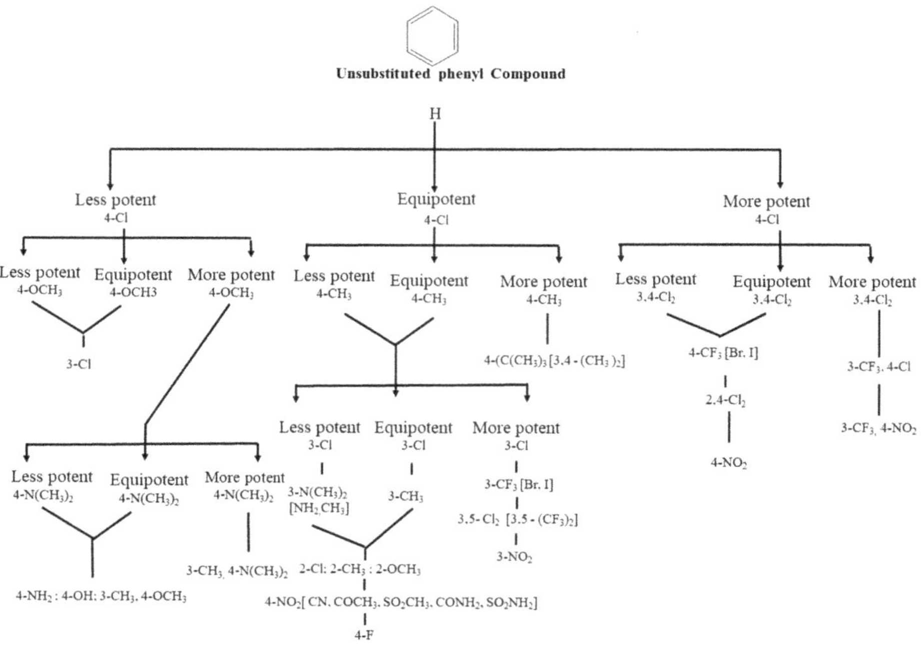

Fig. 4.7: The complete diagram of Topliss scheme I.

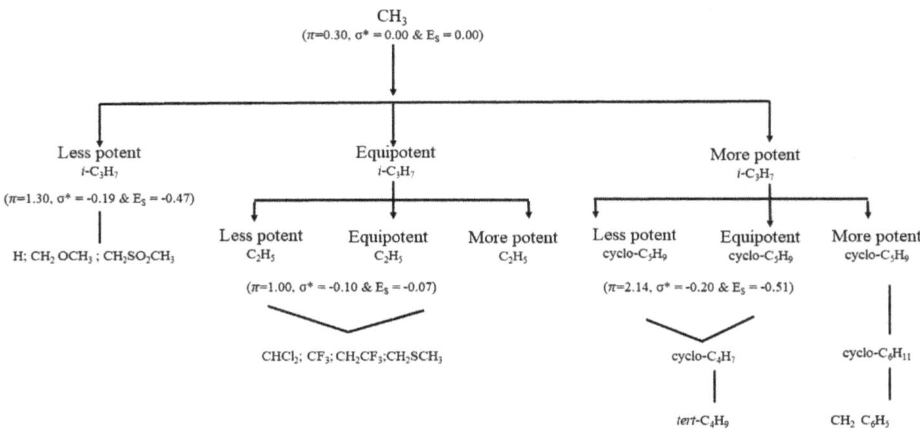

Fig. 4.8: The complete diagram of Topliss scheme II.

and $E_S = -0.51$) substitution will be the next choice as it will increase the π value with a minimum change in the undesired E_S (steric) parameter. An enhanced potency will prompt further evaluation of the potency of cyclohexyl-, benzyl-, and phenyl-substituted compounds.

However, if the cyclopentyl substitution does not enhance the potency, it can be attributed to two reasons. Firstly, π value has increased from its threshold value, and therefore cyclobutyl ($\pi = 1.80$, $\sigma^* = -0.20$ and $E_S = -0.06$) substitution that has a lower π value and also has a low E_S value is recommended. Secondly, biological activity may be inversely related to the σ^* value. Hence, in such case, *tert*-butyl ($\pi = 1.98$, $\sigma^* = -0.30$, and $E_S = -1.54$) substitution is preferred.

4.4.2.2 Isopropyl-substituted compound is equipotent

If the *i*-C_3H_7 substitution does not bring any change in the potency, it suggests that a substituent with less π value is required. To counter the same, substitution with the $-C_2H_5$ group should be checked ($\pi = 1.00$, $\sigma^* = -0.10$, and $E_S = -0.07$). However, after ethyl substitution, if there is a decrease or no change in the biological activity then the search for the further potent compound should be shifted towards the substituent having $+\sigma^*$ values ($CHCl_2$ $\sigma^* = 1.92$, CF_3 $\sigma^* = 2.76$, and CH_2CF_3 $\sigma^* = 0.92$).

4.4.2.3 Isopropyl-substituted compound is less potent

A loss in biological activity with the substitution of *i*-C_3H_7 suggests that either the value of π or σ^* or both are not correlated well with the potency. Hence, in such cases, substitutions with only hydrogen or methoxymethyl, or methylsulfonylmethyl should be investigated.

4.5 Free-Wilson analysis

Free-Wilson analysis (FWA) was proposed by Spencer M. Free Jr. and James W. Wilson in 1964 [20]. FWA stands as a pivotal technique in the realm of drug design, offering a systematic approach to unraveling the intricate relationship between the molecular structure of compounds and their biological activity [21]. This whole model is based on the concept of "additivity" in a series of analogs. This was the first model that correlates the structural features of a compound with its potency, which is totally in contrast with the Hansch analysis where biological activity was correlated with the physicochemical parameters. This is a very effective approach in the initial stage of lead structure optimization. This method, named after the chemist Richard M. Free and the statistician R. A. Wilson, has become an indispensable tool for pharmaceutical researchers aiming to optimize the activity of drug candidates.

The core premise of FWA lies in the strategic modification of specific molecular fragments within a compound while keeping the rest of the molecule constant. By systematically substituting or altering individual fragments, often referred to as "side chains," researchers can decipher the impact of each modification on the compound's overall activity. These fragment-specific activity variations are then analyzed statistically to elucidate the quantitative relationships between different structural features and the observed biological effects.

4.5.1 The classical Free-Wilson model

The approach adopted by Free and Wilson is totally different from that used by Hansch [20]. They proposed that in a series of molecules having a common base compound, the biological activity (BA) will depend on the individual contribution of each substituent and the contribution of all the substituents is additive in nature. The very basic equation of this analysis is as follows:

$$\mathrm{BA} = \Sigma\, a_i + \mu \tag{4.8}$$

where a_i is the contribution of the substituent X_i and μ represents the average of the biological activities of all the compounds measured.

Let us understand the Free-Wilson analysis with a hypothetical model. Let us consider that a simple phenol molecule possesses a certain biological activity and suppose we designed four analogs of phenol that can have two different substitutions, H and CH_3, at *para* and two different substitutions, NH_2 and OCH_3, at the *meta* position. Each of the four analogs will have certain biological activity, which is given in Tab. 4.2.

From the above table, it is understood that the hypothetical biological activity value of the compound with –H and $–NH_2$ substitution is 1.6, with $–CH_3$ and $–NH_2$ substitution is 2.0, with –H and $–OCH_3$ substitution is 2.6, and with $–CH_3$ and $–OCH_3$ substitution is 3.0. Therefore, the average value of the biological activity will be $\mu = 2.3$

Tab. 4.2: A hypothetical model for understanding the Free-Wilson analysis with values of biological activities due to substitutions at different positions.

OH

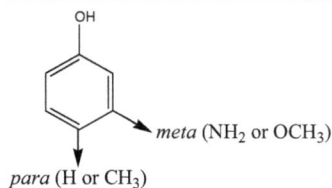

meta (NH_2 or OCH_3)

para (H or CH_3)

	H	CH_3	Average
NH_2	1.6	2.0	1.8
OCH_3	2.6	3.0	2.8
Average	2.1	2.5	$\mu = 2.3$

[(1.6 + 2.0 + 2.6 + 3.0)/4]. From here, we can now calculate the individual contributions of each substituent toward the biological activity of the molecule. The contribution of –H substituent will be the difference between average biological activity value of –H substituent and the overall mean of the biological activity (μ), that is, 2.1–2.3 = −0.2. Similarly, the contribution of $-CH_3$ will be 0.2, the contribution of $-NH_2$ will be −0.5, and that of $-OCH_3$ will be 0.5. Now coming back to the equation of the Free-Wilson model, let us calculate the biological activity of one of the above analogs, say 3-aminophenol (see equation (4.9)):

$$\text{Biological activity} = \text{average}\ (\mu) + \text{effect of } \textit{meta} \text{ substituent} + \text{effect of } \textit{para} \text{ substituent}$$

Putting our values in the above equation (average = 2.3, *meta*-NH_2 = −0.5, and *para*-H = −0.2), the calculated biological activity of 3-aminophenol is 1.6, which is the same as mentioned in Tab. 4.2.

Now the question comes of how this model can be applicable in optimizing the lead structure in drug design. To answer this, let us go back to the above example once again, but this time let us change the scenario that we have the biological activity of only three compounds and the biological activity of one compound, say 4-amino-3-methylphenol is unknown (*X*) (Tab. 4.3). The interesting thing is that by applying the Free-Wilson approach we can calculate the biological activity of the unknown, that is, of *X*.

The values of the contribution of each substituent can be calculated as follows:

$$\text{Contribution of H: } (2.1) - (X+7.2)/4 = (1.2-X)/4$$
$$\text{Contribution of CH}_3\text{: } (X+3)/2 - (X-7.2)/4 = (X-1.2)/4$$
$$\text{Contribution of NH}_2\text{: } (X+1.6)/2 - (X-7.2)/4 = (X-4)/4$$
$$\text{Contribution of OCH}_3\text{: } (2.8) - (X+7.2)/4 = (4-X)/4$$

Tab. 4.3: Determining the biological activity of unknown compound using Free-Wilson analysis.

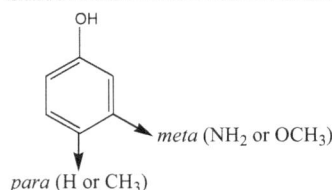

OH

meta (NH_2 or OCH_3)

para (H or CH_3)

	H	CH_3	Average
NH_2	1.6	X	$(1.6 + X)/2$
OCH_3	2.6	3.0	2.8
Average	2.1	$(X + 3)/2$	$\mu = (X + 7.2)/4$

To calculate the value of the biological activity of X of 4-amino-3-methylphenol we have to plug in the values of μ, the contribution of CH_3 and NH_2 groups in the following equation:

$$X = (X + 7.2)/4 + (X - 1.2)/4 + (X - 4)/4 \tag{4.9}$$

By solving the above equation we get the biological activity of the unknown compound as $X = 2.0$.

Hence, from the above example it is clear that by applying FWA, we can calculate the biological values of unknown compounds in a series of compounds. Consider a series of compounds, if substitution is done at two positions, at R_1 and R_2, m numbers of the substituent are proposed at R_1, and n numbers of substituents are proposed at R_2. According to FWA, the minimum number of compounds required to develop the model will be $m + n - 1$. The maximum number of compounds that can be synthesized will be $m \times n$.

4.5.1.1 Limitations of the classical Free-Wilson model

1. The very first limitation of the FWA is that all the molecules in a series should have the same parent structure without any variations.
2. At least substitutions at two different positions in the parent molecule are required for developing a Free-Wilson model. A single-place substitution cannot be studied through this model.
3. Another limitation of the model is that one substitution should not have any influence on the other substitution; otherwise the concept of additivity does not hold well.
4. Prediction of biological activity is possible for only a small number of molecules included in the analysis. The prediction of biological activity is impossible for the molecules that are not included in the analysis.

5. There is no rational explanation behind why one substituent shows a positive effect on biological activity and the other shows a negative effect on biological activity.

The significance of FWA in drug design is manifold. It empowers researchers to customize and fine-tune compound structures with surgical precision, enabling the optimization of desirable properties while minimizing unwanted side effects. Moreover, FWA contributes to a deeper understanding of the mechanisms underlying compound–target interactions, facilitating the design of molecules that effectively interact with specific biological pathways. As drug discovery becomes increasingly intricate, FWA serves as a valuable tool for rational decision-making. By unveiling the individual contributions of molecular fragments to a compound's activity, it offers a roadmap to designing novel therapeutic agents with enhanced efficacy and reduced risks. This analytical approach exemplifies the synergy between chemistry and statistics, underscoring its essential role in modern drug design endeavors.

4.5.2 The Fujita-Ban model (modified classical Free-Wilson model)

The Fujita-Ban model represents a significant advancement in the understanding of molecular interactions and their impact on biological activity [22]. Named after its developers, Japanese chemists Osamu Fujita and Naoto Ban, this model has left a lasting imprint on drug design and computational chemistry. At its core, the Fujita-Ban model delves into the intricate interplay between molecular shapes and their corresponding biological activities. It acknowledges that a molecule's biological response is not solely determined by chemical structure but also influenced by the three-dimensional arrangement of atoms within it. This pioneering perspective takes into account the spatial dimensions and orientation of molecules, allowing for a more nuanced understanding of structure–activity relationships. The model introduces the concept of "shape" as a crucial determinant of molecular interactions with biological receptors. By quantifying molecular shapes and comparing them to known bioactive molecules, the Fujita-Ban Model predicts the potential of untested compounds to elicit specific biological responses. This provides drug researchers with a tool to identify compounds with a higher likelihood of success and optimize molecular structures for desired activities.

The Fujita-Ban model is a slight modification of the Free-Wilson model [23]. Fujito and Ban proposed that the effect of a certain substitution at a position is constant and additive in nature. Based on this assumption they proposed the following equation:

$$\log\left(\frac{A}{A_o}\right) = \Sigma G_i X_i \tag{4.10}$$

where A is the biological activity of the substituted analog, A_0 represents the biological activity of the unsubstituted analog, and G_i is the log of increased or decreased activity of ith substitution when compared to H. X_i is the parameter that represents the value of 1 (if a substituent is present) or 0 (if a substituent is not present). In a series of compounds, the value of log A_0 will always be constant and hence the above equation can be modified as follows:

$$\log A = \Sigma G_i X_i + c \tag{4.11}$$

This equation is also called the modified FWA as the above equation closely resembles the Free-Wilson model with the only difference being that instead of using μ (the average of biological activity of the analogs) the biological activity of the unsubstituted analog is considered, which is more rational.

4.5.2.1 Advantages of the Fujita-Ban model

- The biggest advantage of the Fujita-Ban model is that any of the compounds in the series can be considered as the reference compound whose activity can be considered constant.
- There is comparative ease in generating regression equations in the Fujita-Ban model when compared to the classical Free-Wilson model of analysis.
- It is also possible to add or remove any element from the study that will not affect the quality of the overall equation.
- The substituents that occur together can be considered as one single substituent in the study.

In an era where computational techniques are instrumental in drug discovery, the Fujita-Ban model offers a holistic view of how molecular shapes contribute to biological outcomes. Its insights guide the development of compounds that fit optimally into target binding sites, thus facilitating the creation of more effective and selective pharmaceutical agents. As the paradigm continues to evolve, the Fujita-Ban model remains a cornerstone in the pursuit of improved drug design strategies that capitalize on the geometry of molecules for enhanced therapeutic outcomes.

4.5.3 The relationship between Free-Wilson Analysis (FWA) and Hansch analysis (mixed approach)

FWA and Hansch analysis are both quantitative approaches used in drug design within the field of QSAR studies. They share similarities in their goal of understanding how specific molecular features influence a compound's biological activity, but they differ in their focus and methodology. Hansch analysis focuses on establishing a quantitative relationship between a set of molecular descriptors (often related to electronic, steric, and hydrophobic properties) and biological activity. It uses statistical techniques, such as linear regression, to create a predictive model that relates these descriptors to the observed activity. Hansch analysis is concerned with the overall contribution of various descriptors to the activity of a molecule. It provides insights into the general trends and relationships between structural properties and activity across a set of compounds. FWA, on the other hand, is a more specific approach that focuses on studying the contribution of individual substituents or chemical groups to a molecule's activity. In this method, a parent molecule's activity is compared to the activities of its various derivatives, where each derivative has one specific substituent altered. This allows for a granular understanding of the impact of different substituents on activity. FWA is particularly useful when exploring how variations in a specific part of a molecule influence activity.

While Hansch analysis considers multiple molecular descriptors collectively and their impact on activity across a range of compounds, FWA delves into the individual contributions of substituents within a particular compound. The relationship between these two methods lies in their shared goal of uncovering the structure–activity relationship, but they operate at different levels of detail. In practice, FWA can be seen as a subset of Hansch analysis. The insights gained from FWA can contribute to the development and refinement of the molecular descriptors used in Hansch analysis. By understanding how individual substituents impact activity, one can better select and prioritize the descriptors used in the broader Hansch model. Hansch analysis provides a broader overview of structure–activity relationships across different compounds, while FWA offers a more focused examination of the contributions of specific substituents within a single compound. Both approaches have their place in drug design QSAR, and they can complement each other in uncovering the complexities of molecular interactions and aiding in the optimization of compounds with desired properties.

One major relationship between both the models is that if physicochemical parameters in a linear Hansch equation are constitutive and additive in nature then the values of contributions of each group in the Free-Wilson model can be calculated from the Hansch equation. This can be done with the help of the following equations:

Suppose there is a linear Hansch equation as follows:

$$\log\left(\frac{1}{C}\right) = -k_1\pi + k_2\sigma + k_3E_s + k_4 \tag{4.12}$$

where π is the hydrophobicity parameter, σ is the Hammett electronic parameter, E_s is Taft's steric factor, and k_1 to k_4 are the constants. If all the physicochemical parameters are additive in nature then this equation can be rewritten as follows:

$$\log\left(\frac{1}{C}\right) = \Sigma k_j\Phi_{ij} + C \tag{4.13}$$

where Φ_{ij} is the addition value of all the physicochemical parameters in the model. Now, the value of the substituent contribution (a_i) used in the Free-Wilson model can be calculated by the following equation:

$$a_i = \Sigma k_j \Phi_{ij} \tag{4.14}$$

This relation between the Hansch analysis and the Free-Wilson model was proposed for the first time by Singer and Purcell. This close relationship between both models holds well only if we use the Fujita-Ban model instead of the classical Free-Wilson model. It had also been observed that in certain cases, physicochemical parameters alone were not sufficient to explain the biological activity of the molecules in Hansch analysis. This problem was solved by Hansch again by an "*extra-thermodynamic approach assisted by the Free Wilson method*", popularly known as the mixed approach. In this approach, both the models, Hansch and Free-Wilson models, are combined in one single equation as follows:

$$\log\left(\frac{1}{C}\right) = \Sigma a_i + \Sigma k_j \Phi_{ij} + C \tag{4.15}$$

where Φ_{ij} represents the physicochemical parameters and a_i represents the Free-Wilson type parameters that cannot be parameterized in a physicochemical manner.

4.6 Development of a 2D-QSAR model

We have already discussed that in the initial development phase of QSAR, the biological activity is correlated either with the physicochemical parameters (Hansch analysis) or the contribution of substituents (FMA), or by a mixed approach (combination of both). But over time as the understanding of QSAR increased, there has been a development of more parameters that can be correlated with the biological activity. Topological descriptors are one such parameter that developed in the later stages and the QSAR model generated from correlating the biological activity with these topological descriptors is named as 2D-QSAR models. These models are named 2D-QSAR models as the topological descriptors are derived from the "two-dimensional structures of the molecules."

4.6.1 Topological descriptors

Topological descriptors capture key structural information of molecules using graph theory, focusing on the connectivity and arrangement of atoms and bonds in a two-dimensional representation. The resulting descriptors provide valuable insights into how different molecular features influence a compound's biological activity, aiding in the rational design of new pharmaceutical agents. Topological descriptors are numerical values that quantitatively represent structural aspects of molecules without considering spatial arrangement. They capture information about atom types, bond counts, branching patterns, and cycles in the molecule's chemical structure. These descriptors encode essential structural characteristics that contribute to the interaction of molecules with biological targets. Topological descriptors provide a simplified yet informative representation of molecular structure, enabling the identification of relevant structural motifs associated with specific biological activities. They help uncover trends and relationships between molecular features and observed activity across a dataset of compounds. This information is invaluable for predicting the bioactivity of new compounds, prioritizing molecules for further testing, and optimizing lead compounds. Molecular connectivity indices and atom pair (AP) descriptors are among the two most widely employed topological descriptors for generating a 2D-QSAR model. The most popular topological descriptors used in the 2D-QSAR model generation are given in Tab. 4.4.

Tab. 4.4: Popular topological descriptors employed for the generation of a 2D-QSAR model.

S. no.	Type of descriptor	Description
1	Weiner index (Weiner number)	Oldest and defined as the summation of all the bonds linking all pairs of atoms in the molecule
2	Molecular connectivity indices	Most widely accepted descriptors and calculated from the H-depleted molecular graph
3	2D autocorrelation descriptors	Based on the distance between the atom pairs in terms of number of bonds
4	Eigenvalue-based descriptors	Connectivity and atomic properties are taken into consideration for their calculation

By encapsulating vital structural information in a simplified form, topological descriptors contribute to the systematic analysis of structure–activity relationships and aid in the design of novel compounds with enhanced bioactivity. While topological descriptors offer valuable insights, they might not capture all relevant information in complex biological systems. Advances in machine learning techniques and the integration of other types of descriptors (e.g., physicochemical, 3D) enhance the predictive power of 2D-QSAR models.

4.7 Deriving a 2D-QSAR model

A complete workflow for deriving the 2D-QSAR model is given in Fig. 4.9 and the steps involved in deriving the same are illustrated below:

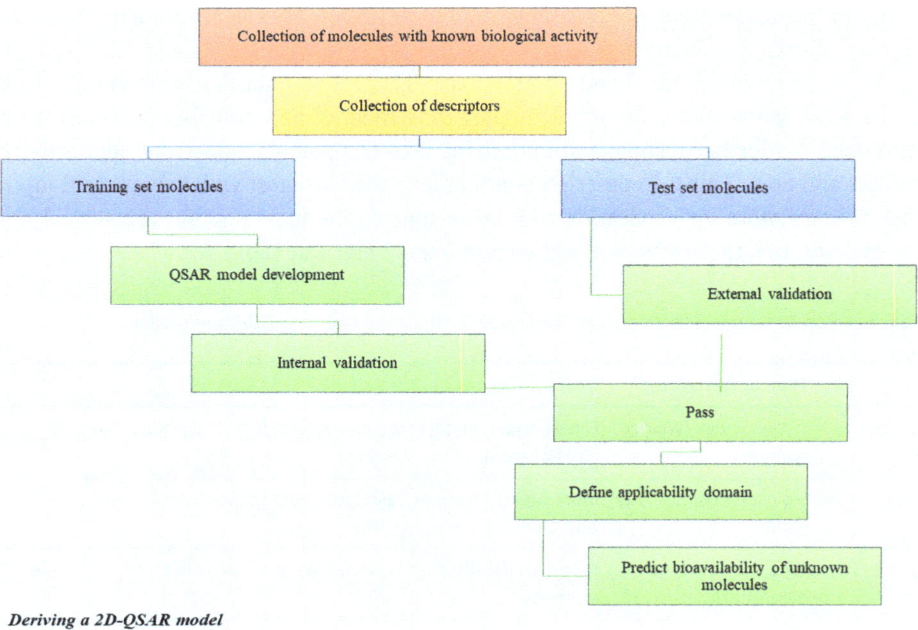

Fig. 4.9: Basic flow diagram of the basic steps involved in the development of a 2D-QSAR model.

a. **Collection of dataset:** The QSAR model is the correlation between the biological activity and the descriptors of the molecules; therefore, in the very first step, molecules whose biological activity is known against a particular target are selected from the literature. These molecules are collectively known as datasets of molecules.
b. **Calculation of descriptors:** This is the second step in the 2D-QSAR model generation. All the two-dimensional molecular descriptors of the compounds, mostly topological descriptors, are calculated. This is done with the help of various open-access software and the most commonly employed software for this purpose include PaDEL, CODESSA, and MOE.
c. **Data pretreatment**: In this step, all the constant and intercorrelated descriptors of the dataset calculated in the previous step are removed. Both constant and intercorrelated descriptors hinder the quality of the generated model and hence it is a prerequisite to remove them before moving further for the QSAR model generation.
d. **Division of dataset:** The dataset of molecules that is collected for the development of the model is randomly divided into two parts, the training dataset, and the test

dataset. The training dataset is employed for the generation of the 2D-QSAR model and the test dataset is employed for the external validation of the generated model. In general practice, 70% of the molecules are used in the training dataset, and the remaining 30% are employed for the test dataset.

e. **2D-QSAR model generation**: After the selection of training and test datasets, a correlation is developed between the biological activity and the molecular descriptors in the form of the following mathematical equation:

$$\text{Biological activity} = k_1D_1 + k_2D_2 + k_3D_3 + \cdots + k_nD_n + C \tag{4.16}$$

where D_1 to D_n are the values of descriptors and k_1 to k_n and C represent the constant values. There are different statistical methods available for the generation of this 2D-QSAR model. The most common method employed in the generation of a 2D-QSAR equation is MLR.

f. **2D-QSAR model validation:** The reliability and prediction ability of the generated 2D-QSAR model is validated in this step. Validation of the 2D-QSAR model is done in two phases:
   i. **Internal validation:** Internal validation is performed on the QSAR equation with the parameters of the training dataset. The most common parameter used is the squared correlation coefficient (R^2) value of the training set. A value of $R^2 > 0.7$ is considered acceptable for the generated QSAR model. Leave-one-out cross-validation method is also employed for the internal validation of the model. Parameter Q^2_{cv} is calculated in this validation with the following equation:

$$Q^2_{cv} = 1 - \frac{\Sigma\,(\text{BA} - \text{BA}_{\text{pred}})^2}{\Sigma\,(\text{BA} - \text{BA}_{\text{mean}})^2} \tag{4.17}$$

   where BA is the observed value of the biological activity, BA_{pred} is the value of biological activity predicted by the model and BA_{mean} is the average of all the biological activities of the training set molecules.
   ii. **External validation**: The quality and robustness of the generated 2D-QSAR model are assessed externally with the help of the test dataset. The most common parameter for assessing the reliability of the generated model is the squared correlation coefficient (R^2_{test}) value of the test set should be greater than 0.6. Another parameter, mean absolute error (MAE) proposed by Kunal Roy is used for external validation of the QSAR model as in the following equation:

$$\text{MAE} = \Sigma\,\frac{(\text{BA}_{\text{obs}} - \text{BA}_{\text{pred}})}{n} \tag{4.18}$$

   where BA_{obs} is the observed value of the biological activity and BA_{pred} is the predicted value of the same by the model of the test set compound, and n is the number of the compounds in the test dataset. A QSAR model, which should not be considered bad, should have the value of $MAE_{95\%}$ less than 0.15, which is calculated after omitting 5% of high residual data.

g. **Defining the applicability domain (AD)**: As per the third principle of the Organization for Economic Co-operation and Development (OECD) it is a must to define the AD for prediction of BA of the new molecules. AD is defined as the chemical structure space in which the developed 2D-QSAR model predicts with the highest accuracy and precision.

Over the last seven decades, we have seen great strides in the development of the QSAR, which is still considered as the most powerful tool of ligand-based drug designing. Although now we see many sophisticated QSAR techniques such as CoMFA, CoMSIA, 4D- or 5D-QSAR, one has to agree that the core concept of the QSAR still lies in the Hansch analysis and Free-Wilson model developed in the sixth decade of the twentieth century. This is the chief reason why Corwin Hansch is considered the father of the modern QSAR methodology.

References

[1] Selassie CD, Mekapati SB, Verma RP. QSAR: Then and now. Current Topics in Medicinal Chemistry, 2002, 2(12), 1357–79.
[2] Du QS, Huang RB, Chou KC. Recent advances in QSAR and their applications in predicting the activities of chemical molecules, peptides and proteins for drug design. Current Protein and Peptide Science, 2008, 9(3), 248–59.
[3] Timmerman H. QSAR and Drug Design: New Developments and Applications. Elsevier, 1995.
[4] Cherkasov A, Muratov EN, Fourches D, et al. QSAR modeling: Where have you been? Where are you going to? Journal of Medicinal Chemistry, 2014, 57(12), 4977–5010.
[5] Kubinyi H. QSAR: Hansch Analysis and Related Approaches. Weinheim, VcH, 1993.
[6] Martin YC. Hansch analysis 50 years on. Wiley Interdisciplinary Reviews: Computational Molecular Science, 2012, 2(3), 435–42.
[7] Jhanwar B, Sharma V, Singla RK, Shrivastava B. QSAR-Hansch analysis and related approaches in drug design. Pharmacologyonline, 2011, 1, 306–44.
[8] Kumar P, Narasimhan B, Sharma D, Judge V, Narang R. Hansch analysis of substituted benzoic acid benzylidene/furan-2-yl-methylene hydrazides as antimicrobial agents. European Journal of Medicinal Chemistry, 2009, 44(5), 1853–63.
[9] Mannhold R, Krogsgaard-Larsen P, Timmerman H. QSAR: Hansch Analysis and Related Approaches. John Wiley & Sons, 2008.
[10] Judge V, Narang R, Sharma D, Narasimhan B, Kumar P. Hansch analysis for the prediction of antimycobacterial activity of ofloxacin derivatives. Medicinal Chemistry Research, 2011, 20, 826–37.
[11] Chaudry MA, James KC. Hansch analysis of the anabolic activities of some nandrolone esters. Journal of Medicinal Chemistry, 1974, 17(2), 157–61.
[12] Patrick GL. An Introduction to Medicinal Chemistry. Oxford University Press, 2023.
[13] Fukuto TR, Metcalf RL. Pesticidal activity and structure, structure and insecticidal activity of some diethyl-substituted phenyl phosphates. Journal of Agricultural and Food Chemistry, 1956, 4(11), 930–5.

[14] Hansch C, Maloney PP, Fujita T, Muir RM. Correlation of biological activity of phenoxy acetic acids with Hammett substituent constants and partition coefficients. Nature, 1962, 194(4824), 178–80.

[15] Hansch C, Muir RM, Fujita T, Maloney PP, Geiger F, Streich M. The correlation of biological activity of plant growth regulators and chloromycetin derivatives with Hammett constants and partition coefficients. Journal of the American Chemical Society, 1963, 85(18), 2817–24.

[16] Hansch C, Fujita T. p-σ-π analysis. A method for the correlation of biological activity and chemical structure. Journal of the American Chemical Society, 1964, 86(8), 1616–26.

[17] Craig PN. Interdependence between physical parameters and selection of substituent groups for correlation studies. Journal of Medicinal Chemistry, 1971, 14(8), 680–4.

[18] Topliss JG. Utilization of operational schemes for analog synthesis in drug design. Journal of Medicinal Chemistry, 1972, 15(10), 1006–11.

[19] Topliss JG. A manual method for applying the Hansch approach to drug design. Journal of Medicinal Chemistry, 1977, 20(4), 463–9.

[20] Free SM, Wilson JW. A mathematical contribution to structure-activity studies. Journal of Medicinal Chemistry, 1964, 7(4), 395–9.

[21] Kubinyi H. Free Wilson analysis. Theory, applications and its relationship to Hansch analysis. Quantitative Structure-Activity Relationships, 1988, 7(3), 121–33.

[22] Fujita T, Ban T. Structure-activity study of phenethylamines as substrates of biosynthetic enzymes of sympathetic enzymes of sympathetic transmitters. Journal of Medicinal Chemistry, 1971, 14, 148–52.

[23] Kubinyi H, Kehrhahn OH. Quantitative structure-activity relationships. 3. A comparison of different Free-Wilson models. Journal of Medicinal Chemistry, 1976, 19(8), 1040–9.

Chapter 5
3D-QSAR in drug designing

> Through the application of 3D-QSAR models, medicinal chemists can prioritize compound synthesis efforts and design molecules with enhanced therapeutic profiles. – Alessandro Pedretti

5.1 Introduction

Traditional quantitative structure–activity relationship (QSAR) models, popularly known as 2D-QSAR models, have allowed us to quantify the relationship between the changes in any of the structural parameters of a molecule with its biological activity [1, 2]. An accumulated understanding of the drug–receptor interaction in the recent past suggests that it is the three-dimensional structure of the molecule that interacts at the binding site of the receptor. So, without considering the three-dimensional (3D) or spatial properties of the molecule, the objective of obtaining a robust QSAR model cannot be achieved [3]. Therefore, 3D-QSAR is considered a natural extension of the traditional QSAR in which biological activity is correlated with the properties/parameters obtained from the 3D structure of the molecule, which is independent of the molecule's topology [4]. By incorporating spatial information and interactions, 3D-QSAR enhances the accuracy of predicting how chemical compounds will interact with biological targets [5]. It utilizes molecular modeling techniques to analyze the alignment and conformation of molecules in a receptor's active site, leading to valuable insights into structure–activity relationships. This approach aids in the design and optimization of potential drug candidates, facilitating more efficient and targeted drug discovery processes [6, 7].

A single bioactive molecule can exist in thousands of different conformations. The calculation of 3D properties, mostly steric and electrostatic, is done from only the bioactive conformation of the molecule, which is a conformation that interacts with the binding site of the receptor [8, 9]. This bioactive conformation can either be obtained experimentally by using techniques like X-ray crystallography, and NMR, or computationally by using molecular mechanics [10]. When compared to the traditional QSAR methodology, the 3D-QSAR study is considered more exhaustive and complex as there is a considerable amount of difference between the 2D- and 3D-QSAR approaches [11]. The basic steps involved in the 3D-QSAR study involve the following:

- **Collection of the dataset:** This is the very first step and like any QSAR modeling technique, it involves the selection of the molecules against a particular biological activity. A dataset is collected for the molecules from which a 3D-QSAR model is generated.
- **Determination of bioactive conformation:** Out of the thousands of available conformations of the molecule, the conformation that interacts with the receptor

https://doi.org/10.1515/9783111434858-005

is selected from all the molecules of the dataset. As discussed above, this can be achieved either experimentally or computationally.
- **Alignment of the molecules**: In this step, alignment of the bioactive conformations is done within a 3D grid space. An interacting partner in the 3D space is also selected which can be water, octanol, or any other solvent.
- **Calculation of the descriptors**: These are the 3D descriptors, mostly steric and electrostatic, which are calculated in this study by probing the bioactive conformer inside the 3D space. The probe is a positively charged atom (carbon or proton) that is used for calculating the steric and electrostatic parameters of the molecule at each point of the lattice of the 3D grid.
- **Generation of the model equation**: After calculating the different descriptors (steric and electrostatic parameters) of the molecules kept within the 3D grid space, the change in the descriptors is correlated with the biological activity of these molecules, and a mathematical equation is generated. The most common statistical tool employed for this is the partial least square (PLS) technique.
- **Validation of the model**: This step is used to assess the reliability and the prediction quality of any of the generated QSAR models. Numerous statistical parameters are employed for this purpose and will be discussed in detail in the later part of the chapter.
- **Interpretation of the results**: This is the last step of the 3D-QSAR model generation. Results obtained from the model are interpreted through various plots and contour maps. Contour maps obtained are the biggest difference between the traditional QSAR and 3D-QSAR models.

5.2 Basic assumptions in 3D-QSAR modeling

To make a 3D-QSAR model as realistic as possible, certain basic assumptions are made before initiating the study [4]. To have complete knowledge of the 3D-QSAR methodology, one should be aware of basic assumptions, which are as follows:
- The drug–receptor interaction is always non-covalent (mostly steric or electrostatic).
- The biological activity of a drug molecule is considered to be directly proportional to its binding with the receptor.
- For all the molecules taken for the study, the binding site of the receptor is considered to be the same.
- The bioactive conformation is the conformation that has the lowest energy of all the possible conformations of the test molecule.
- The biological activity is shown by the drug itself and not by its metabolite.
- Molecules having common structures are assumed to have similar binding modes and similar biological activity.
- The receptor binding site is considered to be rigid in the 3D-QSAR with few exceptions.

- The descriptors calculated of the molecule represent its physical, chemical, and biological parameters.
- It is assumed that loss in the entropy after binding follows the same pattern for all the datasets.

5.3 Classification of 3D-QSAR methodology

3D-QSAR is a broad term comprising all the QSAR approaches in which 3D structures are employed for the generation of descriptors that can be further correlated with biological activity. The classification of 3D-QSAR methodology categorizes approaches that analyze the relationship between molecular structures and biological activities in a 3D context. 3D-QSAR methodologies are categorized into three different types (Fig. 5.1) [4].

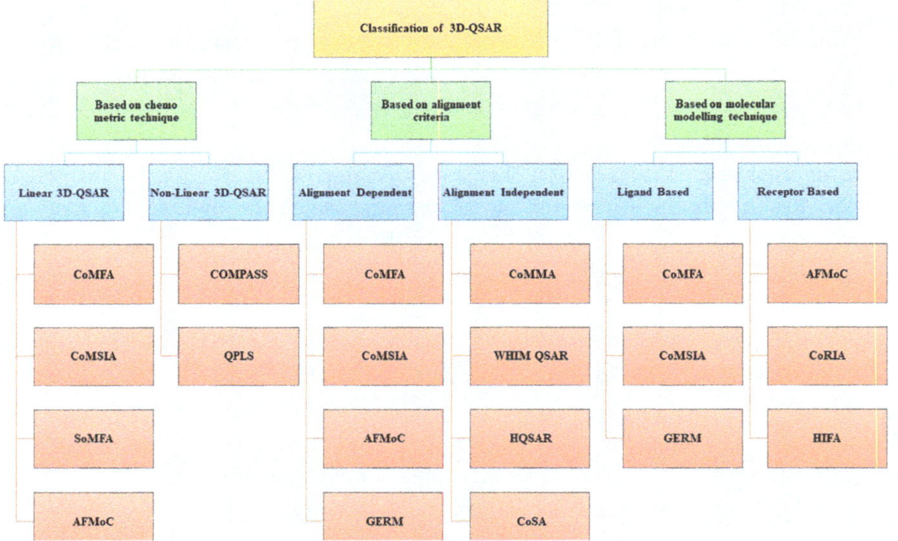

Fig. 5.1: Classification of different 3D-QSAR methodologies based on different techniques.

5.3.1 Based on the chemometric technique used in the development of the QSAR equation

If the correlation between the biological activity and the descriptors is linear then it is classified as linear 3D-QSAR and if the mathematical equation is not linear then it is considered as nonlinear 3D-QSAR. Examples of linear 3D-QSAR include comparative molecular field analysis (CoMFA), comparative molecular similar indices analysis (CoMSIA), consensus adaptation of fields for molecular comparison (AFMoC), and self-organizing molecular field analysis (SoMFA), whereas COMPASS and quadratic PLS are examples of nonlinear 3D-QSAR.

5.3.2 Based on the criterion of alignment

If the bioactive conformers of all the molecules of the dataset are aligned in the 3D grid space for calculations of descriptors then this technique comes under the alignment-based 3D-QSAR model. Examples of alignment-based 3D-QSAR models are CoMFA, CoMSIA, AFMoC, and genetically evolved receptor model (GERM). Likewise, there have been many 3D-QSAR techniques developed that come under alignment independent 3D-QSAR model. These include comparative molecular moment analysis (CoMMA), weighted holistic invariant molecular (WHIM) QSAR, hologram QSAR (HQSAR), and comparative spectral analysis (CoSA).

5.3.3 Based on the molecular modeling technique

This classification is also based on whether the 3D-QSAR model is developed using the information from the ligand or the receptor. If the model is developed based on the known information of the ligands against a particular biological activity then it is classified as ligand-based 3D-QSAR, which includes techniques like CoMFA, CoMSIA, and GERM. If the model is developed based on the known information of the receptor against a particular biological activity then it falls under the receptor-based 3D-QSAR. This includes techniques like AFMoC, comparative residue interaction analysis, and HINT interaction field analysis (HIFA).

5.4 Comparative molecular field analysis (CoMFA)

CoMFA is a pioneering computational method in drug discovery that explores the 3D structure–activity relationship of molecules. CoMFA is one of the most popular techniques of the 3D-QSAR developed in 1988 by Cramer and is considered the prototype of 3D-QSAR [12]. CoMFA transforms molecular structures into grid-based fields representing steric and electrostatic properties. By aligning diverse compounds within a common framework, it analyzes how variations in these fields correlate with differences in biological activity. The method's primary strength lies in generating predictive models that help identify essential structural features contributing to activity, guiding compound optimization. CoMFA's success hinges on its ability to consider molecular shape and electrostatic interactions, offering insights beyond traditional 2D-QSAR. Though it revolutionized structure-based drug design, CoMFA has limitations, such as sensitivity to alignment and grid parameters. Despite this, it remains a foundation for subsequent 3D-QSAR methods and plays a pivotal role in prioritizing molecular candidates, accelerating drug development, and fostering a deeper understanding of structure–activity relationships in the intricate realm of pharmaceutical research. As discussed in the classification above, it is a molecular field-based, ligand-based, alignment-dependent, and linear 3D-QSAR model. The one distinguished property of CoMFA is that it not only provides a statistical correlation between the biological activity and the 3D descriptors of the molecules but also provides a graphical representation of the results, which define the favorable and unfavorable receptor–ligand interactions (contour maps).

5.4.1 Methodology of CoMFA

Various steps involved in the generation of a CoMFA model are depicted in Fig. 5.2 and are discussed in detail below [13].

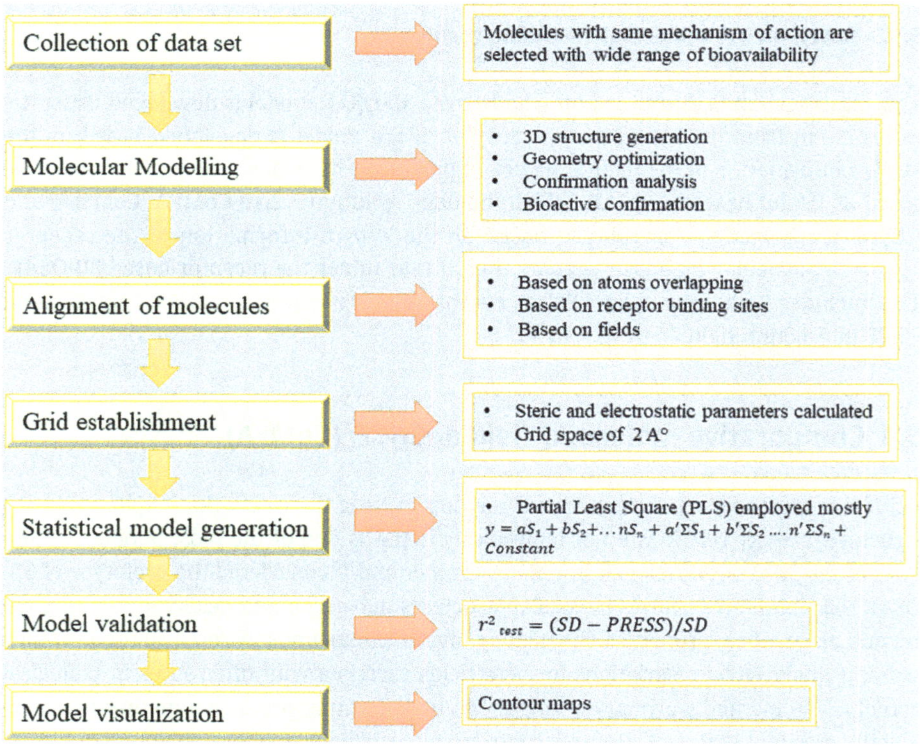
Collection of data set
Molecules with same mechanism of action are selected with wide range of bioavailability
Molecular Modelling
• 3D structure generation
• Geometry optimization
• Confirmation analysis
• Bioactive confirmation
Alignment of molecules
• Based on atoms overlapping
• Based on receptor binding sites
• Based on fields
Grid establishment
• Steric and electrostatic parameters calculated
• Grid space of 2 A°
Statistical model generation
• Partial Least Square (PLS) employed mostly
$y = aS_1 + bS_2 + \ldots nS_n + a'\Sigma S_1 + b'\Sigma S_2 \ldots . n'\Sigma S_n +$ Constant
Model validation
$r^2_{test} = (SD - PRESS)/SD$
Model visualization
Contour maps

Fig. 5.2: Basic workflow for the steps involved in developing a 3D-QSAR model.

5.4.1.1 Collection of the dataset

A dataset is a set of molecules whose biological activities are known and are employed for the generation and validation of any QSAR model. This is a very important step for the successful generation of the CoMFA model. The molecules selected for the development of the model should be of the congeneric series. The experimental protocol applied for the calculation of the biological activity should be the same for all the molecules employed in the study so that their similar mechanism of action can be established. For a robust model generation, it is always desired that the biological activity data of the molecules should be evenly distributed around their mean. Representation of the biological activity of the dataset in the logarithmic scale is preferred to remove the skewness in the data. The structure of the ligands can be obtained from various databases such as Cambridge structural database or protein data bank or they can be drawn in 3D manually using different structure drawing software such as ChemDraw, ChemSketch, and MarvinSketch.

5.4.1.2 Molecular modeling

This step also holds utmost importance in the generation of the CoMFA model and can be further classified into the following steps.

5.4.1.2.1 Optimization of the 3D structure of the molecules

Once the 3D structure of the ligand molecules is obtained as discussed in the above step, their conformational energy minimization is done. Structure optimization of the molecules can be done by any of the following techniques:

a. **Molecular mechanics**: In this method, atoms are taken into consideration for the calculation of energy parameters. This is a fast method and can also be employed for the calculation of the energy of large systems such as enzymes or other large proteins. The total energy of any molecule is assumed to be a total of four different types of energies, stretching energy, bending energy, torsion energy, and nonbonded interaction energy (electrostatic and Van der Waals parameters).
b. **Quantum mechanics**: This is more accurate when compared with the molecular mechanics in the calculation of the energy of the molecule. In quantum mechanics, nuclei and the arrangement of electrons are considered for the calculation of the energy of the molecule. Quantum mechanics methods are also of further two types:
   - **Ab initio methods**: This is a very accurate method but is very time-consuming and is limited to small molecules only. In this method, energy calculations are done from the classical Schrodinger wave equations.
   - **Semiempirical methods**: Semiempirical approach is less accurate when compared to the ab initio method but this method is faster and can be applied easily to large molecules. In these methods, energy calculations are performed by taking some assumptions into in Schrodinger wave equation for

energy calculations. The most common methods included in semiempirical methods are modified neglect of diatomic overlap (MNDO), Austin model 1 (AM1), and parametric method 3 (PM3). This semiempirical approach is the choice of methods in CoMFA for energy calculation and minimization.

5.4.1.2.2 Conformational analysis of the molecules

This step becomes very important as it has been established that most of the molecules do exist in more than one conformation at the physiological temperature. There are several ways through which this conformational analysis can be performed and the choice of the appropriate approach depends on the type of molecules under study in the CoMFA [14, 15].

a. **Systematic search method**: This is the most widely used method in the conformational analysis for the CoMFA model. In this method, different conformations of the molecule are generated by changing the dihedral or torsion angle and keeping the bond length and bond angle constant. This is a very exhaustive technique as all possible conformations of the molecule can be obtained by making small increments in the dihedral angle and the change in this dihedral angle is initiated from 0° and increment is done until 360°. The two major drawbacks of this approach are, firstly, the time for calculating the conformations increases exponentially with a single addition of a rotatable bond, and secondly, it does not account for the ring closure problem in the circular structures.
b. **Molecular dynamics method**: In this method, a molecule whose conformations are to be generated is considered a dynamic structure. The change in the conformation of a molecule is considered due to its kinetic energy and various interaction forces of the neighboring atoms. All possible conformations of the molecule are generated using Newton's second law of motion, that is, force = mass × acceleration ($F = MA$).
c. **Distance geometry method**: This method is of practical use for very large molecules in which systematic search is too time-consuming. A random starting 3D conformer is generated initially and then further calculations are made from here. This is a very rapid method for the calculation of conformations but does not include all the possible conformations of the molecule.
d. **Monte Carlo approach**: In this method, a random change in the initial structure is made and a new conformer is generated. If the energy of the new conformer is lower than the previous one then it is accepted. On the other hand, if the energy of the new conformer is higher than the previous one then a different algorithm is used for accepting or rejecting the new conformer. They are also considered if they do not fall for the local minima in the generation of the conformations.

5.4.1.2.3 Determination of the bioactive conformer

Out of the many possible conformations, it is the bioactive conformer that interacts within the cavity of the receptor, hence making it very important to identify the bioactive conformer before initiating the generation of our CoMFA model. The reliability and quality of any CoMFA model depend highly on how accurately the bioactive conformers are determined in the study [16]. The bioactive conformations can be determined either experimentally or theoretically.

a. **Experimental approaches for determining bioactive conformations**: Different experimental methods employed for bioactive conformation generation include:
   - **X-ray crystallography**: In this method, a drug molecule bound inside the cavity of the receptor is crystallized and the structure is analyzed through X-ray. The conformation in which the drug molecule is bound is considered the bioactive conformation. The major disadvantage of this method is that the solution used for the crystallization is different from the physicochemical solutions, hence, it cannot be established that the ligand present inside the cavity is in its bioactive conformation of not. There is also a high possibility of errors in this method.
   - **NMR spectroscopy**: This is another experimental method for the determination of the bioactive conformation. Unlike X-ray crystallography, here no crystallizing solution is used, instead drug–receptor complex is kept inside the solution and their structure is determined. Here too, there is a high probability that the structure obtained through this method will be the bioactive conformation or close to the same but not fully guaranteed.

b. **Theoretical approaches for determining bioactive conformations**: Besides experimental methods, the bioactive conformation of the molecules can also be calculated through theoretical methods. Calculation done through theoretical approaches is based on the assumption that the global minimum of the different conformations is the bioactive conformation or very close to it. Different theoretical methods used are:
   - **Active analog approach**: This is the most widely used technique for the determination of the bioactive conformer and was developed in 1979 by Garland R. Marshal and coworkers. In this technique, first, the basic pharmacophore around the cavity of the receptor is defined. After that, using a systematic search method, conformations of all possible active analogs are determined.
   - **Ensemble distance geometry approach**: This technique was developed by Sheridan and coworkers in 1986 and is considered an extension of the conventional distance geometry technique. Originally it was developed by determining a common pharmacophore for nicotinic action from four nicotinic agonists, nicotine, cysteine, muscarone, and ferruginine methiodide. In this approach coordinates for the set of molecules for their bioactive conformations are generated. The biggest advantage of this method is that it also handles the cyclic structures where ring closure is a hindrance in calculating the bioactive conformer.

- **Molecular fitting approach**: The molecular fitting technique is also known by the name template forcing approach. This method is only applicable when we have a rigid molecule that binds tightly with the receptor under study. The rigid molecule is taken as a reference compound and certain flexible molecules are mapped onto its surface using molecular mechanics or dynamics approach.

5.4.1.3 Alignment of the molecules

Alignment of the molecules is one of those issues in the generation of the CoMFA model that should be addressed with utmost care. A good predictivity of a CoMFA model is only achieved when exact superimposition is achieved as a small difference in the alignment can vary the result to a very large extent. In the alignment process, generally, the molecule with the least number of conformers or the most active molecule is taken as the template and all other conformations of other molecules are aligned around it. Alignment obtained from a series of molecules is considered as the pharmacophore for that series against the receptor. Different approaches used for the alignment are discussed below:

a. **Atom-based alignment**: Atoms of the molecules are aligned on the atoms of the template molecule. It is best suited for the congeneric series of molecules as it will easily detect drug dissimilarity but this approach cannot be applied to the molecules from different series. This method is also known as the pharmacophore approach.
b. **Shape-based alignment**: Shape-based alignment is the approach in which instead of taking atoms as the starting point for the alignment, a basic shape of the template molecules is taken into consideration for the alignment of the rest of the molecules on it. Although this method of alignment is fast when compared to the atom-based alignment, it is considered less accurate.
c. **Receptor-binding site-based alignment**: This is a more reliable method for the alignment but comes with more complexity due to the enhanced number of degrees of freedom. This type of alignment is achieved through the molecular-docking approach. In this method, active sites of receptors or the residues that interact with ligands are superimposed and then the alignment is done on it.
d. **Field/pseudofield-based alignment**: This is a completely different approach than the abovementioned methods of alignment, as in this approach energy fields (mainly electrostatic or steric fields) of the template molecule are calculated and alignment of the other molecules are done based on their energy fields to minimize the residual mean square difference between the template and other molecules. It is always recommended to first perform the alignment via the abovementioned methods and then do the final alignment through this approach to increase the overall quality of the alignment of the molecules.

5.4.1.4 Setting up the grid and calculation of molecular energy fields

This is the next step after the alignment of the molecules and is divided into two phases:

a. **Setting up of the grid**: In this step, a 3D grid space is identified and aligned molecules are kept in the center of this grid (Fig. 5.3). The default grid space, that is, the distance between the two grid points, is kept at 2 Å, however, it can be varied as per the requirement. Generally low values of the grid space are desirable for a more exhaustive search. Talking about the size of the grid, it should be larger by 3–4 Å compared to the union surface of the aligned molecule. The location of the grid box also plays an important role and it is noticed that if not located properly it can vary the result of the generated CoMFA model.

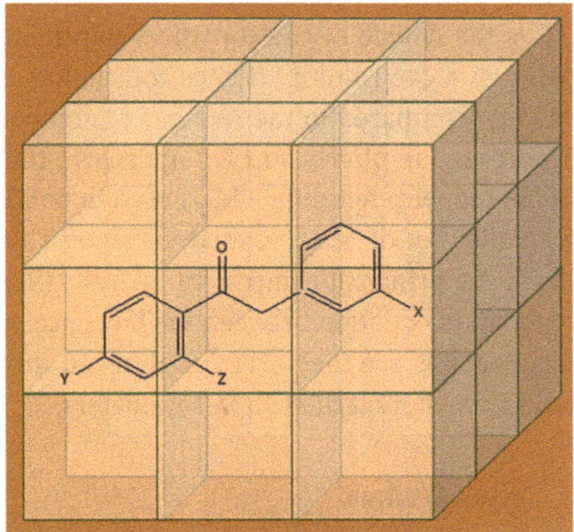

Fig. 5.3: A generated grid box for the generation of the CoMFA model.

b. **Calculations of molecular energy fields**: All the aligned molecules inside the grid box are represented in terms of molecular energy fields (steric and electrostatic fields). A probe atom, generally a sp^3 hybridized carbon or a proton, is used for evaluating these fields. The value of these molecular fields between the probe atom and each molecule is calculated at the regularly spaced point of the grid box. In the CoMFA model, steric fields are represented by the Lennard-jones potential (equation (5.1)), and the electrostatic field are represented by the Columbic potential (equation (5.2)):

$$E_{\text{vdw}} = \Sigma \left(A_{ij} r_{ij}^{-12} - C_{ij} r_{ij}^{-6} \right) \tag{5.1}$$

where E_{vdw} represents the Van der Waals energy, r_{ij} depicts the distance between the molecules at point i and the probe atom at point j, and A_{ij} and C_{ij} are the constants obtained from the Van der Waals radii of the concerned atoms:

$$E_c = \frac{\Sigma\left(q_i q_j\right)}{\epsilon r_{ij}} \tag{5.2}$$

where E_c is the coulomb interaction energy, r_{ij} stands for the distance between the molecule at point i and the probe atom at point j, q_i is the charge of the atom i of the molecule, q_j represents the charge of probe atom, and ϵ is the dielectric constant.

5.4.1.5 Statistical generation of the CoMFA model

Energy field values obtained in the previous step serve as the descriptors in CoMFA analysis for correlating the biological activity values of the compounds. However, the major issue here is that we get thousands of these steric and electric field parameters. Multiple linear regression (MLR), which is the choice of the statistical parameter in traditional QSAR, is not preferred for CoMFA as it can result in chance correlation as the number of descriptors in CoMFA is very large compared to the number of compounds in the study. Hence to overcome this hindrance, another statistical approach, the PLS method developed in 1986 is used for CoMFA model generation. PLS is used in CoMFA to reduce the number of descriptors to a smaller set of uncorrelated variants. This approach is considered a natural extension of the principal component analysis (PCA) of statistics. The model of the CoMFA is a correlation equation between the biological activity and several latent variables (LVs) (Fig. 5.4). LVs are the variables obtained through the linear combinations of the original independent molecular field parameters.

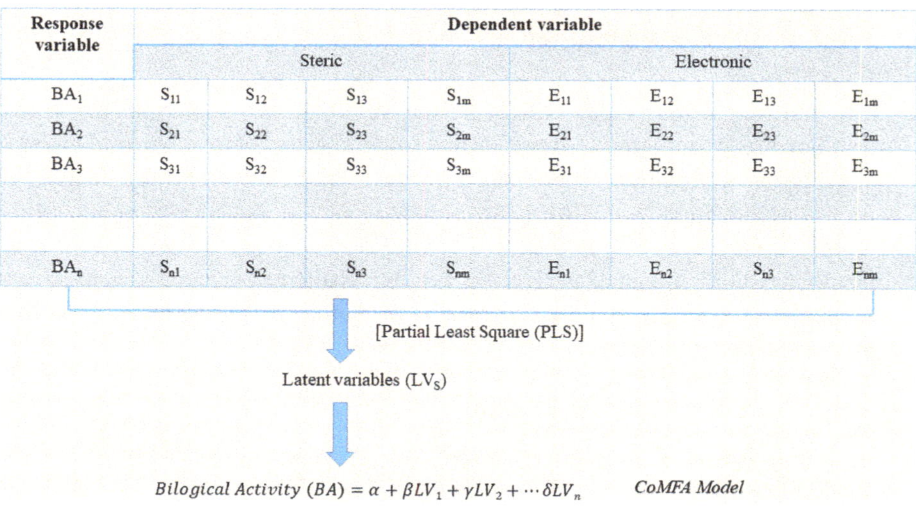

Response variable	Dependent variable							
	Steric				Electronic			
BA_1	S_{11}	S_{12}	S_{13}	S_{1m}	E_{11}	E_{12}	E_{13}	E_{1m}
BA_2	S_{21}	S_{22}	S_{23}	S_{2m}	E_{21}	E_{22}	E_{23}	E_{2m}
BA_3	S_{31}	S_{32}	S_{33}	S_{3m}	E_{31}	E_{32}	E_{33}	E_{3m}
BA_n	S_{n1}	S_{n2}	S_{n3}	S_{nm}	E_{n1}	E_{n2}	S_{n3}	E_{nm}

Fig. 5.4: Development of the CoMFA model equation by using partial least square (PLS) statistical approach.

5.4.1.6 Validation of the CoMFA model

The quality of a CoMFA or any QSAR model is validated through various statistical parameters. CoMFA, like any QSAR model developed, is validated for its predictability and robustness both internally and externally:

a. **Internal validation of the CoMFA model:** In this approach, the quality of the developed model is assessed from the molecules used for its development, that is, from the training dataset. The following two parameters are used for this purpose in the internal validation:
   - **Goodness of fit (R^2):** This is used to assess the quality of the CoMFA model in terms of its reproducibility but does not provide any information regarding the robustness and predicting ability of the generated model. It is calculated by the following equation:

$$R^2 = 1 - \left(\frac{\text{RSS}}{\text{TSS}}\right) \tag{5.3}$$

     where RSS is the residual sum of squares and TSS represents the total sum of squares; both parameters are calculated as shown in the following equations:

$$\text{RSS} = \Sigma\,(Y - Y_{\text{pred}})^2 \tag{5.4}$$

$$\text{TSS} = \Sigma\,(Y - Y_{\text{mean}})^2 \tag{5.5}$$

     where Y represents the observed biological activity value of the molecule, Y_{pred} is the value of the biological activity predicted by the model, and Y_{mean} is the average value of the observed biological activity of all the molecules in the training set. The value of R^2 should be greater than 0.7 for considering the generated model acceptable.
   - **Cross-validation coefficient (Q^2_{cv}):** This is the most popular and widely accepted method for the internal validation of the developed CoMFA model. In this approach, a fixed amount of the molecules are omitted from the training set, which is a leave-one-out method in which one molecule is omitted every time from the training set till all the molecules are omitted one by one. The developed CoMFA model is then utilized to predict the BA values of these omitted molecules and Q^2_{cv} of the model is calculated through the following equation:

$$Q^2_{\text{cv}} = 1\left(\frac{\text{PRESS}}{\text{TSS}}\right) \tag{5.6}$$

     where PRESS represents the predictive error sum of squares and TSS represents the total sum of squares.

     Let us assume that in a CoMFA model, we have n number of molecules in the training set. The CoMFA model is generated from $(n - 1)$ molecules leaving one molecule at each time and predicting the biological activity of the left-out molecule from this generated CoMFA model. If the actual biological activity of the left-out molecule is Y and Y_p is its value predicted by the model then the value of PRESS will be calculated from the following equation:

$$\text{PRSS} = \Sigma\,(Y - Y_{\text{p}})^2 \tag{5.7}$$

PRESS is the total sum of the square of the difference between the actual and predicted values of all the left-out molecules one by one. The Q^2_{cv} value greater than 0.7 indicates that the model developed can be robust and will have good predictability.

b. **External validation of the CoMFA model:** The most widely used parameter for the external validation of the developed CoMFA model is the coefficients of determination (R^2_{pred}). It can be calculated from the following equation:

$$R^2_{\text{pred}} = 1 - \left(\frac{\Sigma\,(Y - Y_{\text{pred}})^2}{\Sigma\,(Y - Y_{\text{mean}})^2}\right) \tag{5.8}$$

where Y is the observed biological activity value of the test molecule, Y_{pred} is the value of biological activity predicted by the model of the test dataset molecule, and the Y_{mean} represents the average value of the observed biological activity of the training set molecules. A value of R^2_{pred} greater than 0.6 is considered acceptable while assessing the robustness of the CoMFA model externally.

There is still a lot of debate amongst the scientific community on whether external validation is required or not in any QSAR model, as some groups of scientists believe that robust internal validation is sufficient for assessing the model, whereas other groups believe that external validation is a must.

5.4.1.7 Visualization and interpretation of the results

The information generated through the CoMFA model is assessed and interpreted through contour maps and the PLS plots. Contour maps are the distinguished features of a CoMFA model that differentiate it from the traditional QSAR model. Contour maps are defined as the 3D space around the molecule, which tells us about the regions of high and low activity due to steric and electrostatic effects (Fig. 5.5). Different colors in the contour maps represent different effects of the groups in the contour map. In the steric contour maps, the green color region represents the region where bulky groups are desired for the biological activity and the yellow colors are the areas where the bulky groups will decrease the biological activity. In the electrostatic contour maps, the blue area represents the requirement of electropositive groups for increasing the biological activity and the red regions emphasize the requirement of the electronegative groups in that region for increasing the biological activity. Besides contour maps useful information is also extracted from the PLS plots such as plots between the predicted and experimental values of the biological activity of training and test datasets (Fig. 5.6).

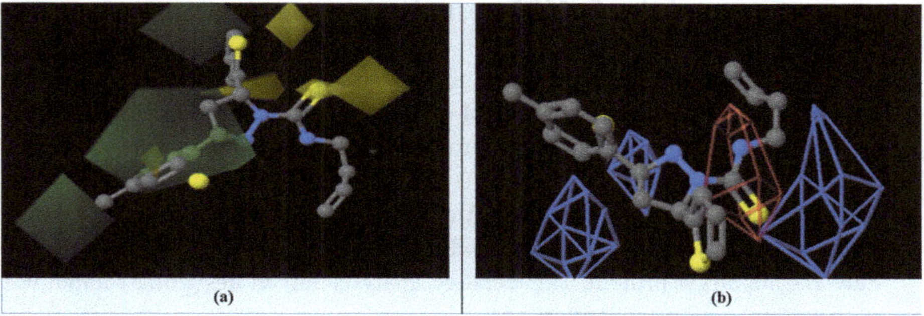

Fig. 5.5: Contour maps obtained from the generated CoMFA model: (a) steric contour map where the green color represents the sterically favorable region and yellow region represents the sterically forbidden region; (b) electrostatic contour map where blue color is the electrostatically favorable region and red region represents the electrostatically forbidden region.

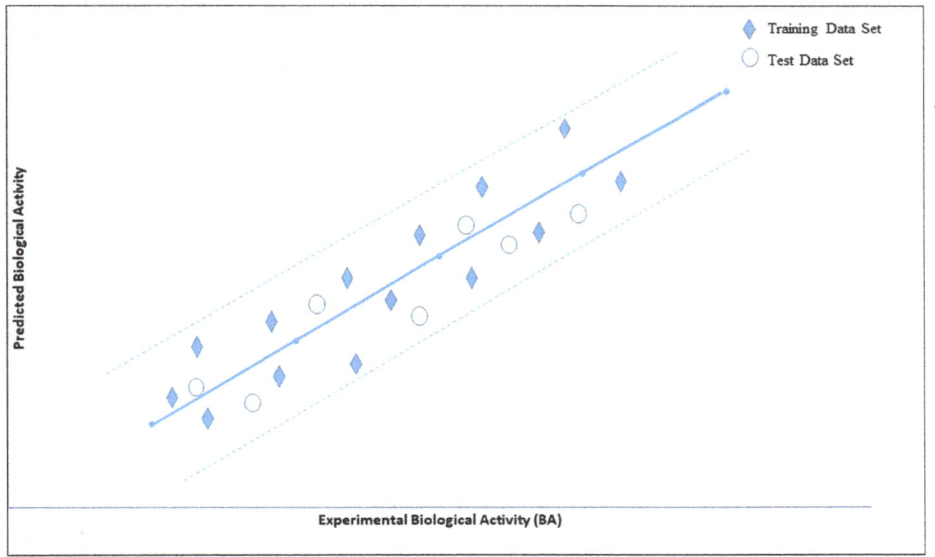

Scatter Plot PLS Plot CoMFA

Fig. 5.6: PLS plot of the CoMFA model plotted between the predicted and the actual biological activity of the compounds employed for the generation of the CoMFA model.

5.5 Advantages of the CoMFA model

- **3D structural information:** CoMFA utilizes three-dimensional structural information of molecules, capturing spatial interactions and alignment within the active site of a protein. This leads to a more realistic representation of molecular interactions compared to traditional 2D-QSAR.
- **Spatial interaction analysis:** CoMFA evaluates steric and electrostatic interactions, which are critical for understanding binding interactions between a molecule and its target protein. This detailed analysis aids in designing compounds with optimal binding characteristics.
- **Predictive power:** CoMFA models can have high predictive accuracy, especially when the alignment and grid parameters are carefully chosen. This accuracy translates into the ability to predict the biological activity of new, untested compounds, aiding in virtual screening and compound prioritization.
- **Structure–activity relationship insights:** CoMFA provides insights into the structure–activity relationships by highlighting how specific regions of a molecule's structure contribute to its biological activity. This information guides medicinal chemists in modifying and optimizing compounds.
- **Visual interpretation:** CoMFA results often include contour maps that visually represent regions of favorable and unfavorable interactions within the active site. These maps help chemists understand where modifications can lead to improved binding affinities.
- **Molecular design:** CoMFA allows for the design of new compounds with desired activity profiles. By altering molecular structures to fit favorable contour regions, chemists can rationally design analogs or derivatives for improved potency.
- **Alignment flexibility:** CoMFA provides a degree of alignment flexibility, accommodating variations in molecule conformations, which is particularly useful for flexible ligands and proteins.
- **Alignment-independent analysis:** CoMFA can be used to analyze datasets with structurally diverse molecules, as the alignment process can help identify common structural features across different molecules.
- **Interpretability:** CoMFA equations contain coefficients for each descriptor, providing a quantitative measure of their contribution to activity. This enhances the understanding of the key molecular features influencing activity.
- **Complementary to experimental methods:** CoMFA can supplement experimental studies by suggesting hypotheses for binding modes and interactions, which can guide subsequent experimental investigations.
- **Time and cost-efficiency:** CoMFA allows researchers to prioritize compounds for synthesis and testing, potentially reducing the number of compounds needing experimental evaluation, and saving time and resources.

5.6 Limitations of the CoMFA model

Besides a lot of significant advantages over traditional QSAR, there are certain limitations of the CoMFA that should be addressed whenever this technique is applied to model generation [16, 17]:

- **Alignment sensitivity:** CoMFA heavily relies on the accurate alignment of molecules within the binding site. Slight variations in alignment can lead to different results, making it sensitive to the chosen alignment method.
- **Grid size and resolution:** The choice of grid size and resolution can impact the results significantly. An inappropriate grid can either miss important interactions or introduce noise into the analysis.
- **Sparse data regions:** CoMFA is sensitive to areas in the active site with sparse data points. If certain regions of the grid have insufficient data, the model's accuracy can be compromised.
- **Overfitting:** Like any QSAR model, CoMFA can be prone to overfitting if not properly validated. Overfitting occurs when the model captures noise in the training data rather than true underlying relationships.
- **Descriptor limitations:** CoMFA uses steric and electrostatic fields as descriptors, which may not capture all the relevant molecular interactions and properties that influence biological activity.
- **Dependence on training set:** The predictive power of a CoMFA model is heavily reliant on the quality and representativeness of the training dataset. A poorly chosen dataset can lead to unreliable predictions.
- **Applicability domain:** CoMFA models are only valid within the applicability domain defined by the training data. Predictions for compounds that fall outside this domain can be unreliable.
- **Assumption of additivity:** CoMFA assumes that the contributions of steric and electrostatic fields are additive, which might not always hold true for complex molecular interactions.
- **Interpretability:** While CoMFA provides insights into which regions of a molecule contribute to its activity, the physical interpretation of the contour maps can be challenging, limiting its usefulness for rational drug design.
- **Variability in biological activity:** Biological activity can be influenced by factors beyond simple steric and electrostatic interactions, such as solvation effects, entropy changes, and dynamic behavior, which CoMFA may not fully capture.
- **Limited scope of interactions:** CoMFA primarily considers steric and electrostatic interactions, potentially overlooking other crucial interactions like hydrogen bonding, π–π stacking, and metal coordination.
- **Computational intensity:** The calculations involved in generating CoMFA models can be computationally intensive and time-consuming, particularly for larger datasets or more complex molecular systems.

5.7 Comparative molecular similar indices analysis (CoMSIA)

CoMSIA model was developed in 1994 by Klebe and coworkers and is considered the natural extension of the CoMFA. CoMSIA is a computational method extensively utilized in drug discovery to elucidate the quantitative relationship between molecular structure and biological activity [18]. CoMSIA extends the principles of CoMFA by incorporating additional physicochemical properties beyond steric and electrostatic fields. It considers fields like hydrophobicity, hydrogen bonding, and lipophilicity, enhancing the predictive accuracy of the model. CoMSIA operates by dividing the three-dimensional space around molecules into grid points and calculating the interactions of molecular probes at these points. These interactions are translated into numerical descriptors that represent various physicochemical properties. Through statistical regression, CoMSIA correlates these descriptors with biological activity data to construct a predictive QSAR model. CoMSIA's ability to encompass multiple molecular properties makes it a valuable tool for understanding complex structure–activity relationships and guiding the design of novel drug candidates with improved potency and selectivity.

Unlike the CoMFA model where we calculate the molecular field-based descriptors (Lennard-Jones and Coulomb potentials), here some arbitrary descriptors are calculated, which are based on the similarity and dissimilarity of the molecules in space. These similarity indices are calculated for each of the aligned molecules using a common probe. Another major advantage of CoMSIA when compared with the CoMFA is that it also incorporates the hydrophobic surface area (which explains the effect of entropy on the binding affinity) and hydrogen bonding parameters in the generation of the model. Therefore, CoMSIA enumerates five parameters using a probe, steric, electrostatic, hydrophobic surface area, hydrogen bond donor, and hydrogen bond acceptor-based descriptors using a probe.

A probe atom, generally having a charge of +1, a radius of 1 Å, hydrophobicity of +1, hydrogen bond acceptor value of 1, and having hydrogen donor value of 1, is used in calculating the similarity indices in the CoMSIA model. Another major difference from CoMFA is the visualization and interpretation of the contour maps. In CoMFA we get those favorable or not favorable regions of the environment where aligned molecules will interact; however, in the CoMSIA model, these regions are occupied by the ligand that will like or dislike the presence of a particular group concerning a particular parameter. The rest of the methodology of CoMSIA, including generation of conformations, alignment of molecules, and statistical approach (PLS), is similar to that described above for the CoMFA model. A brief of steps used in the CoMSIA model generation is given in Fig. 5.7.

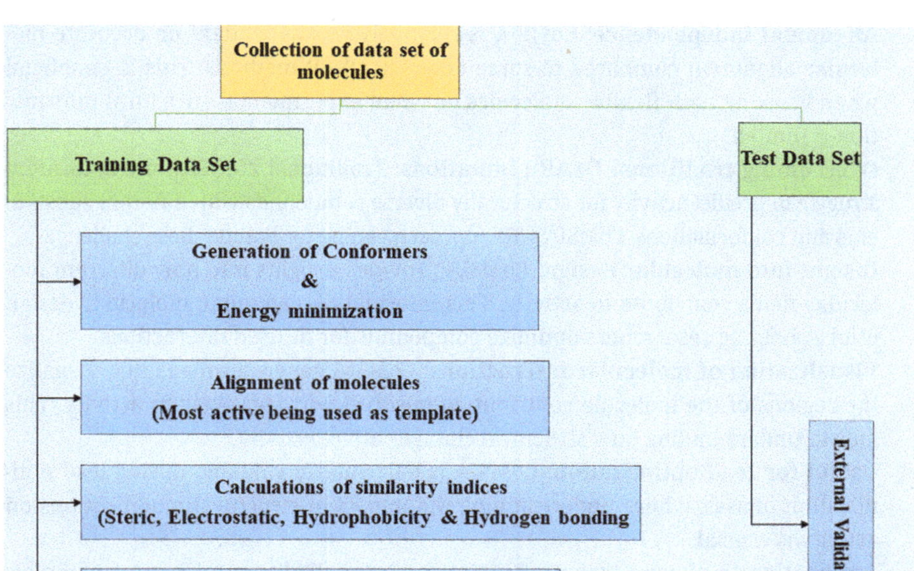

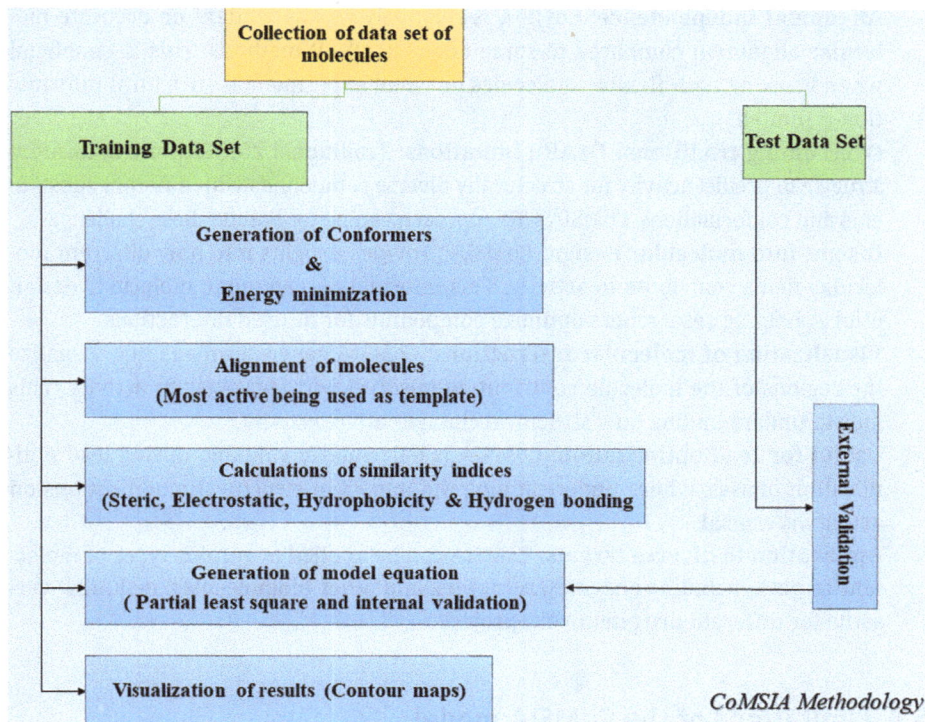

Fig. 5.7: Basic workflow for the steps involved in the generation of a CoMSIA model.

5.8 Advantages of the CoMSIA model

- **Inclusion of multiple molecular fields**: CoMSIA takes into account multiple molecular fields, such as steric, electrostatic, hydrophobic, and hydrogen-bonding properties. This allows it to capture a more comprehensive picture of the interactions between molecules and their targets.
- **Enhanced predictive power:** By incorporating multiple fields, CoMSIA can provide more accurate predictions of molecular activity. Different interactions contribute differently to the overall activity, and CoMSIA can highlight these contributions.
- **Handling of steric effects:** CoMSIA addresses steric hindrance, which is important for accurately predicting molecular interactions. It considers the spatial arrangement of atoms in addition to their electrostatic properties.
- **Flexibility in descriptor calculations:** CoMSIA allows flexibility in selecting and calculating molecular descriptors based on different properties. This versatility enables researchers to tailor the analysis to their specific system.

- **Alignment independence:** CoMSIA is relatively less dependent on accurate molecular alignment compared to some other 3D-QSAR methods. This is beneficial when working with flexible molecules or when experimental structural information is limited.
- **Overcoming traditional QSAR Limitations:** Traditional 2D-QSAR methods often struggle to predict activity for structurally diverse compounds with different substituents and conformations. CoMSIA's 3D approach can better handle these challenges.
- **Insight into molecular design:** CoMSIA provides insights into how different molecular fields contribute to activity. This information can guide molecular design efforts, helping researchers optimize compounds for desired interactions.
- **Visualization of molecular interactions:** CoMSIA generates maps that visualize the regions of the molecule contributing positively or negatively to activity. This aids in understanding how structural changes affect activity.
- **Useful for lead optimization:** CoMSIA is particularly valuable during lead optimization phases, where understanding the impact of structural modifications on activity is crucial.
- **Application to diverse targets:** CoMSIA can be applied to various types of molecular targets, including enzymes, receptors, and other biomolecules, making it versatile for different drug discovery projects.

5.9 Limitations of the CoMSIA model

- **Sensitivity to alignment:** CoMSIA heavily relies on the accurate alignment of molecules in the dataset. Small variations in alignment can lead to significant changes in calculated descriptors, potentially affecting the model's reliability and predictive power.
- **Descriptor selection and influence:** The choice of descriptors used in CoMSIA can significantly impact the results. Selecting an inappropriate set of descriptors or ignoring important ones can lead to biased or inaccurate models. Additionally, the influence of individual descriptors on the overall results may not always be well understood.
- **Sparse data:** CoMSIA models can suffer from sparsity issues, especially when dealing with limited data points or a small number of active compounds. This can lead to overfitting and less reliable predictions.
- **Descriptor interpretation:** While CoMSIA provides molecular fields that correlate with activity, interpreting these fields in a mechanistic and actionable way can be challenging. Understanding how specific field contributions relate to biological interactions is not always straightforward.
- **Dependence on training set:** CoMSIA models are highly dependent on the quality and representativeness of the training set. If the training set does not ade-

quately cover the chemical space or the diversity of the compounds being studied, the model's predictive ability may be compromised.

- **Extrapolation issues:** CoMSIA models are often less accurate when extrapolating predictions to compounds that significantly differ from those in the training set in terms of structure and properties.
- **Assumption of additivity:** CoMSIA assumes that the effects of molecular fields are additive, meaning that the contributions from different fields are independent. This assumption might not always hold true for all molecular interactions.
- **Computational complexity:** CoMSIA calculations can be computationally intensive, especially when dealing with larger datasets or more complex molecules. This complexity can limit the practicality of the method for high-throughput applications.
- **Model overfitting:** Like any QSAR model, CoMSIA models can overfit the noise present in the training data, leading to poor generalization performance on new, unseen compounds.
- **Lack of mechanistic insights:** While CoMSIA can provide correlations between molecular properties and activity, it might not offer deep mechanistic insights into the underlying biological processes.

5.10 Other 3D-QSAR approaches

After the introduction of CoMFA and CoMSIA approaches for drug discovery, which remained the flag-bearer of 3D-QSAR, numerous 3D-QSAR approaches have emerged. A brief description of these approaches is given in Tab. 5.1 for the other alignment-based 3D-QSAR methods.

Tab. 5.1: Description of important alignment-based 3D-QSAR methodologies.

Alignment-based 3D-QSAR models		
S. no.	**Method**	**Description and basic steps involved**
1	Molecular surface analysis (MSA)	– The dataset of the molecules is selected – Conformations of each molecule are generated along with the energy minimization – A bioactive conformer is hypothesized based on the global minima – A shape reference compound is selected, generally the most bioactive one – Pair-wise alignment of the molecules is performed – Calculation of molecular shape-based descriptors is done – Incorporation of other parameters is done if required – Generation of a trial QSAR [19]

Tab. 5.1 (continued)

Alignment-based 3D-QSAR models		
S. no.	**Method**	**Description and basic steps involved**
2	Receptor surface analysis (RSA)	– Molecules are optimized and conformers are generated with the determination of their bioactive conformations – Alignment of the bioactive conformations is done – Generation of receptor surface based on the aligned molecules is performed – Calculation of the descriptors based on the properties of the receptor surface – Development of the QSAR model [20]
3	Self-organizing molecular field analysis (SOMFA)	– The major difference is the calculation of the "mean centered activity" (MCE) of each molecule – Each biological activity is subtracted from the mean of the biological activities, which represents molecules with positive values of MCE as active and those having negative values as inactive – Aligned molecules are placed inside the grid, and steric and electrostatic field values are multiplied by MCE – Calculation of SOMFA descriptors performed and correlated with the biological activity – Instead of contour maps, master grids are generated for the visualization [21]
4	Genetically evolved receptor modeling (GERM)	– Aligned conformation of the molecules is taken as the starting point – The unique thing about the model is it derives a receptor cavity around the aligned conformations using the genetic algorithm approach – Interaction energies of each conformer are calculated around this generated receptor cavity using chemistry at Harvard molecular mechanics (CHARMM) force field – These interaction energies are correlated with the biological activity [22]
5	HINT interaction field analysis (HIFA)	– Considered a hybrid or extension of the traditional CoMFA model – Hydrophobic field parameters of the aligned molecules placed inside the grid and calculated – HINT intermolecular interaction maps are generated to visualize the effect of hydrophobicity on the biological activity [23]

Tab. 5.1 (continued)

Alignment-based 3D-QSAR models		
S. no.	**Method**	**Description and basic steps involved**
6	Adaption of the field for molecular comparison (AFMoC)	– Also known as "reverse CoMFA" as the fields enumerated are derived from the protein environment rather than the ligands – A grid is placed inside the binding site of the protein and "potential fields" are calculated – These potential fields are converted to "interaction field values" by multiplying them by the distance-dependent atom-type properties of the molecules docked inside the cavity – Interaction field values are correlated with the biological activity [24]
7	Molecular quantum similarity measures (MQSM)	– MQSM represents the degree of similarity between the two molecules based on the electron densities – Quantum mechanical calculations are performed for calculating the MQSM between two molecules by using the following integral: $Z_{AB}\,(\Omega_a) = \iint \rho_A(r_1)\,\Omega_a(r_1,r_2)\,\rho_B(r_2)\,dr_1\,dr_2$ where $\rho_A(r_1)$ and $\rho_B(r_2)$ represent the density functions and $\Omega_a(r_1,r_2)$ represents positive definite operator – As the number of MQSM variables obtained is very large, hence, the partial component analysis (PCA) is performed to reduce the number of variables [25]
8	Voronoi field analysis (VFA)	– Alignment of the dataset of the molecules performed – New regions from the aligned molecules space are defined, which are termed as *Voronoi Polyhedrals* that are different from CoMFA where lattice points of grid are defined – The number of variables obtained is less as compared to traditional CoMFA – Steric and electrostatic parameters are obtained from the aligned molecules and steric field parameters are calculated based on the hard-sphere model instead of Lennard-Jones potential [26]

Besides the alignment-based 3D-QSAR models, many other models have been proposed for 3D-QSAR. which are independent of the alignment criteria and are briefly discussed in Tab. 5.2.

Tab. 5.2: Description of important alignment-independent 3D-QSAR methodologies.

Alignment-independent 3D-QSAR models		
S. no.	**Method**	**Description and basic steps involved**
1	Weighed holistic invariant molecular (WHIM) descriptor analysis	– A new set of descriptors named WHIM descriptors is calculated, which defines the 3D structure of the molecule in terms of its size, shape, conformation, and distribution of the atoms – These descriptors are calculated simply from the 3D conformation of the molecule represented by *x*-, *y*-, and *z*-coordinates – Directional WHIM descriptors are calculated using a principal component analysis (PCA) approach – Nondirectional WHIM descriptors are derived from the directional descriptors [27]
2	Molecular hologram QSAR (HQSAR)	– A new set of descriptors known as molecular holograms is calculated, which is based on the fragment fingerprints of the molecules – HQSAR model is generated by using the PLS statistical method for correlating the biological activity with the variables [28]
3	VolSurf	– It is generated with the help of *VolSurf* descriptors, which are obtained by relating the 3D molecular interaction fields of the molecules to their physicochemical properties – This approach is very useful and widely used for ADME prediction [29]
4	Comparative molecular moment analysis (CoMMA)	– Molecular similarity descriptors are enumerated based on the moments of the molecular mass and charge distributions – These descriptors contain information about the shape and charge distribution of the molecules in 3D space [30] – The QSAR model is generated by employing a PLS statistical approach
5	Comparative spectral analysis (CoSA)	– This is the approach in which the spectra of the molecules are employed as descriptors for the establishment of correlation with biological activity – 1H, 13C, infrared, and mass spectra employed for the generation of descriptors – A multivariate regression model (PLS) is used for the generation of the mathematical equation [31]

References

[1] Oprea TI. On the information content of 2D and 3D descriptors for QSAR. Journal of the Brazilian Chemical Society, 2002, 13, 811–5.

[2] Akamatsu M. Current state and perspectives of 3D-QSAR. Current Topics in Medicinal Chemistry, 2002, 2(12), 1381–94.

[3] Katritzky AR, Slavov SH, Dobchev DA, Karelson M. Comparison between 2D and 3D-QSAR approaches to correlate inhibitor activity for a series of indole amide hydroxamic acids. QSAR & Combinatorial Science, 2007, 26(3), 333–45.

[4] Verma J, Khedkar VM, Coutinho EC. 3D-QSAR in drug design-a review. Current Topics in Medicinal Chemistry, 2010, 10(1), 95–115.

[5] Ivanciuc O. 3D QSAR models. QSPR/QSAR Studies by Molecular Descriptors, 2001, 233–80.

[6] Hopfinger AJ, Wang S, Tokarski JS, Jin B, Albuquerque M, Madhav PJ, Duraiswami C. Construction of 3D-QSAR models using the 4D-QSAR analysis formalism. Journal of the American Chemical Society, 1997, 119(43), 10509–24.

[7] Kubinyi H. QSAR and 3D QSAR in drug design Part 1: Methodology. Drug Discovery Today, 1997, 2(11), 457–67.

[8] Oprea TI. 3D QSAR modeling in drug design. Computational Medicinal Chemistry for Drug Discovery, 2004, 571–616.

[9] Thareja S. Steroidal 5α-reductase inhibitors: A comparative 3D-QSAR study review. Chemical Reviews, 2015, 115(8), 2883–94.

[10] Madhavan T. A review of 3D-QSAR in drug design. Journal of the Chosun Natural Science, 2012, 5(1), 1–5.

[11] Hillebrecht A, Klebe G. Use of 3D QSAR models for database screening: A feasibility study. Journal of Chemical Information and Modeling, 2008, 48(2), 384–96.

[12] Cramer RD, Patterson DE, Bunce JD. Comparative molecular field analysis (CoMFA). 1. Effect of shape on binding of steroids to carrier proteins. Journal of the American Chemical Society, 1988, 110(18), 5959–67.

[13] SYBYL 8.1, Tripos International, St. Louis, USA.

[14] Kubinyi H. Comparative molecular field analysis (CoMFA). The Encyclopedia of Computational Chemistry, 1998, 1, 448–60.

[15] Akamatsu M. Current state and perspectives of 3D-QSAR. Current Topics in Medicinal Chemistry, 2002, 2(12), 1381–94.

[16] Hopfinger AJ, Tokarski JS. Three-Dimensional Quantitative Structure-activity Relationship Analysis. Marcel Dekker, Inc., New York, USA, 1997.

[17] Oprea TI. 3D QSAR modeling in drug design. Computational Medicinal Chemistry for Drug Discovery, 2004, 571–616.

[18] Klebe G, Abraham U, Mietzner T. Molecular similarity indices in a comparative analysis (CoMSIA) of drug molecules to correlate and predict their biological activity. Journal of Medicinal Chemistry, 1994, 37(24), 4130–46.

[19] Walters DE, Hopfinger AJ. Case studies of the application of molecular shape analysis to elucidate drug action. Journal of Molecular Structure: THEOCHEM, 1986, 134(3–4), 317–23.

[20] Hahn M. Receptor surface models. 1. Definition and construction. Journal of Medicinal Chemistry, 1995, 38(12), 2080–90.

[21] Robinson DD, Winn PJ, Lyne PD, Richards WG. Self-organizing molecular field analysis: A tool for structure– activity studies. Journal of Medicinal Chemistry, 1999, 42(4), 573–83.

[22] Walters DE, Hinds RM. Genetically evolved receptor models: A computational approach to construction of receptor models. Journal of Medicinal Chemistry, 1994, 37(16), 2527–36.

[23] Semus SF. A novel hydropathic intermolecular field analysis (HIFA) for the prediction of ligand-receptor binding affinities. Medicinal Chemistry Research, 1999, 9(7–8), 535–47.

[24] Gohlke H, Klebe G. DrugScore meets CoMFA: Adaptation of fields for molecular comparison (AFMoC) or how to tailor knowledge-based pair-potentials to a particular protein. Journal of Medicinal Chemistry, 2002, 45(19), 4153–70.

[25] Gironés X, Amat L, Carbó-Dorca R. Using molecular quantum similarity measures as descriptors in quantitative structure-toxicity relationships. SAR and QSAR in Environmental Research, 1999, 10(6), 545–56.

[26] Chuman H, Karasawa M, Fujita T. A novel three-dimensional QSAR procedure: Voronoi field analysis. Quantitative Structure-Activity Relationships, 1998, 17(04), 313–26.

[27] Todeschini R, Gramatica P. New 3D molecular descriptors: The WHIM theory and QSAR applications. Perspectives in Drug Discovery and Design, 1998, 9(0), 355–80.

[28] Lowis DR. HQSAR: A new, highly predictive QSAR technique. Tripos Technical Notes, 1997, 1(5), 17.

[29] Cruciani G, Pastor M, Guba W. VolSurf: A new tool for the pharmacokinetic optimization of lead compounds. European Journal of Pharmaceutical Sciences, 2000, 11, S29–39.

[30] Silverman BD, Platt DE. Comparative molecular moment analysis (CoMMA): 3D-QSAR without molecular superposition. Journal of Medicinal Chemistry, 1996, 39(11), 2129–40.

[31] Bursi R, Dao T, van Wijk T, de Gooyer M, Kellenbach E, Verwer P. Comparative spectra analysis (CoSA): Spectra as three-dimensional molecular descriptors for the prediction of biological activities. Journal of Chemical Information and Computer Sciences, 1999, 39(5), 861–7.

Chapter 6
Statistical approaches used in the QSAR

If your experiment needs statistics, you ought to have done a better experiment – Ernest Rutherford

6.1 Introduction

Statistics is the science of numerical data, which involves collection, analysis, interpretation, representation, and inferring conclusions from these numerical data [1]. Statistics is employed at each step of the quantitative structure–activity relationship (QSAR) model generation and hence, it becomes indispensable for anyone who wishes to learn QSAR to have basic knowledge of statistics [2]. There are different statistical approaches used in the QSAR depending upon the type of QSAR model being generated [3]. Statistical approaches in QSAR encompass a spectrum from linear to nonlinear models, integrating various techniques to model intricate relationships between molecular structures and activities, facilitating the efficient design of new compounds for specific applications [4, 5]. Whether it is a traditional 2D-QSAR or 3D-QSAR, a certain statistical approach is employed in each of them for the generation of the QSAR mathematical equation, which correlates the biological activity with the descriptors of the molecules [6]. A brief overview of the statistical methods used for the generation of the QSAR model is given in Fig. 6.1.

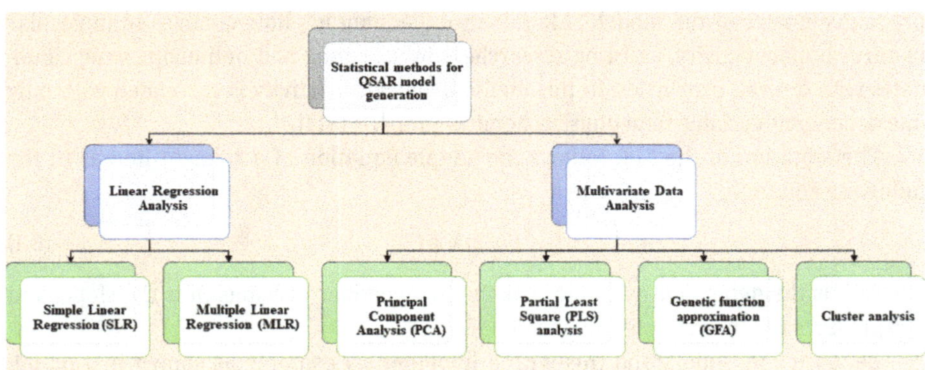

Fig. 6.1: Different statistical approaches employed in the development of different QSAR models.

https://doi.org/10.1515/9783111434858-006

6.2 Linear regression analysis

Linear regression analysis is considered the true mathematical approach for developing a correlation of biological activity with the descriptors of the molecules [7, 8]. In this approach, the biological activity of the molecules is the response, outcome, or dependent variable and the physicochemical parameters or descriptors are termed as predictor, explanatory, or independent variables [9]. The number of independent variables employed for the generation of a mathematical equation of correlation further classifies this approach into simple linear regression (SLR) or multiple linear regression (MLR). Linear regression quantifies this relationship by fitting a linear equation to the data, estimating coefficients that define the influence of molecular descriptors on the activity. Utilizing various molecular features, it predicts the activity of new compounds. QSAR's linear regression enables rational drug design, environmental impact assessment, and toxicity prediction, playing a vital role in optimizing compound properties and accelerating chemical and pharmaceutical research [10].

6.2.1 Simple linear regression (SLR)

In QSAR drug discovery, SLR analysis is a fundamental statistical approach [11]. It aims to establish a relationship between the chemical structure of molecules and their biological activities [12]. By plotting and analyzing the data, this method identifies a linear equation that best fits the relationship between molecular descriptors and activity values [13]. The slope and intercept of the line provide insights into the correlation and predictive power of the model. SLR aids in understanding how changes in molecular features impact activity, assisting researchers in designing and optimizing drug candidates with desired properties. In this method, biological activity is correlated with only one dependent variable (one physicochemical property) [14].

The equation of the SLR follows the simple equation of a straight line as in the following equation:

$$Y = mX + C \tag{6.1}$$

where Y is the dependent variable, X is the independent variable, m is the slope, and C represents the intercept of the line.

Let us try to understand this with a hypothetical example as shown in Tab. 6.1. Suppose there is a series of 10 hypothetical compounds numbered from 1 to 10 along with their hypothetically experimental values of the biological activity and log P. For developing a correlation between biological activity and log P, we will use an SLR approach. Here, our dependent variable will be biological activity (y) and the independent variable will be log P (x).

Tab. 6.1: The hypothetical values of biological activity and the log *P* of 10 hypothetical compounds.

Compound	Biological activity value (negative log of IC_{50})	Log *P* value
1	2.34	4.55
2	2.56	4.78
3	2.63	4.79
4	2.37	4.59
5	2.76	4.88
6	2.14	4.25
7	2.04	4.08
8	2.47	4.66
9	2.31	4.51
10	2.98	4.99

The correlation equation between the biological activity and log P from the SLR should be as in the following equation:

$$\text{Biological activity} = m \log P + C \tag{6.2}$$

where m represents the slope of the line and C represents the intercept of the line. Both of them can be calculated mathematically with the following equation:

$$m = R\left(\frac{\mathrm{SD}_{X\mathrm{var}}}{\mathrm{SD}_{Y\mathrm{var}}}\right) \tag{6.3}$$

where R is the Pearson correlation coefficient and SD represents the standard deviation of X and Y variables. R can be calculated from the following equation:

$$R = \Sigma\left(\frac{(X - X_{\mathrm{mean}})\ (Y - Y_{\mathrm{mean}})}{\sqrt{\Sigma\,(X - X_{\mathrm{mean}})^2\,\Sigma\,(Y - Y_{\mathrm{mean}})^2}}\right) \tag{6.4}$$

where X variable is the value of log P and X_{mean} is its average. Y variable represents biological activity and Y_{mean} is its average value. The values of $\mathrm{SD}_{X\mathrm{var}}$ and $\mathrm{SD}_{Y\mathrm{var}}$ can be calculated with the help of the following equations, respectively:

$$\mathrm{SD}_{X\mathrm{var}} = \sqrt{\Sigma}\left(\frac{(X - X_{\mathrm{mean}})^2}{\sqrt{\Sigma n - 1}}\right) \tag{6.5}$$

$$\mathrm{SD}_{Y\mathrm{var}} = \sqrt{\Sigma}\left(\frac{(Y - Y_{\mathrm{mean}})^2}{\sqrt{\Sigma n - 1}}\right) \tag{6.6}$$

where n depicts the number of observations in this case.

The other parameter C, representing the intercept of the equation, can be calculated from the slope value with the help of the following equation:

Tab. 6.2: Calculated parameters from the values obtained from Tab. 6.1 of 10 hypothetical compounds.

Compound	**Biological activity value (negative log of IC_{50})**	**Log *P* value**	$(Y-Y_{mean})$	$(X-X_{mean})$	$(X-X_{mean})(Y-Y_{mean})$	$(Y-Y_{mean})^2$	$(X-X_{mean})^2$
1	2.34	4.55	−0.12	−0.06	0.0072	0.0144	0.0036
2	2.56	4.78	0.1	0.17	0.0170	0.0100	0.0289
3	2.63	4.79	0.17	0.18	0.0306	0.0289	0.0324
4	2.37	4.59	−0.09	−0.02	0.0018	0.0081	0.0004
5	2.76	4.88	0.3	0.27	0.0810	0.0900	0.0729
6	2.14	4.25	−0.32	−0.36	0.1152	0.1024	0.1296
7	2.04	4.08	−0.42	−0.53	0.2226	0.1764	0.2809
8	2.47	4.66	0.01	0.05	0.0005	0.0001	0.0025
9	2.31	4.51	−0.15	−0.10	0.0150	0.0225	0.0100
10	2.98	4.99	0.52	0.38	0.2204	0.2704	0.1444
	$Y_{mean} = 2.46$	$X_{mean} = 4.61$			$\Sigma = 0.7113$	$\Sigma = 0.7233$	$\Sigma = 0.7056$

$$C = Y_{\text{mean}} - mX_{\text{mean}} \tag{6.7}$$

where Y_{mean} depicts the mean of biological activity (Y), m depicts the slope of the curve, and X_{mean} depicts the mean of the value of descriptor (log P)

Now, we have all the necessary formulas for the calculation of our simple regression model. Table 6.2 depicts the calculated parameters required for the generation of the simple regression model.

The value of Pearson's correlation coefficient (R) (equation (6.8)) can be calculated by plugging in the required values in the following equation:

$$R = \Sigma \left(\frac{0.7113}{\sqrt{(0.7056)(0.7233)}} \right) \tag{6.8}$$

From equation (6.8), the value of R is calculated to be 0.996. Likewise, the slope (m) can be calculated from equation (6.3) and comes out to be 0.98. By using equations (6.5) and (6.6) the values of $\text{SD}_{X\text{var}}$ are calculated to be 0.280 and $\text{SD}_{Y\text{var}}$ is calculated to be 0.283. C, representing the intercept, can be calculated from equation (6.7) and comes out to be −2.05.

We now have all the values of m and C that are required for the generation of an equation for the correlation between the biological activity and log P. From the above values, an SLR equation can be written in the form of the following equation:

$$\text{Biological activity} = 0.98 \log P - 2.05 \tag{6.9}$$

From the above generated simple regression equation, a scatter plot can be derived between the experimental values of biological activity and the calculated values from the equation (Fig. 6.2).

SLR is nowadays seldom used as it is close to impossible to correlate the biological activity of the molecules to a single parameter. With the emergence of numerous descriptors (parameters), the use of other advanced statistical approaches like MLR and PLS has become more prevalent in the QSAR model generation.

6.2.1.1 Advantages of simple linear regression (SLR)

- **Simplicity:** SLR involves only one independent variable (descriptor) and one dependent variable (biological activity). This simplicity makes it easier to interpret and understand the relationship between the descriptor and the activity.
- **Interpretability:** The relationship between the descriptor and the activity is represented by a straight line, making it straightforward to grasp the direction and magnitude of the effect of the descriptor on the activity.
- **Visualization:** The relationship between the descriptor and the activity can be easily visualized through a scatter plot with the regression line, aiding in the intuitive understanding of the data.

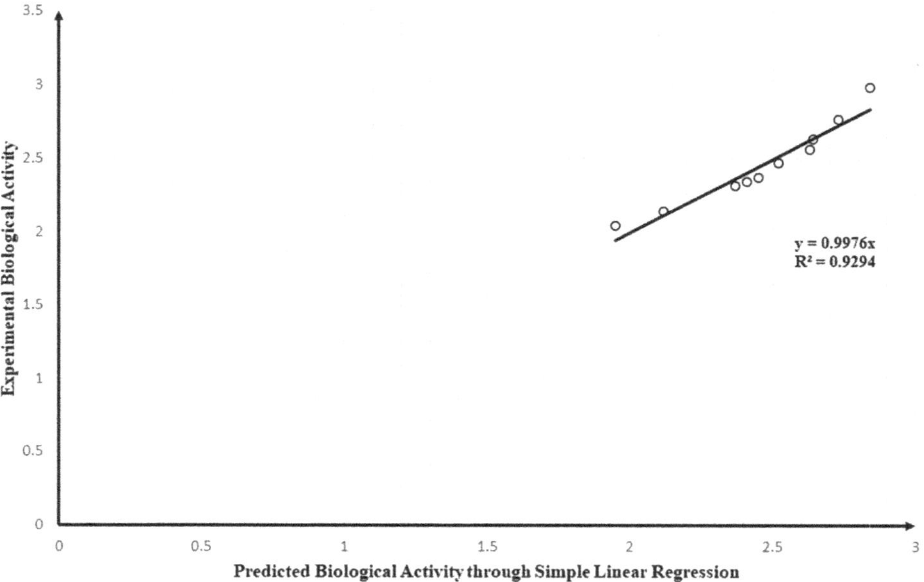

Fig. 6.2: A scatter plot obtained from the simple linear regression between the observed and predicted biological activities.

- **Reduced multicollinearity:** Since only one descriptor is used, issues related to multicollinearity (high correlation between descriptors) are eliminated, leading to more stable coefficient estimates.
- **Feature selection:** SLR can help in the initial screening and selection of descriptors that show the most promising linear relationships with the activity. This can guide more complex modeling efforts.
- **Computational efficiency:** Building and interpreting an SLR model is computationally less intensive than dealing with multiple descriptors in MLR, making it quicker to analyze.
- **Baseline model:** SLR can serve as a baseline model for comparison with more complex models. If the relationship between the descriptor and the activity is close to linear, a simple linear model might provide sufficient accuracy.
- **Initial insight:** SLR can provide an initial insight into the potential influence of a single descriptor on the activity, helping researchers decide whether to explore more advanced modeling techniques.
- **Useful for educational purposes:** SLR is often used to introduce students and newcomers to the concept of regression analysis and its application in QSAR.

However, it is important to note that SLR has its own limitations, particularly when dealing with complex QSAR datasets that involve multiple descriptors with potentially nonlinear relationships. In such cases, SLR might not capture the full complexity of the underlying interactions, and more advanced techniques might be necessary for accurate modeling and prediction.

6.2.1.2 Limitations of simple linear regression (SLR)

- **Simplistic model:** SLR assumes a linear relationship between a single independent variable (descriptor) and the dependent variable (biological activity). This simplicity might not capture the complexity of interactions among multiple descriptors.
- **Multivariate interactions:** Biological activities are often influenced by multiple molecular properties simultaneously. SLR overlooks these multivariate interactions, potentially leading to oversimplified or inaccurate predictions.
- **Limited predictive power:** SLR's single-variable focus might not capture the full variation in the dependent variable, resulting in limited predictive power. QSAR models require more comprehensive approaches to make reliable predictions.
- **Assumption of independence:** SLR assumes that the independent variable is not influenced by any other factors. In QSAR, molecular descriptors can be interrelated, violating this assumption and leading to biased coefficient estimates.
- **Outliers:** Outliers can disproportionately influence SLR models, leading to a skewed fit. QSAR datasets are susceptible to outliers due to experimental errors or unique compound behaviors, affecting the model's accuracy.
- **Data distribution assumptions:** SLR assumes that the residuals (differences between predicted and actual values) are normally distributed. If the data does not meet this assumption, the model's reliability is compromised.
- **Limited descriptor exploration:** SLR only explores the relationship between one descriptor and the activity, potentially missing out on important molecular properties that collectively influence the activity.
- **Overfitting and underfitting:** SLR can suffer from overfitting (capturing noise) or underfitting (oversimplifying the relationship). Balancing model complexity with data fit is challenging, especially in QSAR with limited data.
- **Context dependence:** Similar to multivariate models, SLR might not be universally applicable across diverse compound classes or biological systems, limiting its generalization.
- **Mechanistic insights:** SLR provides a basic correlation analysis but lacks mechanistic insights into the underlying interactions between molecular properties and biological activity.

Given these limitations, researchers often move beyond SLR to embrace more sophisticated techniques such as MLR, nonlinear regression, and machine learning methods. These approaches are better equipped to handle the complexities inherent in QSAR drug discovery, offering improved predictive accuracy and a deeper understanding of molecular interactions.

6.2.2 Multiple linear regression (MLR)

MLR is a pivotal statistical technique within QSAR analysis for drug discovery [15]. It evaluates the relationship between multiple molecular descriptors (features) and the biological activity of compounds [16]. MLR aims to model this connection by fitting a linear equation, considering various descriptors simultaneously [17]. The coefficients in the equation signify the impact of each descriptor on the activity. MLR aids in identifying essential descriptors influencing the activity and predicting activity values for new compounds [18]. It serves as a valuable tool in optimizing drug design by elucidating the quantitative link between molecular properties and biological outcomes, fostering more informed decision-making in drug development [19].

MLR is a simple extension of the SLR in which biological activity (dependent, response, or outcome variable) is correlated with more than one physicochemical parameter or descriptor (independent, predictor, or explanatory variable). MLR is the choice of statistical approach in the traditional QSAR [20].

The general representation of the MLR is depicted in the following equation:

$$Y = m_1X_1 + m_2X_2 + m_3X_3 \ldots\ldots\ldots\ldots\ldots\ldots + m_nX_n + C + \epsilon \tag{6.10}$$

where Y is the dependent variable, m_n is the slope coefficient of each independent variable, X_n represents each independent variable, C is the intercept of the equation, and ϵ represents the error in the calculation of the response variable, also known as residual.

Let us understand the underlying basics of the MLR with an extension of the previous example. In Tab. 6.3 we have given a hypothetical example of 10 compounds with their biological activity data and values of two parameters, log P and polar surface area. The MLR equation to be developed from this data will be in the form of the following equation:

$$\text{Biological activity} = m_1X_1 + m_2X_2 + C \tag{6.11}$$

Tab. 6.3: Hypothetical values of biological activity, log P and polar surface area of 10 hypothetical compounds.

Compound	Biological activity value (negative log of IC_{50}) (Y)	Log P value (X_1)	Polar surface area (X_2)
1	2.34	4.55	17.07
2	2.56	4.78	19.23
3	2.63	4.79	19.34
4	2.37	4.59	16.11
5	2.76	4.88	15.92
6	2.14	4.25	22.32
7	2.04	4.08	15.04

Tab. 6.3 (continued)

Compound	Biological activity value (negative log of IC_{50}) (Y)	Log P value (X_1)	Polar surface area (X_2)
8	2.47	4.66	15.45
9	2.31	4.51	18.12
10	2.98	4.99	19.34

From the above table, we know that we have response variable Y (biological activity) and two independent variables (X_1 and X_2), which are to be correlated using MLR. For generating the MLR equation, we have to calculate the values m_1, m_2, and C that can be determined from the following equations:

$$m_1 = \frac{(\Sigma X_2^2)(\Sigma X_1 Y) - (\Sigma X_1 X_2)(\Sigma X_2 Y)}{(\Sigma X_1^2)(\Sigma X_2^2) - (\Sigma X_1 X_2)^2} \tag{6.12}$$

$$m_2 = \frac{(\Sigma X_1^2)(\Sigma X_2 Y) - (\Sigma X_1 X_2)(\Sigma X_1 Y)}{(\Sigma X_1^2)(\Sigma X_2^2) - (\Sigma X_1 X_2)^2} \tag{6.13}$$

where $\Sigma X_i^2 = \Sigma X_i^2 - \left[(\Sigma X_i)^2/N\right]$, $\Sigma x_i Y = \Sigma X_i Y - [(\Sigma X_i)(\Sigma Y)/N]$, and $\Sigma X_1 X_2 = \Sigma X_1 X_2 - [(\Sigma X_1)(\Sigma X_2)/N]$, and N is the total number of compounds in the study.

The other parameter C, representing the intercept of the equation, can be calculated from the slope value by employing the following equation:

$$C = Y_{\text{mean}} - m_1 X_{1\,\text{mean}} - m_2 X_{2\,\text{mean}} \tag{6.14}$$

Equations (6.12) to (6.14) at first seem to be very complicated but are much easier to understand and calculate, as depicted in Tab. 6.4.

From the parameters obtained from Tab. 6.4, we can easily calculate the values of ΣX_1^2, ΣX_2^2, $\Sigma X_1 Y$, $\Sigma X_2 Y$, and $\Sigma X_1 X_2$. The calculated values are given in Tab. 6.5.

From the values obtained in Tab. 6.5, we can calculate the m_1 and m_2 parameters required for the generation of our MLR equation by using the following equations The calculated values of m_1 and m_2 were found to be 0.48 and −0.20, respectively:

$$m_1 = \frac{(3,193.04)(101.75) - (0.35)(-639.83)}{(210.91)(3,193.04) - (0.35)^2}$$

$$m_2 = \frac{(210.91)(-639.83) - (0.35)(101.75)}{(210.91)(3,193.04) - (0.35)^2}$$

From the values of m_1 and m_2, we can now calculate the intercept value (C) from the following equation. The calculated value of C from this equation is 3.81:

Tab. 6.4: Values calculated from the data of the Tab. 6.3.

Compound	Biological activity (negative log of IC_{50}) (Y)	Log P (X_1)	Polar surface area (X_2)	$(X_1)^2$	$(X_2)^2$	X_1Y	X_2Y	X_1X_2
1	2.34	4.55	17.07	20.70	291.38	10.5	39.94	77.67
2	2.56	4.78	19.23	22.85	369.79	12.24	49.23	91.92
3	2.63	4.79	19.34	22.94	374.03	12.60	50.86	92.64
4	2.37	4.59	16.11	21.07	259.53	10.88	38.18	73.94
5	2.76	4.88	15.92	23.81	253.45	13.47	43.94	77.69
6	2.14	4.25	22.32	18.06	498.18	9.09	47.76	94.86
7	2.04	4.08	15.04	16.65	226.20	8.32	30.68	61.36
8	2.47	4.66	15.45	21.72	238.70	11.51	38.16	71.99
9	2.31	4.51	18.12	20.34	328.33	10.42	41.86	81.72
10	2.98	4.99	19.34	24.90	374.03	14.87	57.63	96.51
	$\Sigma Y = 24.6$	$\Sigma X_1 = 46.08$	$\Sigma X_2 = 177.94$	$(\Sigma X_1^2) = 213.04$	$(\Sigma X_2^2) = 3223.62$	$\Sigma X_1Y = 113.09$	$\Sigma X_2Y = 438.24$	$\Sigma X_1X_2 = 820.3$
	$Y_{mean} = 2.46$	$X_{1mean} = 4.61$	$X_{2mean} = 17.79$					

Tab. 6.5: The values calculated from Tab. 6.4.

S. no.	Parameter	Values obtained
1	$\Sigma X_1{}^2$	213.04 – [(21.25/10)] = 210.91
2	$\Sigma X_2{}^2$	3223.62 – [(305.81/10)] = 3,193.04
3	$\Sigma X_1 Y$	113.09 – [(4.61 × 24.6)/10] = 101.75
4	$\Sigma X_2 Y$	438.24 – [(438.24 × 24.6)/10] = −639.83
5	$\Sigma X_1 X_2$	820.3 – [(46.08 × 177.94)/10] = 0.35

$$C = 2.46 - (0.48)(4.61) - (-0.20)(17.79)$$

From all these calculated values of m_1, m_2, and C, the correlation equation of MLR can be written as in the following equation:

$$\text{Biological activity} = 0.48 \log P - 0.20\ \text{PSA} + 3.81 \tag{6.15}$$

The scatter plot between the experimental biological activity and calculated biological activity can be plotted from the MLR equation (Fig. 6.3).

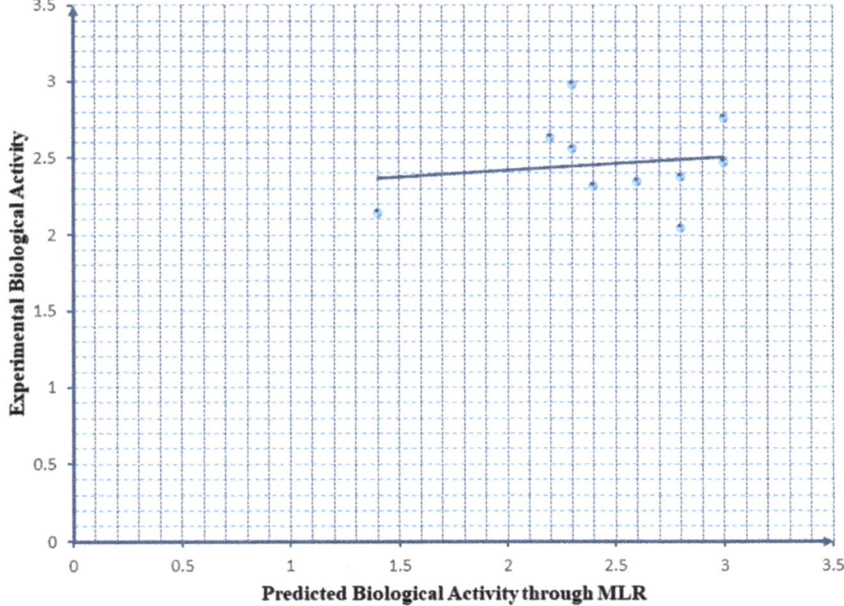

Fig. 6.3: A scatter plot obtained from the multiple linear regression (MLR) between the observed and predicted biological activities.

The MLR approach for the development of the correlation between biological activity and descriptors is widely used in the traditional QSAR. Almost a half-decade has passed since the first QSAR model was developed by Hansch and Fujita and now there has been an emergence of thousands of 2D descriptors. One major hurdle that is observed for the development of a reliable and robust QSAR model is to choose the best descriptors from the pool of descriptors to generate our MLR equation. For the selection of appropriate descriptors, techniques such as best subset selection or genetic algorithm (GA) approach are used. The number of variables to be used for the development of the MLR equation depends upon the number of molecules available for the generation of the model. Generally, it is believed that molecules in the MLR study should be five times or more than the number of descriptors to be used. The major drawback of the MLR is that while selecting the independent variables from the pool of the descriptors, important descriptors could be neglected; hence, more sophisticated statistical approaches are now becoming popular in the selection of the variables.

6.2.2.1 Advantages of multiple linear regression (MLR)

- **Interpretability:** MLR provides coefficients for each descriptor, allowing researchers to interpret the impact of individual molecular properties on the biological activity. This aids in understanding the underlying relationships between chemical structures and activity.
- **Variable selection:** MLR helps identify significant descriptors that contribute most to the activity. This feature selection process guides researchers in focusing on relevant molecular features and excluding irrelevant ones, improving model accuracy.
- **Prediction:** MLR models can predict the activity of new compounds based on their molecular descriptors. This predictive ability accelerates drug discovery by suggesting potential candidates for synthesis and testing, saving time and resources.
- **Simplicity:** MLR is relatively straightforward to implement and comprehend, making it accessible to researchers without advanced statistical expertise. It serves as a foundational technique for those new to QSAR analysis.
- **Insight generation:** By analyzing the coefficients and statistical significance of descriptors, MLR provides insights into the structure–activity relationships, aiding in the design of compounds with desired activity profiles.
- **Model transparency:** MLR models are transparent and offer a clear mathematical relationship between descriptors and activity. This transparency enhances the trustworthiness of the model's predictions and findings.
- **Risk assessment:** MLR can quantify the uncertainty and variability associated with the model's predictions, enabling researchers to assess the reliability and potential risks of using the model in decision-making.

- **Data exploration:** MLR can handle multiple descriptors simultaneously, allowing researchers to explore complex interactions and correlations between molecular features and activity.
- **Foundation for advanced techniques:** MLR serves as a stepping stone for more sophisticated modeling techniques. It helps researchers understand the basic principles of modeling structure–activity relationships before delving into more complex methods.
- **Scientific insights:** MLR aids in formulating hypotheses about the underlying mechanisms responsible for the observed biological activity, contributing to scientific understanding beyond just predictive modeling.

6.2.2.2 Limitations of multiple linear regression (MLR)

- **Assumption of linearity:** MLR assumes a linear relationship between the independent variables (molecular descriptors) and the dependent variable (biological activity). In QSAR, this assumption might not always hold true, as complex interactions between descriptors can lead to nonlinear relationships.
- **Multicollinearity:** When descriptors are highly correlated, multicollinearity occurs. This makes it challenging to discern the individual contribution of each descriptor to the model's outcome, potentially leading to unstable coefficient estimates.
- **Overfitting:** If the model includes too many descriptors relative to the available data points, it might capture noise rather than true relationships, leading to overfitting. This results in poor generalization of new data and unreliable predictions.
- **Limited descriptor space:** QSAR models are constrained by the choice of descriptors. Some relevant molecular properties might not be included, and irrelevant ones could be present, leading to biased or inaccurate predictions.
- **Outliers:** Outliers can significantly influence regression outcomes, pulling the model's fit towards extreme values. QSAR datasets often contain outliers due to experimental errors or rare compound behaviors, affecting the reliability of the model.
- **Sensitive to data distribution:** MLR assumes that the residuals (differences between predicted and actual values) are normally distributed. If the residuals deviate from normality, the model assumptions are violated.
- **Data quality and quantity:** MLR requires a sufficient amount of high-quality data to build robust models. Limited or noisy data can lead to unreliable parameter estimates and poor predictive performance.
- **Lack of mechanistic insight:** MLR provides correlation information between descriptors and activity, but it does not offer mechanistic insights into the underlying biological interactions driving the observed relationships.
- **Extrapolation risk:** MLR models are sensitive to the range of data used for training. Predictions for compounds outside the training data's descriptor space might be unreliable due to extrapolation.

- **Context dependency:** QSAR models might not be transferable across different biological systems or experimental conditions. A model developed for one class of compounds might not perform well for structurally diverse compounds.

To address these limitations, researchers often turn to more advanced techniques such as nonlinear regression, machine learning algorithms, or ensemble methods to build more accurate and robust QSAR models.

6.3 Multivariate data analysis (MVDA)

Multivariate data analysis (MVDA) in QSAR for drug discovery is a pivotal approach that enables the extraction of meaningful insights from complex chemical and biological datasets [21, 22]. MVDA techniques play a crucial role in QSAR by handling datasets with multiple variables, such as molecular descriptors and biological activities, simultaneously [23]. These methods include principal component analysis (PCA), partial least squares (PLS), and cluster analysis, among others [24]. PCA reduces dimensionality by transforming correlated variables into orthogonal components, highlighting dominant patterns in the data [25]. PLS combines the information from both molecular descriptors and biological activities to develop predictive models, effectively identifying key descriptors influencing activity. Cluster analysis categorizes compounds into meaningful groups based on similarity. MVDA not only enhances data exploration and visualization but also guides feature selection, model building, and validation. It helps uncover hidden relationships, identify outliers, and optimize predictive models [26, 27]. Through MVDA, QSAR empowers researchers to prioritize lead compounds, design novel molecules with desired properties, and accelerate drug discovery by narrowing down the chemical space for experimental testing. In essence, MVDA in QSAR offers a systematic and efficient approach to transforming complex multidimensional data into actionable insights, revolutionizing the field of drug development [28].

6.3.1 Principal component analysis (PCA)

PCA is a fundamental technique within MVDA, which holds significant importance in the realm of QSAR for drug discovery. With the burgeoning volume of chemical and biological data, PCA serves as a potent tool to distill the essential information, enabling more efficient decision-making. PCA operates by transforming the original variables (molecular descriptors) into a new set of orthogonal components, termed principal components (PCs). These components capture the maximum variance present in the data while being uncorrelated, simplifying the interpretation of complex relationships. By reducing dimensionality, PCA not only aids in data visualization but also facilitates the identification of critical descriptors that contribute most to the variance in biological activity. PCA also serves to mitigate multicollinearity issues and diminish noise, which often arises in QSAR datasets due to high intercorrelation between descriptors. By retaining a subset of PCs that explain the majority of the variance, PCA streamlines subsequent modeling efforts, resulting in more robust and accurate predictive models. PCA within MVDA for QSAR in drug discovery acts as a transformative process, translating intricate molecular information into a concise representation while preserving the essential characteristics driving biological activity. This technique empowers researchers to uncover patterns, prioritize descriptors, and expedite the identification of promising drug candidates within the vast chemical space [29, 30].

With the evolution of QSAR techniques in the past few decades, the number of descriptors has also increased enormously. Creating MLR from this large pool of independent variables (descriptors) has now become a tedious task due to two main reasons, that is, complexity in the MLR equation generation with a large number of descriptors and colinearity among the descriptors. PCA in simple words is a data reduction statistical approach in which the number of descriptors is reduced rationally without losing the important information among them [31].

To understand this, suppose initially we have a pool of 1,000 descriptors of the molecules. Creating an MLR equation correlating the biological activity of the molecules with these descriptors will be a very tedious task. Also, there will be high colinearity among some of these descriptors. To overcome all this PCA is applied. Each of the 1,000 descriptors will be converted to 1,000 PCs numbered PC_1, PC_2, PC_3, PC_4, . . ., $PC_{1,000}$. Interestingly, these PCs are derived in such a way that the highest information of the data is contained by the PC_1, and as we move from PC_1 to other components information (variation) enclosed by them decreases. Suppose in this case PC_1 contains 80% of the information (variation), PC_2 will have lesser information (say 14%), PC_3 will have even lesser information (say 5%), and moving further the information carried by the PCs will reduce subsequently. Therefore, if we select only the first three PCs from the 1,000 derived for our data analysis, that is, PC_1, PC_2, and PC_3, we will be using 99% of the information of these descriptors. In other words, at the cost of losing 1% accuracy, we can simplify our QSAR model generation by reducing the 1,000 descriptors to 3 useful variables. The choice of the number of PCs to be incorporated solely depends upon how much accuracy is desired.

The exact methodology for the PCA is complex and is done through machine learning. The basic steps involved in the PCA are depicted below:

- **Define the data**: This is the very first step in which original data of the biological activity of all the molecules along with their descriptor values is generated.

- **Creating mean-centering/normalization of data**: The data obtained in the first step is then made mean-centered or normalization of the same is done. Normalization is done to make all the data at the same scale. The purpose of the mean-centered data is to ensure that the PC will be driven toward maximum variance. The general formula for obtaining mean-centered data is given by the following equation:

$$x_i = X_i - X_{\text{mean}} \tag{6.16}$$

where x_i is the mean-centered value of the descriptor X, X_i is the original data value, and X_{mean} is the mean of all the values of descriptor X. For normalization of the data we can use the following equation:

$$z = \frac{x_i}{\text{SD}} \tag{6.17}$$

where z is the normalized data of the descriptor, x_i represents its mean-centered value, and SD is the standard deviation value calculated by the following equation:

$$\text{SD} = \frac{\sqrt{\Sigma(x_i)^2}}{\sqrt{N}} \tag{6.18}$$

where N is the total number of observations.

The calculation of mean-centered data and normalization is illustrated in Tab. 6.6 with a hypothetical example.

Tab. 6.6: Example of calculation of mean-centered data and normalization.

S. no.	Descriptor 1 (*X*1)	Descriptor 2 (*X*2)	Descriptor3 (*X*3)	x1 mean-centered	x2 mean-centered	x3 mean-centered	z1	z2	z3
1	10	100	2	−08	−100	−2	−1.07	−1.12	−1.42
2	20	300	4	02	100	0	0.27	1.12	0
3	20	100	3	02	−100	−1	0.27	−1.12	−0.71
4	10	200	5	−08	0	1	−1.07	0	0.71
5	30	300	6	12	100	2	1.60	1.12	1.42
	Mean = 18	Mean = 200	Mean = 4						

– **Creation of the covariance matrix**: The covariance matrix is a square symmetrical matrix of order n (n is the number of rows and columns). The purpose of creating the covariance matrix is to identify the correlation among the descriptors. A negative covariance between two descriptors means a negative correlation whereas a positive covariance between them demonstrates a positive correlation between them. As in the hypothetical example given in Tab. 6.6, a covariance matrix of order 3 will be obtained from the normalized values of the descriptors as illustrated below:

$$\begin{bmatrix} \text{Cov}_{z1z1} & \text{Cov}_{z1z2} & \text{Cov}_{z1z3} \\ \text{Cov}_{z2z1} & \text{Cov}_{z2z2} & \text{Cov}_{z2z3} \\ \text{Cov}_{z3z1} & \text{Cov}_{z3z2} & \text{Cov}_{z3z3} \end{bmatrix}$$

Individual covariance between the two variables (suppose x and y) can be calculated by the following equation:

$$\text{Covarience}_{(xy)} = \frac{\Sigma\,(x_i - x_{\text{mean}})(x_i - x_{\text{mean}})}{N} \tag{6.19}$$

where x_i represents the individual value of the x variable, y_i is the value of the individual y variable, $x_{\text{mean},}$ and y_{mean} represent their respective mean values, and N represents the total number of molecules in the study.

– **Calculations of eigenvalues and eigenvectors:** Eigenvalues and eigenvectors are calculated from the covariance matrix given above. These two terms are obtained from linear algebra and are used in computing the PCs. Eigenvectors tell us about the direction toward which maximum information of the variables (variance) is available that constitutes our PC and eigenvalues tell us about the magnitude of that PC. It must be noted that both these terms exist in pairs and are equal to the number of variables in our study.

Consider an example where three variables are in the initial study. We will get three eigenvalues denoted by $\lambda1$, $\lambda2$, and $\lambda3$, where $\lambda1$ will be the largest value and $\lambda3$ is the smallest value. Eigenvectors associated with them will be denoted by $V1$, $V2$, and $V3$, respectively. Now suppose if the value of $\lambda1$ is 10, $\lambda2$ is 3, and $\lambda3$ is 1, then the total of these three values is 14 and it implies that PC1 corresponding to $\lambda1$ and $V1$ will have 71.42% [(10/14) × 100] of the total information (variance). Similarly, PC2 will have 21.42% of the information, and the remaining 7.14% of the information will be possessed by the PC3.

– **Formation of the feature vector**: As of now, we already know that the number of PCs identified is equal to the number of initial variables in the study. The next step in the PCA is to keep the N number of the PC from all the computed ones. The choice of the number of PCs to be retained and to discard the rest of them totally depends upon the user and can either be done statistically or manually. Generally, PCs that explain 95% of the variance are retained in the study. The next step is the formation of the feature vector, which is simply a matrix containing the eigenvectors in columns. Each column of the vector matrix is made of an eigenvector and the number of columns is equal to the number of PCs we want to retain in the study.

– **Reducing the dimensionality of the dataset**: This is the final step of the PCA in which the dataset is rearranged as per the new PCs computed. The values of PCs are calculated simply by multiplying the transpose of the abovementioned feature vector matrix and the transpose of the original dataset.

Here one must understand that PCA is just a data mining technique in which an initial large pool of descriptors is reduced to a few numbers that contain the maximum information of the variables. The PCs, denoted as PC1, PC2, PC3, . . ., PCn, are then employed for the generation of regression equations for the development of the correlation between the biological activity and these newly calculated variables, that is, PCs. This method is known as PC regression (PCR). The general equation obtained from PCR is given as in the following equation:

$$\text{Biological activity} = m_1\text{PC}_1 + m_2\text{PC}_2 + m_3\text{PC}_3 + \cdots + m_n\text{PC}_n + C \qquad (6.20)$$

where PC1, PC2, . . . , PCn represent the PCs obtained from PCA, m_1, m_2,. . ., m_n represent the respective coefficients of PCs, and C is the intercept.

6.3.2 Partial least square (PLS) analysis

PLS analysis is a pivotal technique within MVDA that serves as a potent tool to unravel intricate relationships between molecular descriptors and biological activities, facilitating the predictive modeling of a compound's pharmacological behavior [32]. In the context of QSAR, where datasets entail numerous variables, PLS efficiently tackles the challenge of multicollinearity and high dimensionality [33]. This method constructs latent variables or components that encompass both descriptor and activity information, emphasizing the most influential variables driving compound-activity correlations. By simultaneously considering all variables, PLS captures shared variance and retains predictive power, even in scenarios with noisy or incomplete data [34].

PLS operates by iteratively extracting components that maximize the covariance between descriptors and activities, consequently constructing a model that not only describes the data's structure but also enables accurate activity predictions for new compounds [35]. PLS's ability to handle complex relationships and its capacity to manage "small sample, high dimension" scenarios make it a staple in QSAR applications. PLS analysis within the framework of MVDA in QSAR offers an efficient means to unravel intricate molecular-activity associations, thereby revolutionizing the process of drug development through data-driven insights [36].

PLS is considered as the extension of the PCA and is an MVDA approach used for the reduction of variables when we have a large pool of variables and ultimately to develop a regression model that correlates the biological activity with the new variables. It basically satisfies two conditions: first variables chosen are highly correlated with biological activity and secondly, these variables contain the maximum amount of variance among them [37, 38].

In PCA, PCs are calculated based on covariance among the variables. However, one major drawback of this approach is that small PCs that are obtained sometimes hold a very high correlation with the output variable, that is, with the biological activity, and therefore, should not be neglected (Fig. 6.4). To overcome this, the PLS approach is applied in which, instead of PCs, latent variables are computed based on the covariance between the independent variables and the dependent variables.

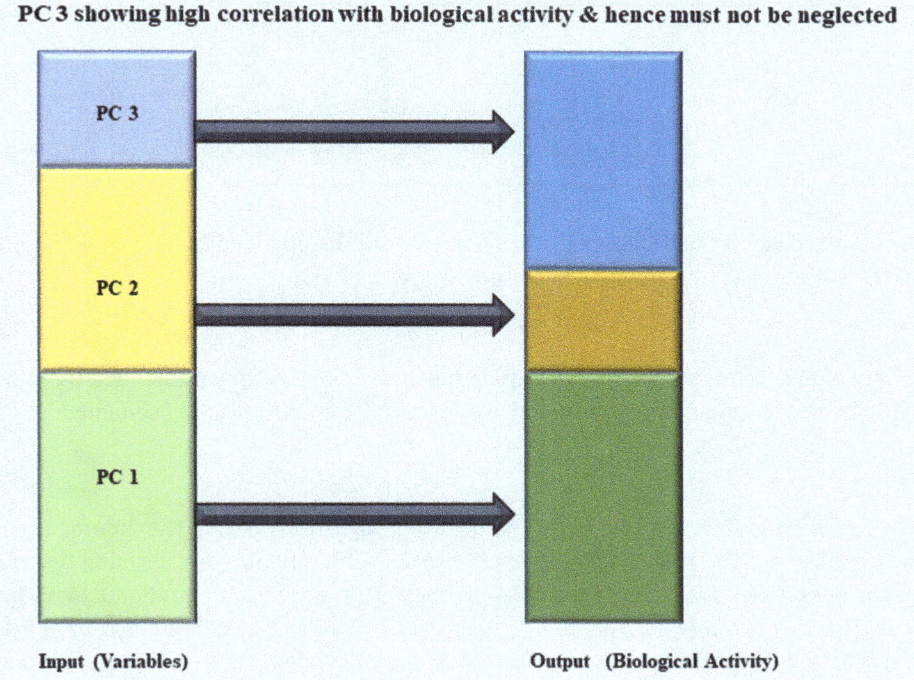

Fig. 6.4: The principal component sometimes has a small weightage but it has a high contribution to the biological activity, and hence must be included in the model generation.

Let us understand this with a hypothetical example given in Tab. 6.7. Assume we have six molecules along with their biological activity and descriptor values. In this example, if we go for traditional MLR for the development of the correlation between the biological activity and two variables (*X*1 and *X*2), we will find that there is very strong colinearity between these two variables. Therefore, it will hinder the quality of the generated QSAR model. By applying PCA, we can get PC1 but how it is correlated with the BA is not taken into consideration. Hence a better approach PLS is applied to calculate the latent variable. This is a combination of *X*1 and *X*2 in such a way that it is correlated with the biological activity rationally.

Tab. 6.7: Example of six hypothetical compounds along with their biological activity (*Y*) and descriptor values (*X*1 and *X*2).

Compound	Biological activity (*Y*)	*X*1	*X*2
1	120	126	38
2	125	128	40
3	130	128	42
4	121	130	42
5	135	130	44
6	140	132	46

A new variable, termed the latent variable (LV), will be obtained by the combination of the variables *X*1 and *X*2 and can be represented as follows:

$$\mathrm{LV} = \alpha_1 X1 + \alpha_2 X2 \tag{6.21}$$

where α_1 and α_2 are the weights corresponding to the concerned variables and the sum of their square should always be equal to one as shown in the following equation:

$$\alpha_1^2\alpha_2^2 = 1 \text{ or } \alpha_2 = \sqrt{1} - \alpha_1^2 \tag{6.22}$$

We can have numerous combinations of α_1 and α_2 that will give the sum of their squares as one. The role of PLS comes into play in this situation for deciding the appropriate combination. Suppose, α_1 is 0.5, then from the above equation α_2 will be 0.87. The new variable, LV, can be calculated as LV = 0.5*X*1 + 0.87*X*2 and the covariance between the biological activity and this LV can be calculated. These covariance values obtained from the different combinations of α_1 and α_2 are plotted against the α_1 (Fig. 6.5). The value of α_1 at which covariance is highest is selected. The α_2 is calculated for the α_1 by employing equation (6.21). In this example, α_1 came out as 0.52 and the corresponding α_2 value comes as 0.85. The LV1 obtained can be depicted as shown in the following equation:

$$\mathrm{LV1} = 0.52X1 + 0.85X2 \tag{6.23}$$

The next step will be the calculation of LV for the molecules from the above equation given in Tab. 6.8.

Now we can apply SLR on the biological activity and the LV1 values. The correlation equation we will get will be expressed as follows:

$$\text{Biological activity} = m\,\mathrm{LV1} + C \tag{6.24}$$

where *m* is the slope coefficient and *C* is the intercept of the line. In our example, this equation can be written as follows:

$$\text{Biological activity} = 1.96\,\mathrm{LV1} - 72.8 \tag{6.25}$$

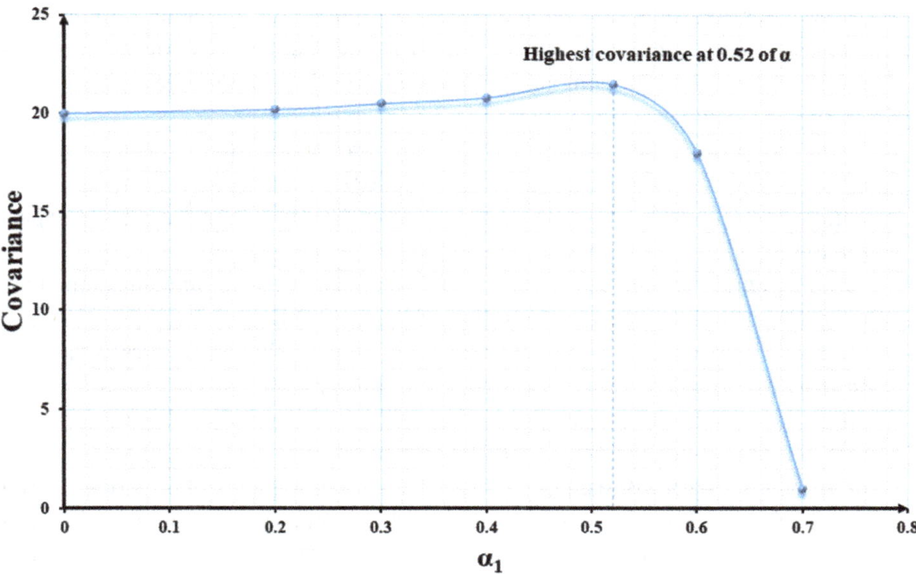

Fig. 6.5: The plot between the covariance and one of the weight components (α_1) in the generation of the partial least square (PLS)-based model.

Tab. 6.8: Calculation of the latent variable from the data obtained from Tab. 6.7.

Compound	Biological activity (*Y*)	*X1*	*X2*	LV1
1	120	126	38	97.8
2	125	128	40	100.6
3	130	128	42	102.3
4	121	130	42	103.3
5	135	130	44	105
6	140	132	46	107.7

Replacing its value in equation (6.23), we will get the correlation equation as follows:

$$\text{Biological activity} = 1.96\ (0.52X1 + 0.85X1) - 72.8 \tag{6.26}$$

6.3.2.1 Advantages of PLS analysis

- **Handling multicollinearity:** PLS effectively addresses multicollinearity, where predictor variables (molecular descriptors) are highly correlated. It creates latent variables (components) that are linear combinations of the original descriptors, reducing collinearity issues and enhancing model stability.

- **Optimal feature selection:** PLS identifies the most informative descriptors contributing to the observed biological activities. This feature selection process simplifies the model by focusing on the most relevant information, leading to more interpretable and accurate models.
- **Predictive power:** PLS models have demonstrated strong predictive capabilities, allowing researchers to make reliable predictions about the activity of new compounds based on their structural characteristics. This is crucial for guiding experimental efforts and prioritizing compounds for further evaluation.
- **Handling small sample sizes:** In drug discovery, datasets with limited compounds but numerous descriptors are common. PLS copes well with small sample sizes, making it suitable for scenarios where traditional regression techniques might fail.
- **Robustness to noisy data:** PLS is robust in the presence of noise and outliers, which are common in real-world datasets. It identifies patterns despite the presence of measurement errors or unusual data points.
- **Multivariate relationships:** PLS considers both the descriptor-response relationship and interrelationships among descriptors simultaneously. This ability to capture complex multivariate relationships improves the understanding of structure–activity relationships.
- **Overfitting prevention:** PLS includes a regularization parameter that helps prevent overfitting, maintaining the model's generalization capabilities even when dealing with high-dimensional data.
- **Visualization and interpretation:** PLS generates score plots and loading plots that enable visualization of compound clusters, trends, and influential descriptors. This aids in data exploration and hypothesis generation.
- **Iterative improvement:** PLS allows for iterative model refinement by adding or removing descriptors, enhancing the model's performance progressively.
- **Guiding compound design:** PLS-based QSAR models offer insights into the structural features that impact compound activity, guiding the rational design of new molecules with desired properties.

6.3.2.2 Limitations of PLS analysis

- **Model complexity:** PLS models can become quite complex, especially when dealing with a large number of descriptors. This complexity can make it difficult to interpret the relationships between descriptors and biological activities.
- **Sensitivity to outliers:** PLS is sensitive to outliers in the data. Outliers can disproportionately influence the model, leading to potentially inaccurate predictions.
- **Multicollinearity:** PLS assumes that the predictor variables (descriptors) are not highly correlated with each other. If there is significant multicollinearity among the descriptors, it can lead to unstable and unreliable model estimates.
- **Interpretability:** While PLS can predict activity, it might not provide direct insight into the underlying mechanisms of the molecular interactions. The relationships

between descriptors and biological activities can be challenging to interpret, especially when dealing with a large number of descriptors.

- **Selection of components:** PLS requires the selection of the number of components to retain in the model. Selecting an inappropriate number of components can impact the model's performance.
- **Data quality:** The accuracy and reliability of PLS models heavily depend on the quality of the input data. Inaccurate or noisy data can lead to biased model outcomes.
- **Applicability domain:** PLS models might not perform well outside the range of the training data, as they rely on capturing the relationships observed in the training set. Extrapolation can be problematic.
- **Dimensionality reduction:** While PLS aims to reduce dimensionality, it might not eliminate all irrelevant features. The inclusion of irrelevant or redundant descriptors can degrade model performance.
- **Human expertise:** PLS requires domain knowledge for proper model interpretation and validation. Lack of expertise can lead to misinterpretation of results or inadequate model assessment.

6.3.3 Genetic function approximation (GFA)

Genetic function approximation (GFA) represents a significant advancement in QSAR modeling for drug discovery. GFA, a machine learning technique, enhances this predictive capacity by identifying complex relationships between molecular features and activity. At its core, GFA employs a GA to select and refine relevant molecular descriptors or features from a large pool of potential variables. These descriptors encode various structural, electronic, and physicochemical characteristics of molecules. By iteratively selecting, combining, and evolving descriptors, GFA uncovers intricate correlations that traditional QSAR methods might overlook. GFA's strength lies in its ability to handle high-dimensional data and capture nonlinear relationships between molecular features and biological activity. This enables it to provide more accurate predictions, aiding in the prioritization of potentially active compounds during drug discovery. Moreover, GFA offers insights into the underlying mechanisms of molecular interactions, facilitating a deeper understanding of structure–activity relationships. In QSAR for drug discovery, GFA stands as a versatile tool that bridges the gap between complex chemical information and actionable insights. Its integration has the potential to accelerate the identification of promising drug candidates while minimizing costly and time-consuming experimental efforts [39–41].

GFA is slightly a different approach in the MVDA. The inspiration behind the GFA is from Holand's GA developed in 1975 [42] and Friedman's multivariate adaptive regression splines (MARS) algorithm developed in 1991 [43]. Unlike approaches discussed earlier (MLR, PCA, and PLS), where only one model is generated, in the GFA approach many models are generated out of which the best one is selected. In this approach, initially, a population of the parent models is generated, and then based on the "parent models" further models are constructed which are considered the "offspring models." The offspring models generated are equivalent to or better than the parent models constructed by applying the principles of genetic evolution and the terminology used for it is GA. In the QSAR model building, the GFA or GA approach is often used in combination with the MLR or PLS, and this combined approach (GA-MLR or GA-PLS) is always considered superior to other approaches for the generation of a robust QSAR model [44–46].

6.3.4 Pattern recognition or multivariate data classification procedures or cluster analysis

The one major advantage of the MVDA is that it provides lots of information about the data but also comes with a hindrance to comprehension of this large data [47]. Sometimes, the approaches discussed above are not sufficient enough to handle such large multivariate data in statistics. Hence, multivariate data classification procedures are introduced to classify the large datasets into smaller groups having similar characteristics or patterns (Fig. 6.6). The two main objectives of these classification procedures are the simplification of the data and development of a model having better predictability. In these approaches, each independent variable (descriptors) of the molecule is assigned a certain score value; then a plot is plotted for score values against variables for each molecule (Fig. 6.7). Before plotting, the score value of each

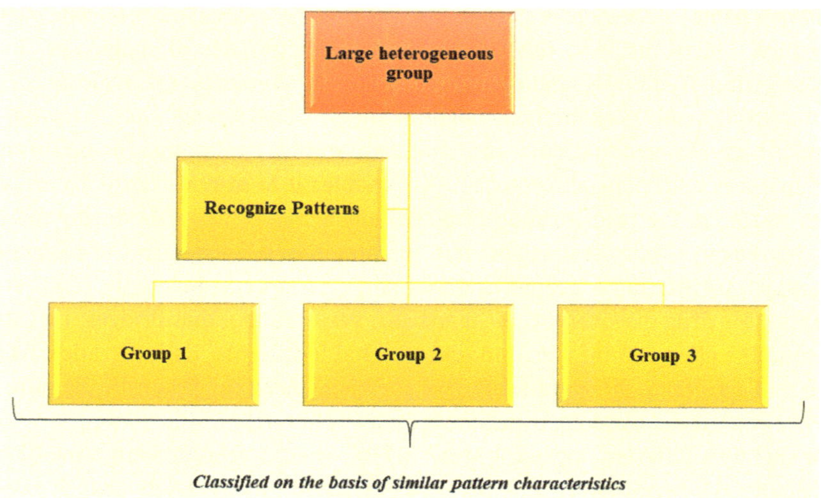

Fig. 6.6: Variables in the multivariate variable models are grouped into a set of variables based on their similarity in the patterns.

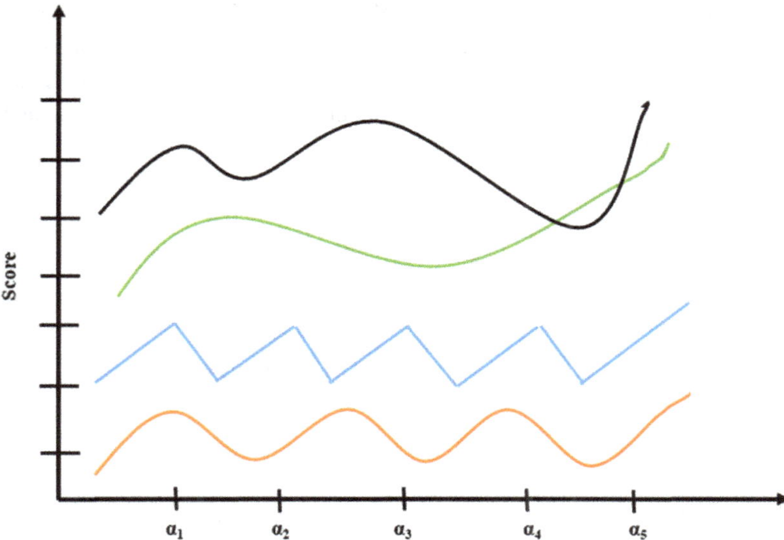

Fig. 6.7: The plot between the score value of each variable and the variable itself.

variable is normalized to a common scale to avoid variability among the values. The height of the plot represents the magnitude of the score possessed by each variable and the shape of the plot represents the pattern of the variables. The variables employed for the cluster analysis are assumed to be independent and uncorrelated [48].

6.3.4.1 Measurement of proximity

Proximity measurement is used to define the parameters based on which clustering or grouping will be done in the cluster analysis. Two types of such parameters are used for this purpose: similarity indices and dissimilarity indices. The high value of similarity indices between the molecules indicates they have common characteristics and the same will be true if the value of dissimilarity indices is low between the molecules. In other words, molecules with high similarity indices or low values of dissimilarity indices will be placed in the same group.

6.3.4.2 Algorithms of cluster analysis

There are many algorithms defined for carrying out cluster analysis. The three most widely used approaches are described further.

6.3.4.2.1 Hierarchical algorithm

This is a popular means of performing the cluster analysis, which is further divided into two types [49]:

– Agglomerative hierarchical clustering: This is a bottom-up clustering approach in which initially all the data points are considered an individual cluster. Then, we start merging the two nearest related data points and the process is continued till all the data points are merged into one single cluster. In the end, the single cluster is divided into small clusters as per the requirement (Fig. 6.8).

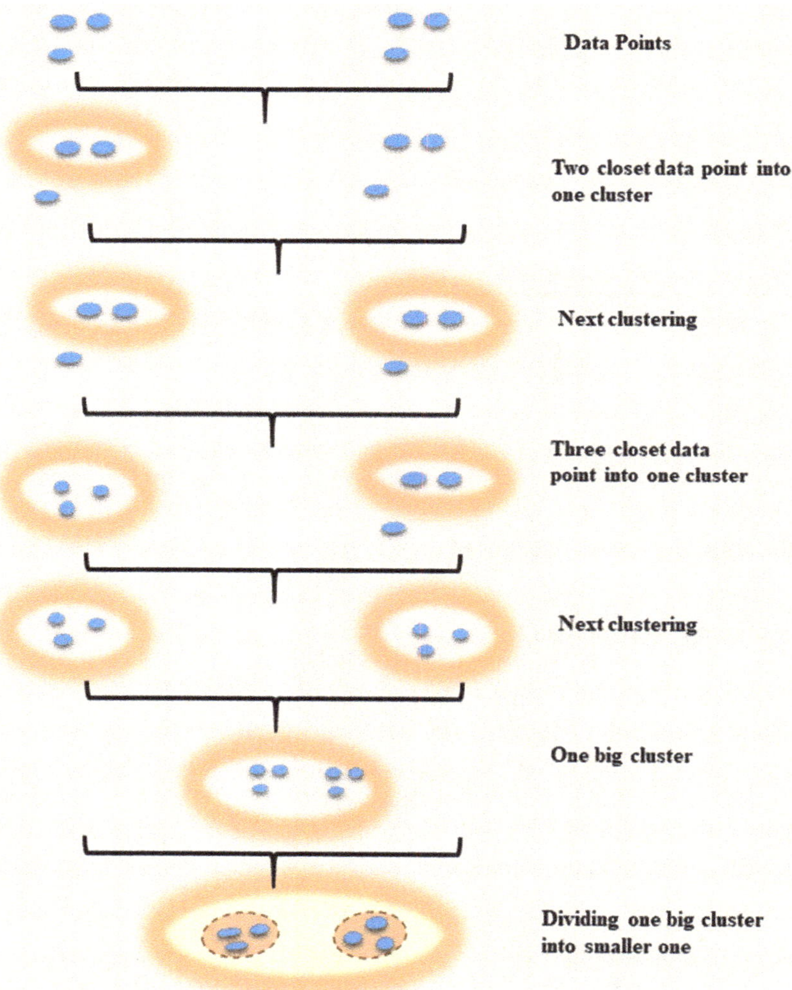

Fig. 6.8: Agglomerative hierarchical clustering of the cluster analysis.

– Divisive hierarchical clustering: This is also known as the top-down approach. This approach is totally opposite to the agglomerative hierarchical clustering, as in this method, all data points are considered as one single cluster and then dividing of the same into smaller clusters is performed depending upon the requirement (Fig. 6.9).

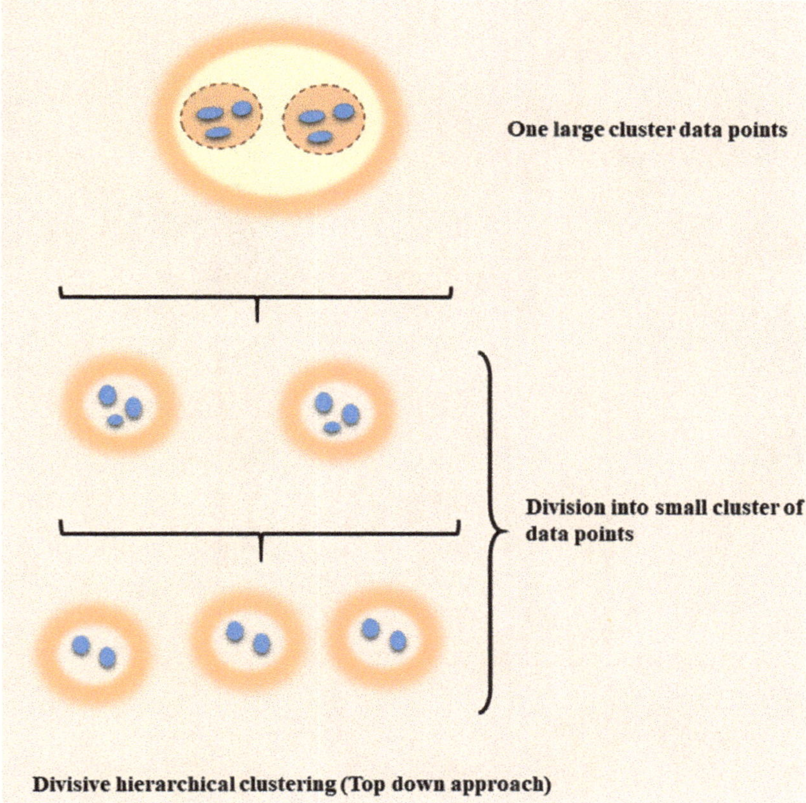

Fig. 6.9: Divisive hierarchical clustering (top-down approach) of the cluster analysis.

6.3.4.2.2 Artificial neural network (ANN)-based clustering

This is another popular classification or clustering approach. The methodology is analogous to the functioning of the human neurons present in the brain. A simple model representing an ANN model (McCulloh-Pitts model) is given in Fig. 6.10. The basic idea behind this type of clustering is that an external input (stimulus) is provided in the form of variables, and then with a combination of external weights (at synapses), the desired output is obtained depending on the activation factor (linear, hyperbolic, Gaussian, etc.) [50].

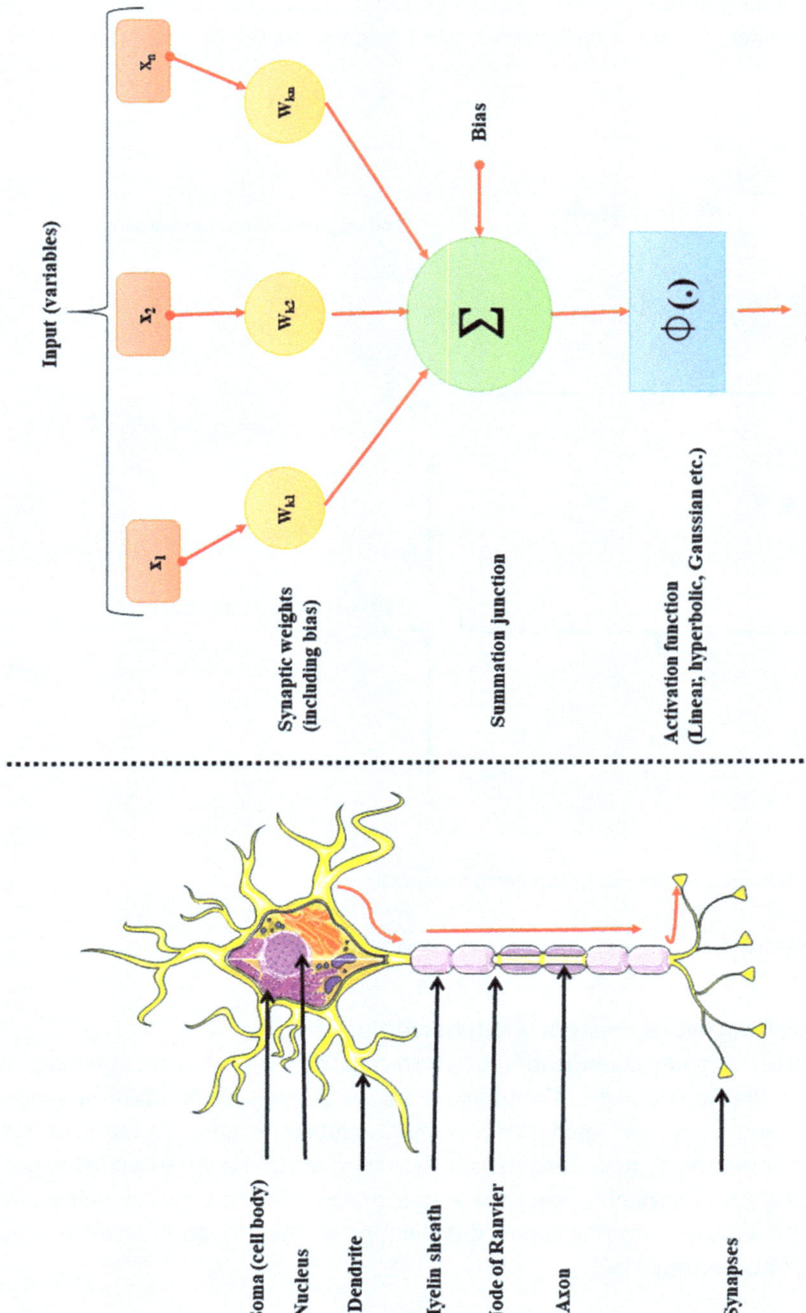

Fig. 6.10: Artificial neural network (ANN)-based clustering.

6.3.4.2.3 K-nearest neighbor (KNN) clustering

K-nearest neighbor-based clustering is a nonparametric approach for the classification [51]. Here, *K* stands for the total number of neighbors required for clustering. In the example given in Fig. 6.11, suppose there are two groups present in the study, for example, group A represented by circles and group B represented by squares. If a new data point *x* is entered in the model, it can be grouped in either of the groups based on the value of *K*. If *K* is 3, then the nearest three data points from *x* will be studied and it will be clustered in group B. But, if we consider *K* as 5, then it will be grouped in group A.

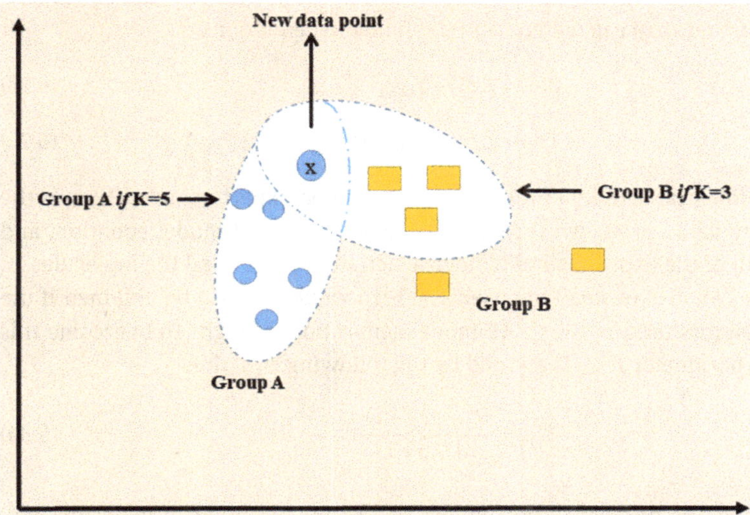

Fig. 6.11: K-nearest neighbor (KNN) clustering.

6.4 Statistical parameters for the validation of the QSAR models

Irrespective of the QSAR model, whether developed traditionally or through more sophisticated approaches such as 3D-QSAR, they are to be validated for assessing their robustness and predictability. Certain common internal and external validation parameters are used for checking the quality of the QSAR model developed [52, 53].

6.4.1 Internal validation parameters for the QSAR model

In the internal validation of the QSAR model, an assessment of the QSAR model is done from the molecules that are used for the development of that model [54]. It is done through the following parameters:

– **Goodness of fit (R^2):** Goodness of fit, popularly known as the R^2 value, is one of the most commonly employed parameters for the internal assessment of the QSAR model. The R^2 value tells us about the reproducibility of the QSAR model but provides little or no information regarding its robustness and predictability. An R^2 value greater than 0.7 is considered acceptable in the QSAR modeling and R^2 calculation is done by employing the following equation:

$$R^2 = 1 - (\mathrm{RSS}/\mathrm{TSS}) \tag{6.27}$$

where RSS represents the residual sum of squares and TSS is the total sum of squares. Both of the parameters are calculated from the following equations:

$$\mathrm{RSS} = \Sigma\,(Y - Y_{\mathrm{pred}})^2 \tag{6.28}$$

$$\mathrm{TSS} = \Sigma\,(Y - Y_{\mathrm{mean}})^2 \tag{6.29}$$

where Y is the biological activity of the molecule obtained experimentally, Y_{pred} is the biological activity value predicted through the developed QSAR model equation, and Y_{mean} is the mean of the experimental biological activity values of all the molecules.

The R^2 value can also mislead the user as a high value can also be obtained if the number of descriptors used in the QSAR model generations is high. To overcome this situation, a new parameter R^2_{adj} is defined by the following equation:

$$R^2_{\mathrm{adj}} = \frac{R^2 - p(n-1)}{n-p+1} \tag{6.30}$$

where p represents the number of descriptors and n is the number of molecules employed for the QSAR model generation. For a reliable and robust QSAR model, the difference between R^2 and R^2_{adj} should not be greater than 0.3.

– **Cross-validation coefficient (Q^2_{cv})**

The internal validation of the QSAR model is also performed by another parameter known as the cross-validation coefficient (Q^2_{cv}). The basic methodology behind this type of cross-validation approach is that a specific amount of molecules are omitted from the training set every time the QSAR model is built. Then the generated QSAR model is used for the prediction of biological activity of the omitted molecules. This step is repeated until all the molecules get omitted once. The parameter Q^2_{cv} is calculated by using the following equation:

$$Q^2_{\mathrm{cv}} = 1 - (\mathrm{PRESS}/\mathrm{TSS}) \tag{6.31}$$

where PRESS is the predictive error sum of squares and TSS is the total sum of squares. The PRESS can be calculated from

$$\mathrm{PRESS} = \Sigma\,(Y - Y_{\mathrm{p}})^2 \tag{6.32}$$

where Y is the actual biological activity value of the omitted molecule and Y_{p} is its value of biological activity predicted by the generated QSAR model. An acceptable QSAR model should have a Q^2_{CV} value greater than 0.7.

6.4.2 External validation of the QSAR model

External validation is performed to assess the robustness of the QSAR model by using an external test set of molecules that are not used for the generation of the QSAR model. Parameters used for external validation are:

6.4.2.1 Coefficients of determination (R^2_{pred})

Similar to the R^2 parameter employed for the internal validation of the QSAR model, the parameter used for the external validation of the QSAR model is coefficients of determination (R^2_{pred}), which is calculated similarly to R^2 as follows:

$$R^2_{\mathrm{pred}} = 1 - [\Sigma\,(Y - Y_{\mathrm{pred}})^2 / \Sigma\,(Y - Y_{\mathrm{mean}})^2] \tag{6.33}$$

Here, Y is the experimentally observed biological activity of the molecule used in the test dataset, Y_{pred} is its value obtained by the QSAR model, and Y_{mean} is the mean value of the biological activity of the training set molecules. For external validation of a QSAR model, a value greater than 0.6 is considered acceptable.

6.4.2.2 Mean absolute error (MAE)

This is another parameter given by Kunal Roy and his team for the external validation of a QSAR model [55]. It is mathematically expressed as follows:

$$\mathrm{MAE} = \Sigma\,\left(Y_{\mathrm{obs}} - Y_{\mathrm{pred}}\right)/n \tag{6.34}$$

where Y_{obs} and Y_{pred} represent the experimentally obtained and predicted value of the test set molecule, and n represents the number of molecules in the test set. For the external validation of the QSAR, $\mathrm{MAE}_{95\%}$ parameter is calculated by omitting the top 5% high residual molecules from the test set. A value of $\mathrm{MAE}_{95\%}$ less than 0.15 is desired for rating a generated QSAR model as “not bad.”

Combining all the above approaches, Golbraikh and Tropsha proposed statistical parameters for the internal and external validation of a QSAR model [56] (Tab. 6.9).

Tab. 6.9: Golbraikh and Tropsha parameters for the internal and external validation of the QSAR model.

S. no.	Parameter	Threshold value
1	Q^2_{CV}	Threshold value $Q^2 > 0.5$
2	R^2	Threshold value $R^2_{train} > 0.6$
3	$\lvert R^2 - R^2_{adjusted} \rvert$	Threshold value $\lvert R^2 - R^2_{adjusted} \rvert < 0.3$
4	k or k'	$0.85 < k < 1.15$ and $0.85 < k' < 1.15$
5	$R^2_{pred} - R^2/R^2_{pred}$	Threshold value $R^2_{pred} - R^2/R^2_{pred} < 0.1$

All the parameters mentioned are already explained above except k or k', which represent the regression slopes of the QSAR equation generated, which passes through the origin.

References

[1] Campbell JM, Herzinger CV. Statistics and Single Subject Research Methodology. Single Subject Research Methodology in Behavioral Sciences. Routledge, 2009, 417–53.

[2] Todeschini R, Consonni V, Gramatica P. Chemometrics in QSAR. Comprehensive Chemometrics. Elsevier, 2009, 4, 129–72.

[3] Gramatica P. On the development and validation of QSAR models. Computational Toxicology, 2013, 2, 499–526.

[4] Polanski J, Bak A, Gieleciak R, Magdziarz T. Modeling robust QSAR. Journal of Chemical Information and Modeling, 2006, 46(6), 2310–8.

[5] Tropsha A. Best practices for QSAR model development, validation, and exploitation. Molecular Informatics, 2010, 29(6-7), 476–88.

[6] Cherkasov A, Muratov EN, Fourches D, et al. QSAR modeling: Where have you been? Where are you going to? Journal of Medicinal Chemistry, 2014, 57(12), 4977–5010.

[7] Montgomery DC, Peck EA, Vining GG. Introduction to Linear Regression Analysis. John Wiley & Sons, 2021.

[8] Seber GA, Lee AJ. Linear Regression Analysis. John Wiley & Sons, 2012.

[9] Papa E, Dearden JC, Gramatica P. Linear QSAR regression models for the prediction of bioconcentration factors by physicochemical properties and structural theoretical molecular descriptors. Chemosphere, 2007, 67(2), 351–8.

[10] Duchowicz PR. Linear regression QSAR models for polo-like kinase-1 inhibitors. Cells, 2018, 7(2), 13.

[11] Krieger AM, Pollak M, Yakir B. Surveillance of a simple linear regression. Journal of the American Statistical Association, 2003, 98(462), 456–69.

[12] Kim HY. Statistical notes for clinical researchers: Simple linear regression 3–residual analysis. Restorative Dentistry & Endodontics, 2019, 44(1).

[13] Marill KA. Advanced statistics: Linear regression, part I: Simple linear regression. Academic Emergency Medicine, 2004, 11(1), 87–93.

[14] Pak SI, Oh TH. Correlation and simple linear regression. Journal of Veterinary Clinics, 2010, 27(4), 427–34.

[15] Afantitis A, Melagraki G, Sarimveis H, et al. A novel simple QSAR model for the prediction of anti-HIV activity using multiple linear regression analysis. Molecular Diversity, 2006, 10, 405–14.

[16] Bhhatarai B, Garg R, Gramatica P. Are mechanistic and statistical QSAR approaches really different? MLR studies on 158 cycloalkyl-pyranones. Molecular Informatics, 2010, 29(6–7), 511–22.
[17] Akbari S, Zebardast T, Zarghi A, Hajimahdi Z. QSAR modeling of COX-2 inhibitory activity of some dihydropyridine and hydroquinoline derivatives using multiple linear regression (MLR) method. Iranian Journal of Pharmaceutical Research: IJPR, 2017, 16(2), 525.
[18] Hajimahdi Z, Safizadeh F, Zarghi A. QSAR analysis for some 1, 2-benzisothiazol-3-one derivatives as caspase-3 inhibitors by stepwise MLR method. Iranian Journal of Pharmaceutical Research: IJPR, 2016, 15(2), 439.
[19] Rajathei DM, Parthasarathy S, Selvaraj S. QSAR analysis of multimodal antidepressants vortioxetine analogs using physicochemical descriptors and MLR modeling. Current Computer-Aided Drug Design, 2019, 15(4), 294–307.
[20] Noolvi MN, Patel HM, Bhardwaj V. A comparative QSAR analysis of quinazoline analogues as tyrosine kinase (erbB-2) inhibitors. Medicinal Chemistry, 2011, 7(3), 200–12.
[21] Ferreira M. Multivariate QSAR. Journal of the Brazilian Chemical Society, 2002, 13, 742–53.
[22] Pirhadi S, Shiri F, Ghasemi JB. Multivariate statistical analysis methods in QSAR. Rsc Advances, 2015, 5(127), 104635–65.
[23] Eriksson L, Johansson E. Multivariate design and modeling in QSAR. Chemometrics and Intelligent Laboratory Systems, 1996, 34(1), 1–9.
[24] Miyashita Y, Li Z, Sasaki SI. Chemical pattern recognition and multivariate analysis for QSAR studies. TrAC Trends in Analytical Chemistry, 1993, 12(2), 50–60.
[25] Wold S, Dunn III WJ. Multivariate quantitative structure-activity relationships (QSAR): Conditions for their applicability. Journal of Chemical Information and Computer Sciences, 1983, 23(1), 6–13.
[26] Schmidli H. Multivariate prediction for QSAR. Chemometrics and Intelligent Laboratory Systems, 1997, 37(1), 125–34.
[27] Wold S, Sjöström M, Andersson PM, et al. Multivariate design and modelling in QSAR, combinatorial chemistry, and bioinformatics. Molecular Modeling and Prediction of Bioactivity, 2000, 27–45.
[28] Freitas MP. Multivariate QSAR: From classical descriptors to new perspectives. Current Computer-Aided Drug Design, 2007, 3(4), 235–9.
[29] Yoo C, Shahlaei M. The applications of PCA in QSAR studies: A case study on CCR5 antagonists. Chemical Biology & Drug Design, 2018, 91(1), 137–52.
[30] Hemmateenejad B, Miri R, Elyasi M. A segmented principal component analysis – Regression approach to QSAR study of peptides. Journal of Theoretical Biology, 2012, 305, 37–44.
[31] Goodarzi M, Dejaegher B, Heyden YV. Feature selection methods in QSAR studies. Journal of AOAC International, 2012, 95(3), 636–51.
[32] T Stanton D. QSAR and QSPR model interpretation using partial least squares (PLS) analysis. Current Computer-Aided Drug Design, 2012, 8(2), 107–27.
[33] Baroni M, Costantino G, Cruciani G, Riganelli D, Valigi R, Clementi S. Generating optimal linear PLS estimations (GOLPE): An advanced chemometric tool for handling 3D-QSAR problems. Quantitative Structure-Activity Relationships, 1993,12(1), 9–20.
[34] Olah M, Bologa C, Oprea TI. An automated PLS search for biologically relevant QSAR descriptors. Journal of Computer-Aided Molecular Design, 2004, 18, 437–49.
[35] Varmuza K, Filzmoser P, Dehmer M. Multivariate linear QSPR/QSAR models: Rigorous evaluation of variable selection for PLS. Computational and Structural Biotechnology Journal, 2013, 5(6), e201302007.
[36] Eriksson L, Gottfries J, Johansson E, Wold S. Time-resolved QSAR: An approach to PLS modelling of three-way biological data. Chemometrics and Intelligent Laboratory Systems, 2004, 73(1), 73–84.
[37] Johnels, D., Gillner, M., Nordén, B., Toftgård, R. and Gustafsson, J.Å. Quantitative structure-activity relationship (QSAR) analysis using the partial least squares (PLS) method: The binding of polycyclic

aromatic hydrocarbons (PAH) to the rat liver 2, 3, 7, 8-tetrachlorodibenzo-P-dioxin (TCDD) receptor. Quantitative Structure-Activity Relationships, 1989, 8(2), 83–9.

[38] Long HX, Wang YQ, Lin Y, Lin ZH. QSAR study on ACE inhibitors by using OSC-PLS algorithm. Journal of the Chinese Chemical Society, 2010, 57(3A), 417–22.

[39] Sivakumar PM, Babu SK, Mukesh D. QSAR studies on chalcones and flavonoids as anti-tuberculosis agents using genetic function approximation (GFA) method. Chemical and Pharmaceutical Bulletin, 2007, 55(1), 44–9.

[40] Roy K, Roy PP. Exploring QSAR and QAAR for inhibitors of cytochrome P450 2A6 and 2A5 enzymes using GFA and G/PLS techniques. European Journal of Medicinal Chemistry, 2009, 44(5), 1941–51.

[41] Elsamman A, Khaled KF, Halim SA, Abdelshafi NS. Development of QSAR based GFA predictive model for the effective design of a new bispyrazole derivative corrosion inhibitor. Journal of Molecular Structure, 2023, 1293, 136230.

[42] Holland JH. Genetic algorithms and adaptation. Adaptive control of ill-defined systems, 1984, 317–33.

[43] Friedman JH. Multivariate adaptive regression splines. The Annals of Statistics, 1991, 19(1), 1–67.

[44] Riahi S, Ganjali MR, Norouzi P, Jafari F. Application of GA-MLR, GA-PLS and the DFT quantum mechanical (QM) calculations for the prediction of the selectivity coefficients of a histamine-selective electrode. Sensors and Actuators B: Chemical, 2008, 132(1), 13–9.

[45] Nekoei M, Goudarzi N, Nekoei S, Mohammadhosseini M. QSAR study of arylsulfonylpiperazine inhibitors of 11β-HSD1 by GA-MLR, GA-PLS and GA-ANN. Analytical Chemistry Letters, 2014, 4(1), 14–28.

[46] Absalan G, Hemmateenejad B, Soleimani M, Akhond M, Miri R. Quantitative structure–micellization relationship study of gemini surfactants using genetic-PLS and genetic-MLR. QSAR & Combinatorial Science, 2004, 23(6), 416–25.

[47] Kadam RU, Roy N. Cluster analysis and two-dimensional quantitative structure–activity relationship (2D-QSAR) of Pseudomonas aeruginosa deacetylase LpxC inhibitors. Bioorganic & Medicinal Chemistry Letters, 2006, 16(19), 5136–43.

[48] Pirhadi S, Shiri F, Ghasemi JB. Multivariate statistical analysis methods in QSAR. Rsc Advances, 2015, 5(127), 104635–65.

[49] Teles HR, Ferreira LL, Valli M, Coelho F, Andricopulo AD. Hierarchical clustering and target-independent QSAR for antileishmanial oxazole and oxadiazole derivatives. International Journal of Molecular Sciences, 2022, 23(16), 8898.

[50] Salt DW, Yildiz N, Livingstone DJ, Tinsley CJ. The use of artificial neural networks in QSAR. Pesticide Science, 1992, 36(2), 161–70.

[51] Zheng W, Tropsha A. Novel variable selection quantitative structure– property relationship approach based on the k-nearest-neighbor principle. Journal of Chemical Information and Computer Sciences, 2000, 40(1), 185–94.

[52] Veerasamy R, Rajak H, Jain A, Sivadasan S, Varghese CP, Agrawal RK. Validation of QSAR models-strategies and importance. International Journal of Drug Design & Discovery, 2011, 3, 511–9.

[53] Gramatica P. Principles of QSAR models validation: Internal and external. QSAR & Combinatorial Science, 2007, 26(5), 694–701.

[54] Kubinyi H. Validation and predictivity of QSAR models. InQSAR & molecular modelling in rational design of bioactive molecules. In: Proceedings of the 15th European Symposium on QSAR & Molecular Modelling, Istanbul, Turkey, 2004, 30–3.

[55] Roy K, Das RN, Ambure P, Aher RB. Be aware of error measures. Further studies on validation of predictive QSAR models. Chemometrics and Intelligent Laboratory Systems, 2016, 152, 18–33.

[56] Tropsha A, Golbraikh A. Predictive quantitative structure-activity relationships modeling. Handbook of Chemoinformatics Algorithms, 2010, 33, 211.

Chapter 7
Molecular and quantum mechanics in drug designing

If Quantum mechanics hasn't profoundly shocked you, you haven't understood it yet
– Neil Bohr

7.1 Introduction

Computer-aided drug designing (CADD) has become an integral part of any drug discovery process [1, 2]. Not only does it reduce the cost and time of the research but it also provides us valuable insights into the structural, physical, and chemical properties of potential drug molecules. Molecular mechanics (MM) and quantum mechanics (QM) play a pivotal role in the drug discovery process, revolutionizing the way new pharmaceuticals arė designed and developed [3, 4]. MM provides insights into the structural properties and interactions of molecules, while QM delves into the fundamental behaviors of atoms and subatomic particles [5, 6]. In drug designing, MM aids in simulating the interactions between molecules, enabling researchers to predict how a potential drug compound interacts with its target protein or receptor [7]. This information guides the modification of molecular structures to enhance binding affinity and specificity, thereby increasing the chances of therapeutic success. Additionally, MM simulations elucidate the stability of drug molecules and their potential side effects, critical for optimizing drug candidates [8, 9]. QM, on the other hand, delves into the electronic structure of molecules, offering a deep understanding of molecular behavior at the atomic level [10]. QM calculations provide valuable insights into molecular energetics, reaction mechanisms, and electronic properties, aiding in the prediction of how a drug molecule undergoes chemical transformations within the body. This knowledge is indispensable for designing drugs with desired reactivity and metabolic pathways [11, 12].

The synergy between MM and QM expedites the drug discovery process by guiding the rational design of compounds with improved pharmacological properties, reduced toxicity, and enhanced efficacy [13]. By providing a comprehensive understanding of molecular interactions and reactivity, these computational tools accelerate the identification of promising drug candidates, offering a more efficient and cost-effective approach to revolutionizing healthcare and advancing the field of pharmaceutical research [14]. Mechanics is a simple terminology of physics that deals with the study of the motion of objects and their subsequent effect on the environment in which they are present [15]. In the context of CADD, it is the application of principles of mechanics on potential drug candidates and target receptor molecules to understand their different aspects, which include:

- Different conformations of drug molecules
- Structure and conformations of target receptor molecules

https://doi.org/10.1515/9783111434858-007

- Drug-receptor binding energy calculations
- Calculations of the chemical and physical properties of molecules
- The most energetically stable molecule under different conditions
- Effect of solvents on the drug-receptor interactions
- Other calculations that involve energy terms of the molecules, including drug and target receptor

Molecular modeling is the combined terminology used in CADD for these types of studies involving the principles of mechanics (Fig. 7.1). The two most widely used approaches of molecular modeling for the purpose are MM and QM.

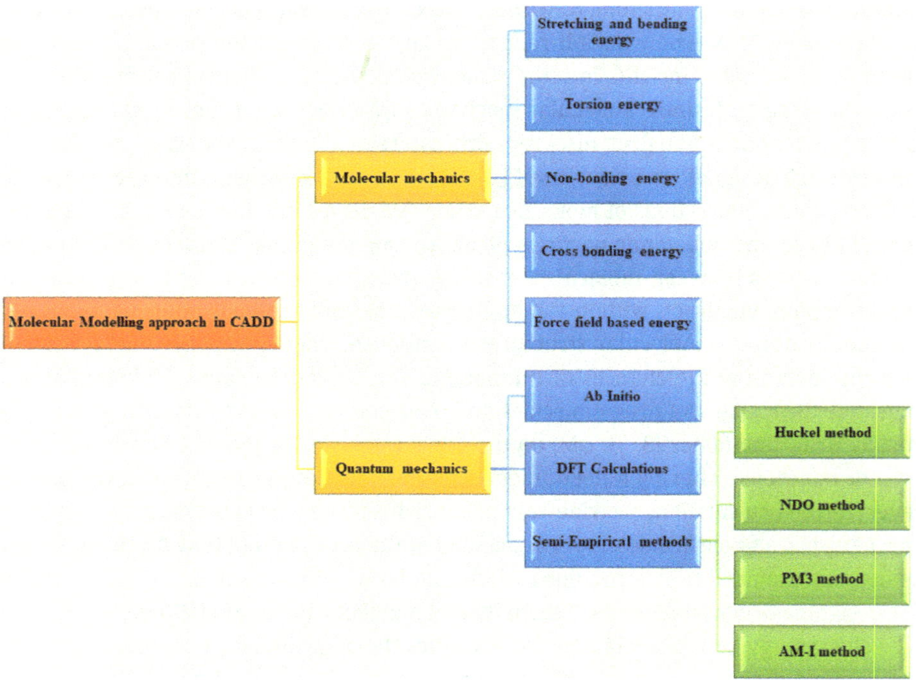

Fig. 7.1: Different molecular modeling approaches used in computer-aided drug designing (CADD).

7.2 Molecular mechanics (MM) in drug design

MM is the simpler approach which employs the use of "classical Newtonian mechanics" principles for all the calculations [16]. In this approach, any molecule, whether drug or receptor, is treated as made up of atoms, which are joined by various bonds that exhibit spring-like properties [17]. Another major assumption in the MM approach is that it considers the additive nature of different parameters employed in the calculations [18]. This is of concern, particularly, when calculating the potential energy of any molecule. Total potential energy possessed by any molecule, whether drug or target receptor, is the summation of different individual energy terms of the molecule, namely stretching energy, bending energy, torsion energy, and the non-bonding energy parameters (Fig. 7.2) [19]. Mathematically this total energy possessed by the molecule can be expressed by the following equation:

$$E_{\text{total}} = \Sigma E_{\text{bend}} + \Sigma E_{\text{str}} + \Sigma E_{\text{torsion}} + \Sigma E_{\text{nonbonding}} \tag{7.1}$$

where E_{total} is the total potential energy of the molecule, ΣE_{bend}, ΣE_{str}, $\Sigma E_{\text{torsion}}$, and $\Sigma E_{\text{nonbonding}}$ are individual energy terms due to stretching, bending, torsion, and non-bonding interactions, respectively, of the molecule.

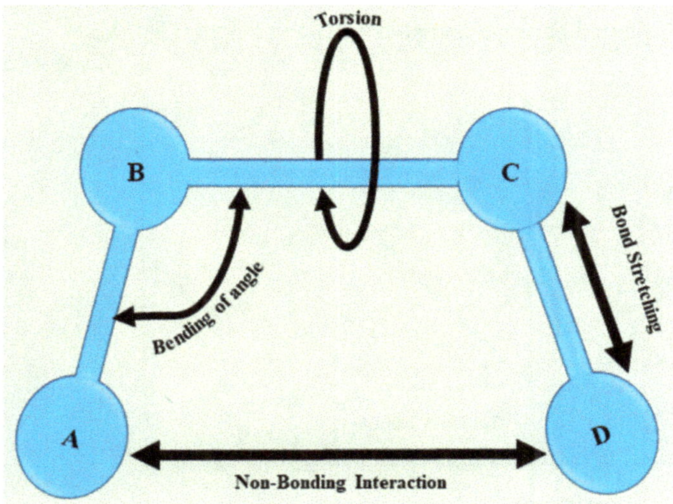

Fig. 7.2: Different types of interactions occurring inside a molecule that constitute its potential energy.

7.2.1 Stretching and bending energy

The change in stretching energy in the molecule is represented by ΣE_{str} of equation (7.1) and denotes the changes occurring due to the changes in the bond length of the molecules, whereas, ΣE_{bend} of equation (7.1) represents the bending energy changes due to changes in the bond angles inside the molecule. Both of these parameters follow the simple Hooke's law and are considered to follow the simple harmonic motion. Mathematically, these are calculated by using the following two equations:

$$\Sigma E_{str} = \Sigma k_b (r - r_0)^2 \tag{7.2}$$

$$\Sigma E_{bend} = \Sigma k_\theta (\theta - \theta_0)^2 \tag{7.3}$$

where k_b and k_θ are the force constants of Hook's law, r is the current bond length, r_0 is the initial bond length, θ is the current bond angle, and θ_0 is the initial bond angle.

Both of the above equations employed are quadratic in nature and a parabola is obtained when the energy is plotted against the bond angles or bond lengths (Fig. 7.3). It is important to note that the energy changes due to bond stretching are quite large and cannot be calculated accurately using simple Hooke's law assumptions. A more accurate way of calculating this energy is through Morse potential by using the following equation:

$$\Sigma E_{str} = \Sigma D_e (1 - e^{-a(r - r_0)})^2 \tag{7.4}$$

where D_e and a are constants defining the depth and width of the potential well, respectively, when a plot is plotted between potential and bond length (Fig. 7.4).

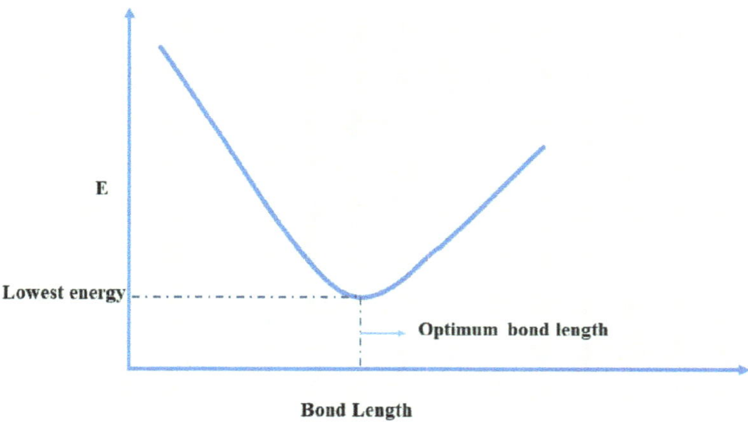

Fig. 7.3: The energy plot diagram between the energy of the molecule and the bond angle inside the molecule.

7.2.2 Torsion energy

Torsion energy is represented by $\Sigma E_{torsion}$ in equation (7.1) and it comes into play when the molecule contains more than three atoms. These energies play a role due to the rotation of bonds. To understand this, consider an example of *n*-butane; when we rotate the molecule along the carbon–carbon bond we obtain different conformations, which differ in the energy (Fig. 7.5). The most stable conformation in terms of energy is the anti-form having a dihedral angle between the CH_3–CH_3 groups of 180°. Similarly fully eclipsed conformation will have a dihedral angle of 0° between the CH_3–

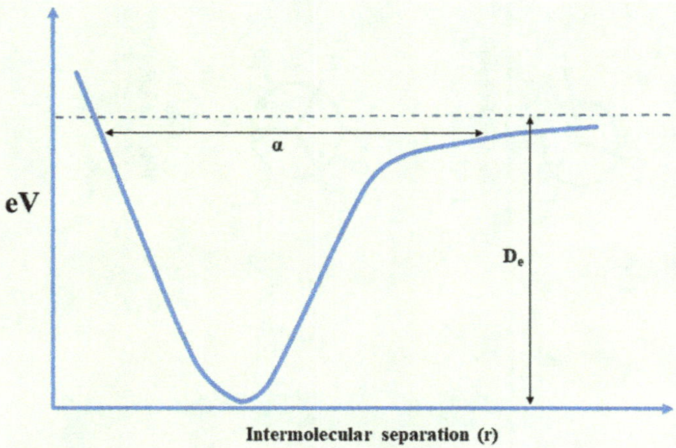

Fig. 7.4: The plot between the potential energy of the molecule and the interatomic distance inside the molecule.

CH_3 groups and is the least stable in energy terms due to high repulsion between the bulky methyl groups. Likewise, other conformations having different energies exist between these two extreme conformations.

Mathematically, this energy term is calculated by using the following equation:

$$\Sigma E_{\text{torsion}} = \Sigma A[1 + \cos(n\tau + \phi)] \tag{7.5}$$

where A is the constant term, τ is the current dihedral angle, and ϕ is the initial dihedral angle of the molecule.

7.2.3 Nonbonding energy terms

Nonbonding energy terms is represented by $\Sigma E_{\text{nonbonding}}$ in equation (7.1). It depicts the changes in the energy due to electrostatic, Van der Waals, and hydrogen-bonding type of interactions inside the molecule. Mathematically, it can be denoted by the following equation:

$$\Sigma E_{\text{nonbonding}} = \Sigma E_{\text{electrostatic}} + \Sigma E_{\text{Van der Waals}} + \Sigma E_{\text{H-bond}} \tag{7.6}$$

To calculate total nonbonding energy changes in a molecule, electrostatic, Van der Waals, and H-bonding type energy changes are calculated separately and then added. The electrostatic energy changes in the molecule can be calculated through Coulomb's law and can be calculated using the following equation:

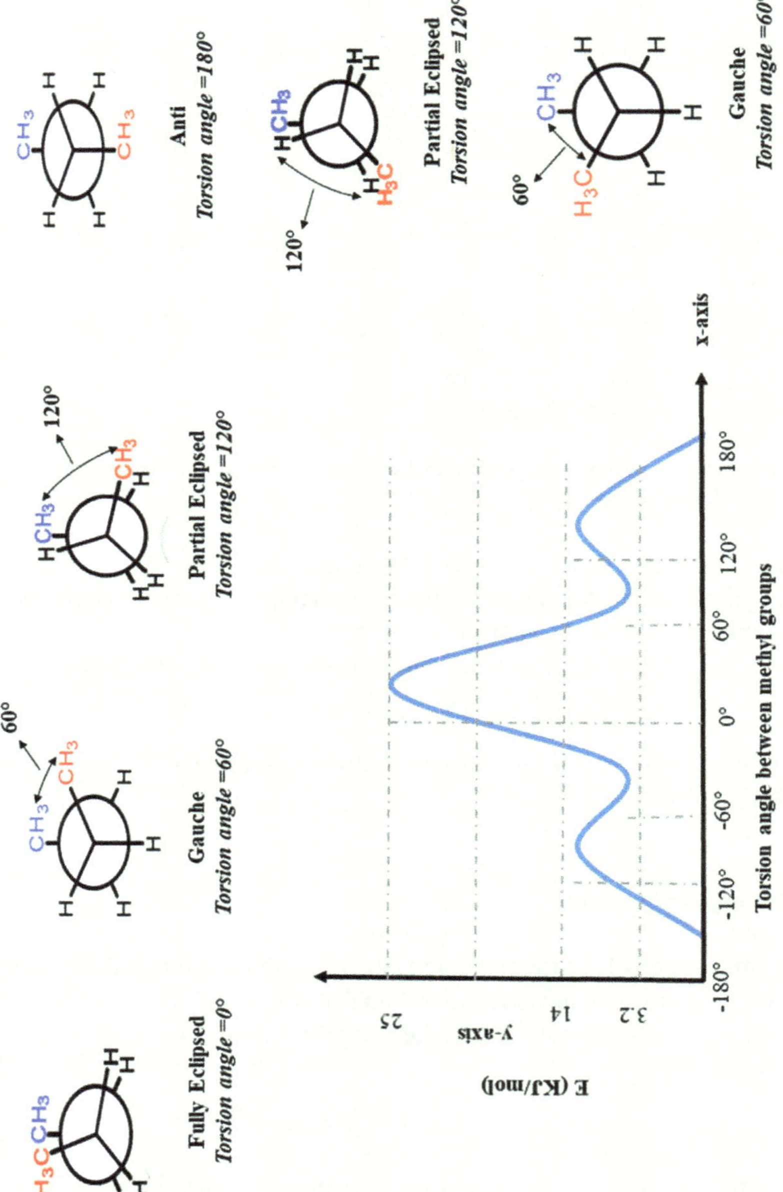

Fig. 7.5: The energy plot for the energy of the different conformations of the *n*-butane.

$$\Sigma E_{\text{electrostatic}} = \Sigma(q_1 q_2/\epsilon\, r_{12}) \tag{7.7}$$

where q_1 and q_2 are individual charges on atom 1 and atom 2, r_{12} is the distance between these two atoms, and ϵ represents the dielectric constant. Similarly, the Van der Waals interactions can be calculated from the following equation of Lennard-Jones potential:

$$\Sigma\, E_{\text{Van der Waals}} = \Sigma\left[\left(-A_{12}/r_{12}^{6}\right) + \left(B_{12}/r_{12}^{12}\right)\right] \tag{7.8}$$

where A and B are the constants known as the degree of stickiness and degree of hardness, respectively, and r_{12} represents the distance between atoms 1 and atom 2. H-bonding type of interaction also plays an important role in the nonbonding interactions and can be calculated by using the following equation:

$$\Sigma\, E_{H\text{-bond}} = \Sigma\left[\left(A_{12}/r_{12}^{12}\right) - \left(B_{12}/r_{12}^{10}\right)\right] \tag{7.9}$$

where r_{12} and constants A and B represent the same meaning as discussed above in the Lennard-Jones potential equation (7.8).

7.2.4 Cross-bonding energy terms

A single molecule consists of many atoms connected through several bonds. In these atoms stretching, bending, and rotation of the bonds occurs simultaneously inside the molecule (Fig. 7.6). Therefore, for calculating the energy of the whole molecule, it is indispensable to consider these cross interactions separately and energy related to them should be calculated separately. Mathematically they can be calculated by using the following equations:

$$\Sigma\, E_{\text{str-str}} = \Sigma\, k_{12}(r_1 - r_{10})(r_2 - r_{20}) \tag{7.10}$$

$$\Sigma\, E_{\text{str-bend}} = \Sigma\, k_{12}(r_1 - r_{10})(\theta_2 - \theta_{20}) \tag{7.11}$$

$$\Sigma\, E_{\text{bend-bend}} = \Sigma\, k_{12}(\theta_1 - \theta_{10})(\theta_2 - \theta_{20}) \tag{7.12}$$

where r_1 and r_2 represent the current bond length and r_{10} and r_{20} represent the initial bond length between atoms, respectively. θ_1 and θ_2 are current bond angles and θ_{10} and θ_{20} are their initial bond angles, respectively.

7.2.5 Concept of force field in molecular mechanics (MM)

The concept of a force field within MM is a fundamental tool that enables the simulation and prediction of molecular interactions and behaviors. A force field is a set of mathematical functions and parameters that describes the energy and forces acting between atoms and molecules in a system. It guides the computation of potential energy

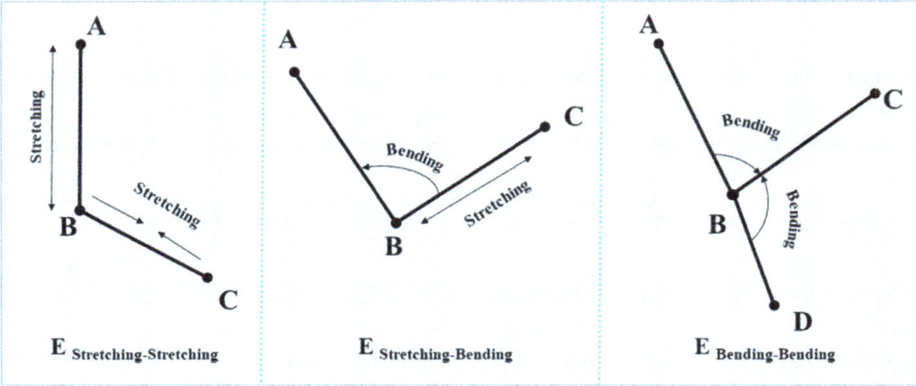

Fig. 7.6: Different types of the cross-bonding energy terms inside the molecule, that is, stretching–stretching, stretching–bending, and stretching–bending interactions.

surfaces and molecular dynamics trajectories, offering insights into how molecules interact and change over time. Force fields account for various types of interactions, including bond stretching, angle bending, dihedral rotation, Van der Waals forces, and electrostatic interactions. These interactions collectively shape the overall energy landscape of a molecule, influencing its stability, conformational changes, and binding affinities. In drug discovery, force fields are employed to model the interactions between drug candidates and their target proteins or receptors. By calculating the potential energy of different conformations and configurations, researchers can predict the most stable and energetically favorable binding modes. This information guides structural modifications of drug candidates to enhance their binding affinity and selectivity. However, it is important to note that force fields have limitations, such as their empirical nature and potential inaccuracies in capturing specific interactions. Therefore, they are often parameterized and validated against experimental data to improve their accuracy. In essence, the concept of a force field in MM serves as a bridge between theoretical calculations and real-world molecular behaviors, facilitating the rational design of drug molecules with the desired properties for effective and targeted therapies.

While studying CADD approaches, especially structure-based drug designing, the force field is one of the most common terms we come across. For making any sort of calculation in the MM we must be familiar with the meaning of force field. The simplest definition of the force field is that it is a certain set of defined formulae or functions that are used for the calculations in MM. At present, there are numerous force fields available for computational scientists and each of them has its own algorithm for the energy calculation of the molecules. The choice of force fields to be employed depends upon the type of computational study being conducted, that is, for the energy minimization of a drug molecule; the choice of force field will be different from that for the force field employed for the molecular dynamics simulations. Force fields like

CHARMM, AMBER, OPLS, MMFF, Dreiding, and UFF are a few of the most commonly employed force fields in MM.

Classification of the force fields is done based on their development. The first generation of the force fields, also known as class I force fields, are the initial force fields that were developed and they use the basic equation (7.1) mentioned above for the calculations of energy. The second generation of the force fields, also known as class II force fields, are the ones that are more advanced than class I and also incorporate the cross-binding energy terms for the energy calculations of the molecule. The latest in this class is the class III force fields, which were developed in the 1990s. These force fields also take into account the electronic polarization factors while calculating the energy of the molecules. Examples of each class of the force field are given in Tab. 7.1.

Tab. 7.1: Classification of various force fields applied in molecular mechanics.

Classification of force fields		
S. no.	**Types of force field**	**Examples**
1	Class I force fields	AMBER, CHARMM, COSMIC, DREIDING, ECEPP, ENCAD, ESFF, GROMOS, MM1, OPLS, PFF, TRIPOS, YETI
2	Class II force fields	CFF, CVFF, EFF, MM2, MM3, MM4, MMFF, QMFF, SDFF, UFF
3	Class III force fields	Allinger's MM4, Accelrys CFF, AMOEBA, DRUDE FF

With time, continuous evolution is seen in the development of various force fields; some of the class I and II force fields like OPLS, CHARMM, MM, and AMBER have also introduced their versions that take into account the phenomenon of polarization.

7.3 Quantum mechanics (QM) in drug designing

QM has emerged as a transformative force in the realm of drug designing, enabling a deeper understanding of molecular interactions and behaviors at the atomic and subatomic levels [20]. QM, a fundamental theory in physics, describes the behavior of matter and energy at the smallest scales, providing insights into the electronic structure and properties of molecules [21]. In drug designing, QM calculations offer unparalleled precision in predicting molecular properties, reaction mechanisms, and energetics [22]. By simulating the quantum behavior of electrons and nuclei within molecules, QM allows researchers to explore intricate details of chemical reactions, such as bond formation and breaking, which are crucial in drug metabolism and interactions with target proteins. QM approaches enable the accurate computation of electronic energies, helping the determination of molecular stability and reactivity. Additionally, QM studies shed light on the electronic properties of drug candidates, aiding in the identification of electron-rich or electron-deficient regions crucial for binding to protein-active sites [23, 24]. Furthermore, QM computations facilitate the analysis of molecular spectroscopy, offering insights into vibrational frequencies, electronic transitions, and NMR spectra. These spectroscopic data aid in characterizing drug structures and interactions. While computationally intensive, QM provides invaluable insights into the behavior of molecules that cannot be fully captured by classical methods. By integrating QM calculations with other techniques, such as MM and molecular dynamics simulations, drug designers can make more informed decisions in optimizing drug candidates for enhanced efficacy and reduced toxicity [25]. In the QM methodology, electrons and nuclei are considered as separate entities for all the calculations, which is contrary to the principle of MM. There is no consideration made for the inclusion of "chemical bonds" in the QM-based calculations [26]. Basically in the QM approach, a solution of the time-independent Schrodinger wave equation is sought for making all the calculations. Solutions of the Schrodinger wave equation provide us with information regarding the position and energy of the electron in space. Based on the methodology, QM approaches used in CADD can be divided into ab initio methods, density functional theory (DFT) calculations, and semiempirical approaches.

7.3.1 Ab initio methods in the CADD

The meaning of the Latin term ab initio is "from the beginning." Hence, all the calculations involved in this methodology are derived mathematically from the first principle and there is no inclusion of any experimental data for the approximation purpose [27]. All calculations in the ab initio approaches are dependent on the solutions of the time-independent Schrodinger wave equation:

$$\left[\left(\frac{d^2}{dx^2}\right)+\left(\frac{d^2}{dy^2}\right)+\left(\frac{d^2}{dz^2}\right)\right]\Psi+\left[(8\pi^2 m)(E-V)\frac{\Psi}{h^2}\right]=0 \qquad (7.13)$$

The above is the simplest depiction of the complex time-independent Schrodinger wave equation where the $[(d^2/dx^2)+(d^2/dy^2)+(d^2/dz^2)]$ term represents the partial derivative of the particle with respect to all three direction coordinates in the space, Ψ is the wave function, m is the mass of the particle, E is the total energy of the particle, V represents its potential energy, and h is the Planck's constant [28].

Ab initio methods in CADD play a pivotal role in unraveling the intricate molecular interactions that drive drug discovery. These methods are rooted in QM and offer an atomistic perspective on molecular behavior, sidestepping empirical parameters. By solving the Schrödinger equation, ab initio methods provide accurate insights into electronic structure, energy landscapes, and reaction pathways. They are used to predict properties critical for drug design, such as binding affinities, reaction energies, and molecular geometries. Additionally, ab initio methods excel in exploring noncovalent interactions and transition states, enhancing our understanding of drug-receptor interactions. However, ab initio methods are computationally demanding, limiting their application to small molecules and specific scenarios. Hybrid approaches, which combine ab initio methods with less computationally intensive techniques, strike a balance between accuracy and efficiency, making them valuable tools in rational drug discovery efforts.

The most common ab initio approach used in CADD is Hartree-Fock calculations, which is also known as Hartree-Fockself consistent field calculations, for multiparticle systems. In the year 1927 Douglas Hartree, an English scientist, proposed the Hartree equations for the distribution of electrons in an atom and proposed the self-consistent field method for solutions of the Schrodinger wave equation. Based on these equations, Douglas Hartree proposed the Hartree theory, "Final field calculated for the electrons from the solutions of Schrodinger wave equation is self-consistent with the initially assumed field."

7.3.1.1 Approximations in the Hartree-Fock calculations

The electrostatic interactions of the electrons with other electrons and the nucleus are of utmost importance and approximations used for its calculations are termed as first approximations [29, 30]. The two basic assumptions made in the first approximations are:

- Coulomb interactions between all the electrons inside the atom must be considered.
- All electrons inside the atoms move independently of each other.

All other approximations in the Hartree-Fock calculations are less important and are done after the first approximation is performed and are collectively known as second approximations.

A major approximation in Hartree-Fock calculations is the assumption that electrons move independently within an averaged electron density, ignoring electron correlation effects. This simplification, known as the "Hartree-Fock approximation," limits the accuracy of predictions for systems with strong electron correlation. Another critical approximation is the use of basis sets to describe the wave functions of molecules. These are finite sets of functions that approximate the true electron distribution. Choosing an appropriate basis set is crucial, as it affects both accuracy and computational cost. While these approximations might sacrifice some precision, they remain essential for efficiently predicting molecular behavior in drug discovery. Researchers continually seek ways to balance accuracy and computational feasibility, often combining ab initio methods with other approaches to yield more reliable insights into drug-receptor interactions and molecular properties.

7.3.1.2 Methodology of Hartree-Fock calculations

The complete workflow for the methodology employed in these calculations is given in Fig. 7.7. The basic steps involved in these calculations include:

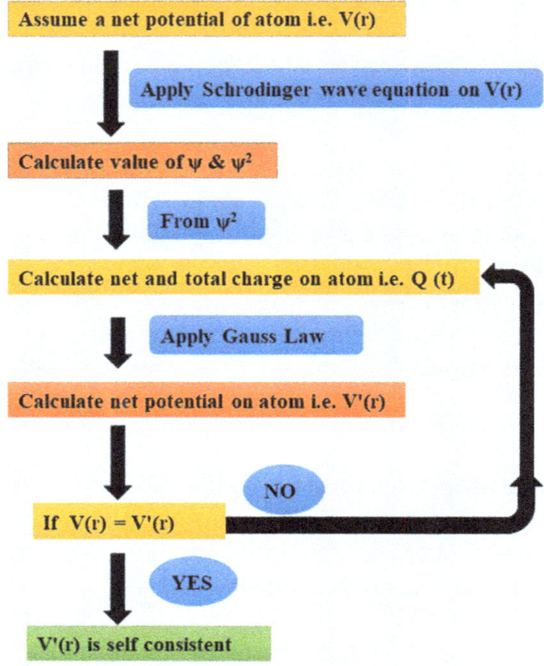

Fig. 7.7: Workflow for the different steps involved in Hartree-Fock calculations.

- Initially, a net potential of the atom *V*(*r*) is assumed.
- From the value of *V*(*r*), the time-independent Schrodinger wave equation is solved to get the values of the wave function (Ψ) for each electron. It is established that Ψ wave function in the Schrodinger wave equation does not have any physical significance; however, Ψ^2 has its physical significance as it tells us about the probability of finding the electron in space (orbital).
- From the value of Ψ^2, which is the orbital of the electron, the net charge on the electron is calculated.
- The total charge distribution of the atom (Q_t) is calculated thereafter from the individual charge distribution of each of the electrons in the atom.
- From the value of Q_t, the net potential of the atom $V'(r)$ is calculated by applying the Gauss law calculations as shown in the following equations:

$$\Phi_{\mathrm{E}} = \frac{Q_{\mathrm{t}}}{\varepsilon_0} \tag{7.14}$$

$$E = \frac{\Phi_{\mathrm{E}}}{\mathrm{Area}} \tag{7.15}$$

$$V'(r) = \int E.dr \tag{7.16}$$

where Φ_E is the electric flux, Q_t is the total charge, E represents the electric field, and $V'(r)$ is the net potential to be calculated.

- Now, if the $V(r) = V'(r)$, the net potential is self-consistent and the whole procedure is repeated until we get the self-consistent net potential value.
- From this value of the self-consistent net potential, we perform our energy calculations.

The use of ab initio approaches including Hartree-Fock calculations is limited to small molecules as it is a very rigorous and time-consuming approach. Applying these principles to the large biomolecules is quite challenging and requires more development in this field.

The major uses of this approach in CADD include:

- It is the most accurate method for calculating the 3D molecular arrangement of the small drug molecules that can be further employed for ligand- and structure-based drug designing approaches.
- For understanding the molecular interactions, ab initio approaches can be efficiently employed for calculating the important parameters such as distribution of electrons, dipole moments, and molecular orbital energies.
- To assess the stability of different conformations of a drug molecule, ab initio methods are very useful and accurate. These stable conformations obtained can be further employed for further CADD approaches such as molecular docking, QSAR model generation, and molecular dynamics simulations.
- The molecular descriptors calculated through ab initio methods are highly accurate and can be used for the generation of very robust and reliable QSAR model generation. These types of QSAR models are gaining popularity in the CADD and are known as QM-QSAR models.
- Ab initio methodologies can also be employed for the generation of spectral properties of the molecules such as UV absorption, IR spectra, and NMR spectra of the drug molecules.

7.3.2 Density functional theory (DFT) in CADD

DFT has emerged as a pivotal tool in CADD, revolutionizing the way researchers understand molecular interactions and properties. DFT is a quantum mechanical approach that enables accurate predictions of electronic structure and properties of molecules [31]. In CADD, it finds applications in predicting molecular geometries, calculating electronic properties, and understanding the energetics of molecular interactions. DFT's ability to balance accuracy and computational efficiency has made it a valuable asset in rational drug design. By elucidating molecular structures and energies, DFT aids in identifying potential binding sites, optimizing ligand conformations, and predicting interaction strengths between drug candidates and target biomolecules. Additionally, DFT allows the exploration of key physicochemical properties, such as electrostatic potentials, charge distributions, and polarizabilities, providing insights into the intermolecular forces governing ligand-receptor interactions. In the pursuit of drug discovery, where experimental approaches can be time-consuming and costly, DFT offers a cost-effective alternative for screening and prioritizing potential drug candidates. Despite its successes, DFT does have limitations, particularly in accurately capturing dispersion forces and accurately modeling highly correlated systems. However, advancements like hybrid DFT methods and dispersion corrections have ameliorated some of these issues. Its ability to provide valuable insights into molecular properties and interactions at a reasonable computational cost makes it an indispensable tool in the quest for novel and effective pharmaceutical agents [32].

The Hartree-Fock method discussed above does not consider the correlations between the electrons in an atom while performing the calculations and is a very time-consuming methodology. Both of these issues were addressed by the introduction of DFT calculations for computational chemistry in 1998. Therefore, DFT methods not only offer high accuracy but also perform all the calculations in less time. Another advantage of the DFT calculations is that it avoids the very complex Schrodinger wave equation for all the calculations [33].

The DFT is based on the Hohenburg-Kohn theorem, which states that all ground state properties of a system are the functional of its electron density (ρ). In other words, if we know the electron density of the system then we can calculate its total energy and other molecular properties related to the ground state. Mathematically it can be represented by the following equation:

$$E_0(\rho) = T(\rho) + V(\rho) + E_{XC}(\rho) \tag{7.17}$$

where $E(\rho)$ depicts the value related to the total electron density function, $T(\rho)$ depicts the total energy of the system (kinetic and potential energy), $V(\rho)$ depicts the electrostatic energy that arises due to the interaction between the nuclei and electrons and between the electrons themselves, and E_{XC} depicts the energy due to exchange and correlation of the electrons.

This type of interaction due exchange and correlation of the electrons is neglected in Hartree-Fock methods due to the complexity of the Schrodinger wave equation. In the above equation, $E(\rho)$ and $V(\rho)$ are calculated to the best of known accuracy; however, E_{XC} cannot be calculated accurately and hence approximations are used for this calculation. In all the E_{XC} calculations, the total energy is divided into two parts, energy due to the exchange of electrons (E_X) and energy due to the correlation between the electrons (E_C). Mathematically, it is depicted by the following equation:

$$E_{XC}(\rho) = E_X(\rho) + E_C(\rho) \tag{7.18}$$

In DFT, generally, three types of algorithms are employed for E_{XC} calculations, which include local density approximations, generalized gradient approximations, and hybrid approximations. Hybrid approximations are the ones that are most widely used and considered to be most accurate in performing the DFT calculations in CADD. The most widely accepted applications of DFT in CADD are as follows:

- DFT is widely used for studying the spatial arrangement of drugs and receptor molecules including their bond lengths, bond angles, conformations, and torsion strains.
- The basic use of the DFT is to perform the energy calculation of the molecules and hence it can be employed for finding out the most stable conformer of the drug molecules.
- DFT-based QSAR has also gained popularity nowadays and uses on the calculations of molecular descriptors based on DFT for the generation of a QSAR model.
- DFT methodology is also a very helpful tool employed for the calculations of binding energy between the drug and receptor molecule. DFT is very successful in determining the ionic, covalent, and hydrogen bonds between the drug and receptor; however, weaker bonds such as ion–dipole and dipole–dipole interactions cannot be determined through DFT.
- Organometallic compounds play a very important role in biological systems as there are numerous known metal-containing drugs available, and also, thousands of metal-containing enzymes are known that act as targets for many diseases. DFT technique is used successfully for the modeling of these organometallic compounds for the drug designing process.

7.3.3 Semiempirical methods in the CADD

The semiempirical approach is a hybrid of both MM and ab initio methods in which mathematical approximation along with some predefined data from the experiments is used for our calculations. As compared to ab initio and DFT methods, it is faster but less accurate and can also be employed for larger molecules. To begin the calculations, semiempirical methods also initiate from the Hartree-Fock equations but proceed through many approximations instead of finding the exact solutions to these equations, which accounts for their hastiness and lower accuracy. One major reason for the fast calculations through semiempirical approaches is that they only take into consideration the valence electrons and neglect the core electrons [34]. There are several semiempirical approaches employed in the CADD methodology (Fig. 7.8), which include the following.

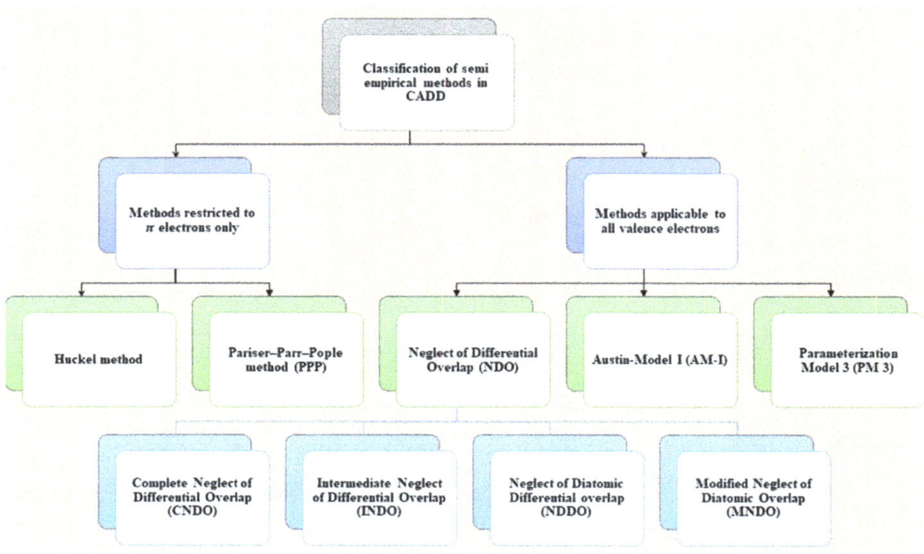

Fig. 7.8: Different semiempirical approaches employed in quantum mechanics (QM).

7.3.3.1 Huckel and extended Huckel methods

The Huckel theory was proposed in the early 1930s and is considered the "mother" of all the approximate molecular orbital methods [35]. The existence of the Huckel theory even in the current era when highly powerful computational tools for the calculation of molecular energies are available says enough about its significance. The one major drawback of the Huckel theory-based calculations is that it is limited to the calculation of the molecular shape and energies of the only π system present in a molecule.

The extension of the Huckel theory to the sigma orbital was done by Roald Hoffman in 1963, who named it as the extended Huckel's method [36]. Both of these methods have the advantage that they can be applied to the whole of the periodic table without many complications. Both of these methods are of seldom use in the current era of semiempirical calculations due to their lack of inaccuracy.

7.3.3.2 Pariser-Parr-Pople (PPP) method

Pariser-Parr-Pople (PPP) method is considered as the further improvement of Huckel's and extended Huckel's method [37]. Similar to the Huckel method, this approach is also applicable to the calculations of the molecular shape and energies of the only π system present in a molecule. The improvement made in this method is that it incorporates the electronic repulsion calculations between the π and valence electrons present in the molecule. The most important use of the PPP method in molecular modeling studies is that it is used for defining the aromaticity in the compounds.

7.3.3.3 Neglect of differential overlap (NDO) method

This is considered as one of the most widely employed semiempirical approaches and was developed by John Pople [38, 39]. There are groups of NDO methods that are used and all have one common principle behind them, which is the modification of the Hartree-Fock equations by modifying the complex integral part of the same into a simpler one using some assumptions. The most basic assumption, which is done in NDO, is that the overlap between a pair of any two different orbitals in a molecule is set to zero. If the differential overlap is between the two orbitals of the same atom then it is termed a monoatomic differential overlap and on a similar note, if the overlap is between the two orbitals of different atoms of the molecule then the differential overlap is called the diatomic differential overlap.

Based on the degree of the neglect of overlap between the two orbitals inside the molecule, NDO methods are further modified.

7.3.3.3.1 Complete neglect of differential overlap (CNDO) method

- Very helpful in the prediction of the conformations of molecules.
- Vibration frequencies and bond lengths are not calculated very accurately.
- Energies and spectral information of the molecules are calculated poorly.
- Calculation of the distance between the diatomic molecules is also done in lesser time.

7.3.3.3.2 Neglect of diatomic different overlap (NDDO) method

- The overlap between the two different orbitals is neglected only if it is between the orbitals of the different atoms of the molecule.
- The basic assumption in this method is that the molecular orbitals of two different atoms are perpendicular, that is, orthonormal to each other.

7.3.3.3.3 Intermediate neglect of differential overlap (INDO) method

- This is an improvement over the CNDO method.
- This method takes into consideration an important parameter that the interaction between the two electrons also depends upon their relative spin.
- This interaction is of considerable amount if the electrons are present on the same atom, for example, methylene molecule (Fig. 7.9).
- If the spin of the electrons is parallel it results in a lowering of the energy in contrast to the energy between the electrons of paired spin.

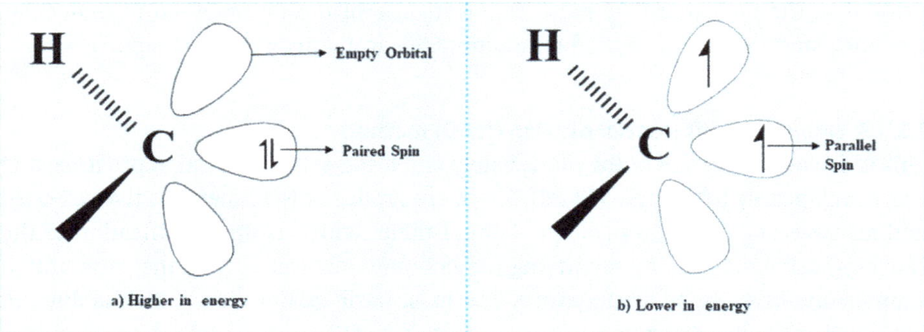

Fig. 7.9: Methylene molecule in two different orientations, that is the methylene molecule having paired spin possesses high energy, whereas the methylene molecule having parallel spin possesses low energy.

7.3.3.3.4 Modified neglect of diatomic overlap (MNDO) method

- This method is a natural extension of the NDDO method.
- Was developed to perform the calculations, especially for the molecules which contain lone pair of electrons.
- Ground state properties such as heat of formation and geometries of the molecules are taken into consideration for all the calculations in this method.
- Additional reference parameters such as ionization potentials and dipole moments are employed to enhance the accuracy of the result.
- Not suitable for estimating the hydrogen bond interactions and conjugated system in the molecule.

7.3.3.4 The Austin model 1 (AM1) method

The Austin model 1 (AM1) method was developed by Dewar and coworkers to provide more accuracy and overcome the limitations of the NDO methods [40]. This method is very similar to the MNDO method except for the calculations involved for the nonbonded repulsions. The introduction of the Gaussian function in the AM1 method has improved the accuracy of energy calculations many folds, especially for the organic molecules. Although this method was not able to resolve the lowest energy conformation of a dimer of water molecule accurately, it is very helpful in calculating certain other parameters such as heat of formation. Despite being a very effective semiempirical method for the calculations, this approach has certain limitations also which include:

- Calculations involving the steric effects are always overvalued if assessed through the AM1 method.
- Whenever AM1 is employed for the five-membered ring, it always predicts the rings are more stable than their actual stability.
- If a molecule contains too much charge localization, then the AM1 method will always calculate its heat of formation inaccurately.

7.3.3.5 The parametric method 3 (PM3)

The parametric method 3 (PM3) method was developed by James Stewart and this approach also uses a similar methodology to AM1 but with different parameters [41]. The accuracy of the PM3 is also considered slightly superior when compared to the AM1 or any other semiempirical method. One biggest difference with the AM1 method is that in the PM3 method parameterization is done automatically, whereas in AM1 it is done manually. The Gaussian functions employed in the PM3 approach use only two Gaussian terms per atom in the calculations. PM3 methods are best known for the calculations involving hypervalent molecules and the calculations of the overall heat of formation. Despite its high accuracy, it does not predict the rotational barrier in the molecules having an amide group. PM3 is also known for imparting pyramidal hybridization to the nitrogen atoms most of the time, so it is not a preferred method for calculation for nitrogen-containing molecules where hybridization is an important criterion.

References

[1] Hung CL, Chen CC. Computational approaches for drug discovery. Drug Development Research, 2014, 75(6), 412–8.

[2] Nascimento IJ, de Aquino TM, da Silva-Júnior EF. The new era of drug discovery: The power of computer-aided drug design (CADD). Letters in Drug Design & Discovery, 2022, 19(11), 951–5.

[3] Niazi SK, Mariam Z. Computer-aided drug design and drug discovery: A prospective analysis. Pharmaceuticals, 2023, 17(1), 22.

[4] Leelananda SP, Lindert S. Computational methods in drug discovery. Beilstein Journal of Organic Chemistry, 2016, 12(1), 2694–718.

[5] van der Kamp MW, Mulholland AJ. Combined quantum mechanics/molecular mechanics (QM/MM) methods in computational enzymology. Biochemistry, 2013, 52(16), 2708–28.

[6] Hehre WJ. A Guide to Molecular Mechanics and Quantum Chemical Calculations. Wavefunction, Irvine, CA, 2003.

[7] Sousa SF, Ribeiro AJ, Neves RP, Brás NF, Cerqueira NM, Fernandes PA, Ramos MJ. Application of quantum mechanics/molecular mechanics methods in the study of enzymatic reaction mechanisms. Wiley Interdisciplinary Reviews: Computational Molecular Science, 2017, 7(2), e1281.

[8] Atkins PW, Friedman RS. Molecular Quantum Mechanics. Oxford University Press, 2011.

[9] Zheng M, Waller MP. Adaptive quantum mechanics/molecular mechanics methods. Wiley Interdisciplinary Reviews: Computational Molecular Science, 2016, 6(4), 369–85.

[10] Kollman P. Theory of complex molecular interactions: Computer graphics, distance geometry, molecular mechanics, and quantum mechanics. Accounts of Chemical Research, 1985, 18(4), 105–11.
[11] Woods CJ, Manby FR, Mulholland AJ. An efficient method for the calculation of quantum mechanics/ molecular mechanics free energies. The Journal of Chemical Physics, 2008, 128(1).
[12] Gao J. Methods and applications of combined quantum mechanical and molecular mechanical potentials. Reviews in Computational Chemistry, 1996, 119–85.
[13] Sun Q, Chan GK. Exact and optimal quantum mechanics/molecular mechanics boundaries. Journal of Chemical Theory and Computation, 2014, 10(9), 3784–90.
[14] Zhang Y, Lee TS, Yang W. A pseudobond approach to combining quantum mechanical and molecular mechanical methods. The Journal of Chemical Physics, 1999, 110(1), 46–54.
[15] Bakowies D, Thiel W. Hybrid models for combined quantum mechanical and molecular mechanical approaches. The Journal of Physical Chemistry, 1996, 100(25), 10580–94.
[16] Pissurlenkar RR, Shaikh MS, Iyer RP, Coutinho EC. Molecular mechanics force fields and their applications in drug design. Anti-Infective Agents in Medicinal Chemistry (Formerly Current Medicinal Chemistry-Anti-Infective Agents), 2009, 8(2), 128–50.
[17] Bekono BD, Sona AN, Eni DB, Owono LC, Megnassan E, Ntie-Kang F. Molecular mechanics approaches for rational drug design: Force fields and solvation models. Physical Sciences Reviews, 2021, 8(3), 457–77.
[18] Vanommeslaeghe K, Guvench O. Molecular mechanics. Current Pharmaceutical Design, 2014, 20(20), 3281–92.
[19] Sears A, Batra RC. Macroscopic properties of carbon nanotubes from molecular-mechanics simulations. Physical Review B, 2004, 69(23), 235406.
[20] Raha K, Peters MB, Wang B, Yu N, Wollacott AM, Westerhoff LM, Merz Jr KM. The role of quantum mechanics in structure-based drug design. Drug Discovery Today, 2007, 12(17–18), 725–31.
[21] Zhou T, Huang D, Caflisch A. Quantum mechanical methods for drug design. Current Topics in Medicinal Chemistry, 2010, 10(1), 33–45.
[22] Arodola OA, Soliman ME. Quantum mechanics implementation in drug-design workflows: Does it really help? Drug Design, Development and Therapy, 2017, 2551–64.
[23] Mucs D, Bryce RA. The application of quantum mechanics in structure-based drug design. Expert Opinion on Drug Discovery, 2013, 8(3), 263–76.
[24] Heifetz A, (ed). Quantum Mechanics in Drug Discovery. Humana Press, New York, 2020.
[25] Manathunga M, Götz AW, Merz Jr KM. Computer-aided drug design, quantum-mechanical methods for biological problems. Current Opinion in Structural Biology, 2022, 75, 102417.
[26] Kotev M, Sarrat L, Gonzalez CD. User-friendly quantum mechanics: Applications for drug discovery. Quantum Mechanics in Drug Discovery, 2020, 231–55.
[27] Zhou T, Huang D, Caflisch A. Quantum mechanical methods for drug design. Current Topics in Medicinal Chemistry, 2010, 10(1), 33–45.
[28] Segall MD, Payne MC, Ellis SW, Tucker GT, Boyes RN. *Ab initio* molecular modeling in the study of drug metabolism. European Journal of Drug Metabolism and Pharmacokinetics, 1997, 22, 283–9.
[29] Petke JD, Whitten JL, Douglas AW. Gaussian lobe function expansions of Hartree–Fock solutions for the second-row atoms. The Journal of Chemical Physics, 1969, 51(1), 256–62.
[30] Medwick PA. Douglas Hartree and early computations in quantum mechanics. Annals of the History of Computing, 1988, 10(2), 105–11.
[31] Tandon H, Chakraborty T, Suhag V. A brief review on importance of DFT in drug design. Research in Medical & Engineering Sciences, 2019, 39, 791–5.
[32] Patel MA, Deretey E, Csizmadia IG. Will ab initio and DFT drug design be practical in the 21st Century?: A case study involving a structural analysis of the β2-adrenergic G-protein coupled receptor. Journal of Molecular Structure: Theochem, 1999, 492(1 –3), 1–8.

[33] Chermette H. Density functional theory: A powerful tool for theoretical studies in coordination chemistry. Coordination Chemistry Reviews, 1998, 178, 699–721.

[34] Pople JA, Beveridge DL. Molecular Orbital Theory. NY, 1970.

[35] Hückel E, Hückel W. Theory of induced polarities in benzene. Nature, 1932, 129(3269), 937–8.

[36] Hoffmann R. An extended Hückel theory I hydrocarbons. The Journal of Chemical Physics, 1963, 39(6), 1397–412.

[37] Michl J. Use of semiempirical models for calculation of B terms in MCD spectra. II. The Pariser-Parr-Pople (PPP) model: A consequence of the pairing symmetry in π systems with alternant skeletons. The Journal of Chemical Physics, 1974, 61(10), 4270–3.

[38] Pople JA, Santry DP, Segal GA. Approximate self-consistent molecular orbital theory. I. Invariant procedures. The Journal of Chemical Physics, 1965, 43(10), S129–35.

[39] Pople JA, Segal GA. Approximate self-consistent molecular orbital theory. III. CNDO results for AB2 and AB3 systems. The Journal of Chemical Physics, 1966, 44(9), 3289–96.

[40] Dewar MJ, Zoebisch EG, Healy EF, Stewart JJ. Development and use of quantum mechanical molecular models. 76. AM1: A new general purpose quantum mechanical molecular model. Journal of the American Chemical Society, 1985, 107(13), 3902–9.

[41] James JS, Stewart J. Optimization of parameters for semiempirical methods II applications. Journal of Computational Chemistry, 1989, 10, 221–64.

Chapter 8
Energy minimization techniques in molecular modeling

Energy minimization is the alchemist's crucible of molecular modeling, transforming raw data into the golden structures of knowledge. – Anonymous

8.1 Introduction

Energy minimization techniques play a pivotal role in molecular modeling by enabling the exploration of energetically favorable conformations and structures of molecules [1]. These methods seek to find the lowest energy state of a molecular system by iteratively adjusting atomic positions and optimizing bond lengths, angles, and dihedral angles while considering steric clashes and energy interactions [2, 3]. One commonly used technique is the steepest descent algorithm, where the system's energy is successively reduced by moving each atom along the gradient of the energy surface [4]. While effective, this method can be slow and prone to convergence issues. Conjugate gradient methods, another approach, consider the history of atomic movements to efficiently navigate the energy surface, converging to the minimum energy configuration more rapidly [5]. More advanced methods like the L-BFGS (limited-memory Broyden-Fletcher-Goldfarb-Shanno) algorithm combine the benefits of both steepest descent and conjugate gradient approaches, accelerating convergence while conserving memory [6].

Energy minimization techniques are vital in tasks such as protein–ligand docking and structure refinement [1]. In drug discovery, these methods help predict the stable binding geometry between a drug candidate and its target protein, guiding the design of molecules with improved binding affinity and selectivity. While these techniques simplify complex molecular interactions, they provide valuable insights into molecular energetics and aid in the rational design of novel therapeutic agents. Stability of any molecule whether a drug or a receptor, can be assessed only by studying its energetic state. It is always the stable conformation of any drug or receptor molecule that is of interest in any drug-designing process. This search for stable conformation led to the investigation of different approaches applicable to energy minimization in the drug discovery process. Whenever a molecule is kept in space its energy is defined with respect to different internal coordinates or Cartesian coordinates [7–9].

For a molecule with having "*x*" number of atoms, the Cartesian coordinates required to define its energy will be $3x$, and the internal coordinates required to define its energy will be $3x - 6$. Considering an example of a methane molecule (CH_4; four atoms), for defining its energy in space and for the calculation of its minimum energy we require 6 $((3 \times 4) - 6)$ internal coordinates and 12 Cartesian coordinates. The internal coor-

https://doi.org/10.1515/9783111434858-008

dinates are used for the energy minimization when employing quantum mechanics methods, whereas, Cartesian coordinates are employed while using the molecular mechanics approach for the energy minimization. This was a simple example of a four atom-containing molecule; however, for the larger molecules, energy calculation is quite complicated as it involves many atoms that are to be considered [10].

The energy of any molecule in space is defined mainly based on these four parameters:

- Energy is associated with the bond lengths and bond angles
- Energy is associated with the torsions (dihedral angles)
- Van der Waals steric energy terms
- Electrostatic energy terms

The goal of any energy minimization method is to find the best combinations of these four terms to have the least energy or the most stable conformer of that particular molecule. The energy referred to here is the potential energy of the molecule. The graph between the energy and different coordinates of the molecule is termed as "potential energy surface."

8.2 Concept of the local and global minima

Molecular modeling techniques, such as energy minimization and molecular dynamics simulations, aim to explore and characterize the energy landscape by navigating through various conformational states. By understanding the distribution of local and global minima, researchers can gain valuable insights into the potential binding modes, stability, and reactivity of drug molecules. This knowledge aids in the rational design of drug candidates with improved properties for successful therapeutic interventions [11–13].

In molecular modeling for drug discovery, the concepts of local and global minima play a significant role in understanding the energy landscapes of molecules. These minima represent points of lowest potential energy and are crucial for predicting molecular stability, reactivity, and interactions. A local minimum refers to a configuration in which a molecule is at a lower energy state compared to its immediate surroundings, but it might not be the lowest-energy configuration possible for the entire molecular system. In the context of drug discovery, local minima correspond to different conformations and arrangements that a molecule can adopt. These conformational changes can influence how a drug candidate interacts with its target protein or receptor. Properly accounting for local minima is vital to accurately predict molecular properties and behaviors. A global minimum, on the other hand, is the most stable configuration across the entire energy landscape of a molecule. It represents the most energetically favorable arrangement among all possible conformations. Identifying the global minimum is critical in drug design, as it provides insights into the most stable and relevant

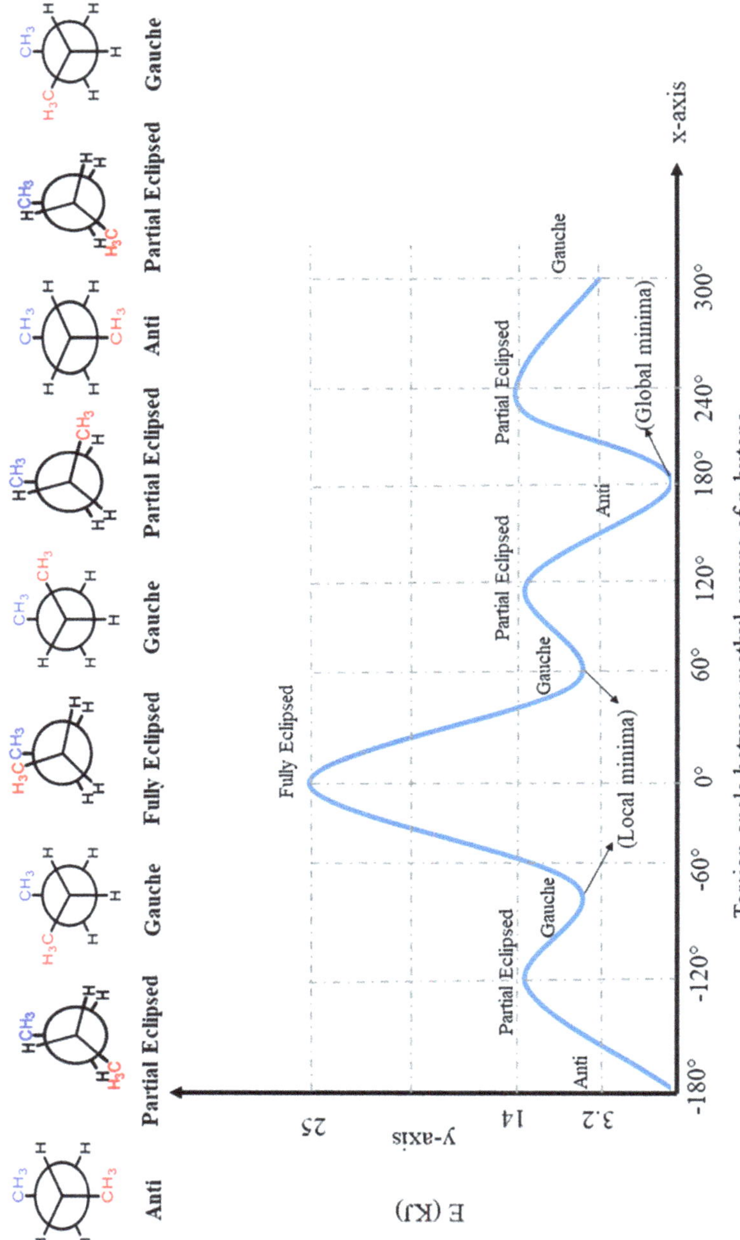

Fig. 8.1: Concept of global and local minima through different conformations of *n*-butane.

structure for predicting binding interactions and other molecular behaviors. It is always more tedious to compute the global minima of a molecule as compared to the local minima. To understand the concept of local and global minima, consider an example of conformations of *n*-butane (Fig. 8.1). *n*-Butane molecule mainly exists in four conformations, that is, fully eclipsed, anti-, gauche, and partially eclipsed conformations. Out of these conformations gauche conformer corresponds to the local minima whereas the anti-conformer is the global minima [11–13].

There are many methods used for the calculation and minimization of the energy of any system. These methods are broadly classified into two types (Fig. 8.2):

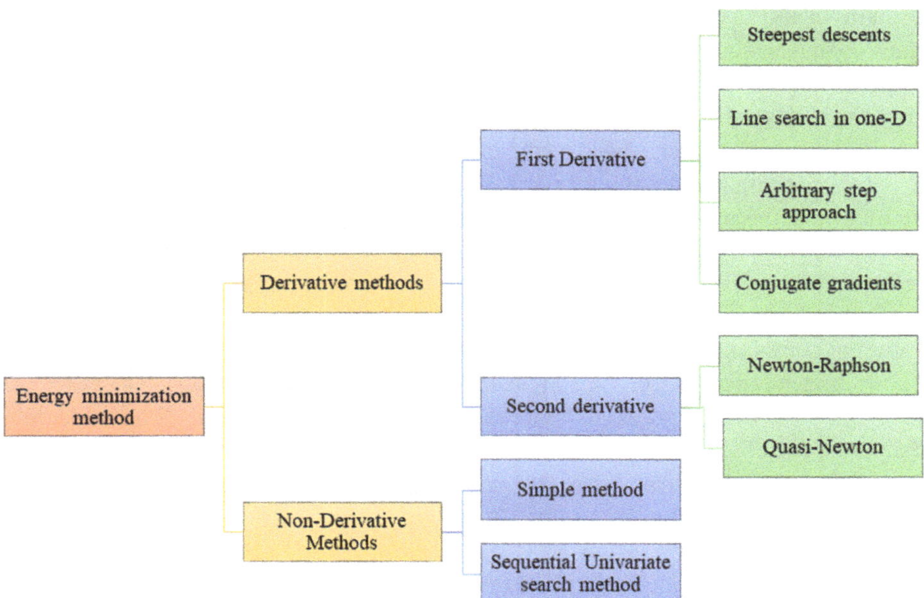

Fig. 8.2: Classification of different methods used for the energy minimization of the molecules in computer-aided drug designing (CADD).

- Methods in which derivatives of energy are calculated with respect to the coordinates for energy minimization.
- Methods that are independent of the calculations of derivatives for the energy minimization.

8.3 Derivative energy minimization methods

Derivative-based energy minimization methods are indispensable techniques within molecular modeling for predicting stable molecular structures. These methods optimize molecular geometries by iteratively adjusting atomic positions to find the configuration with the lowest potential energy. These methods utilize information from energy gradients, which are derivatives of the energy with respect to atomic coordinates [1]. Gradient descent algorithms, like the steepest descent and conjugate gradient methods, exploit these gradients to guide the iterative search for energy minima. The algorithms work by moving atoms along the negative gradient direction, effectively descending the energy surface. This process continues until a local energy minimum is reached, signifying a stable molecular structure. Derivative-based energy minimization methods offer efficiency and accuracy, particularly for moderately sized molecules. They are widely used in various molecular modeling applications, including protein–ligand docking, conformational analysis, and molecular dynamics simulations. By harnessing the power of energy gradients, these methods provide crucial insights into molecular behavior and interactions, enhancing our understanding of complex biological systems and aiding drug discovery efforts.

As the name suggests, these methods use the derivative of energy concerning the different coordinates for the minimization of the energy of the system. Derivative methods are the most popular with high acceptability in the research fraternity for energy minimization. The basic principle behind the derivative methods is that the energy (E) of the molecule is considered as the function of different independent variables x_1, x_2,. . ., x_n ($E = f(x_n)$). The main objective of any derivative method is to find values of these independent variables that result in the minimum value of E [14].

To understand this, consider an example of one independent variable system (x) for minimizing the energy (E) of a molecule. The first derivative of E with respect to x should be zero for obtaining a minimum value of energy ($dE/dx = 0$); whereas if we take the second derivative of the same with respect to x we can come across the three situations (Fig. 8.3):

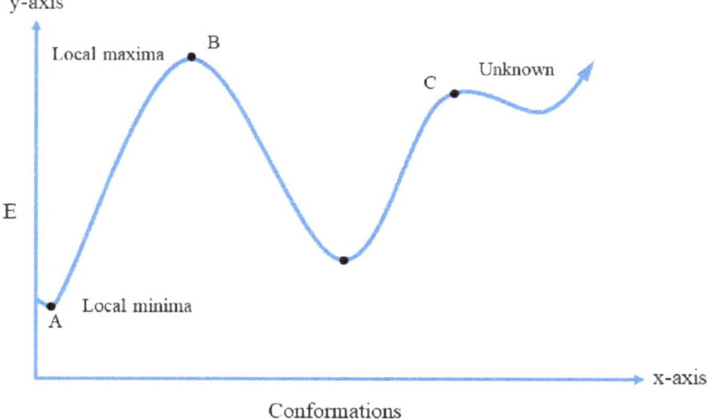

Fig. 8.3: Understanding the concept of local and global minima through the derivative energy minimization method.

- $d^2E/dx^2 > 0$, corresponds to the local minima (point A in the figure)
- $d^2E/dx^2 < 0$, corresponds to the local maxima (point B in the figure)
- $d^2E/dx^2 = 0$, corresponds to an unknown point in terms of energy (point C in the figure)

We can understand the above statement with an example. Consider there is an energy function (E) as depicted in the following equation:

$$E = 2x^2 - 8x + 12 \tag{8.1}$$

The following equations can hence be derived:

$$\frac{dE}{dx} = 4x - 8 \tag{8.2}$$

$$\frac{d^2E}{dx^2} = 4 \tag{8.3}$$

As discussed above, for a local minima $dE/dx = 0$, and putting $4x - 8 = 0$, we can get the first minima of minimum energy at $x = 2$. Also, $d^2E/dx^2 > 0$, therefore, this point is the local minima (Fig. 8.4).

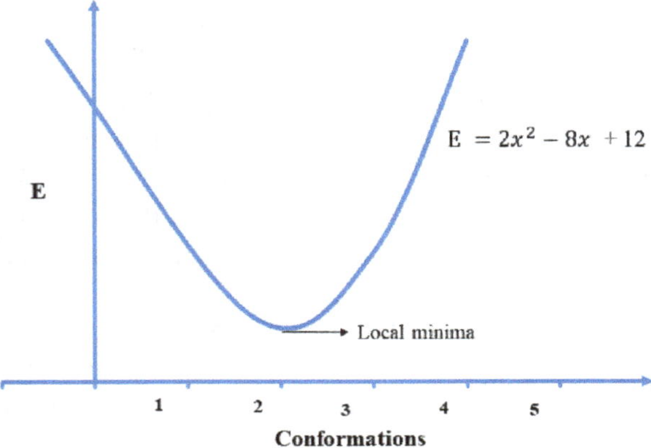

Fig. 8.4: Illustration of derivative energy minimization method for calculating the local minima.

Based on the highest derivative employed for energy minimization, these derivative methods are further classified into first-order and second-order methods.

8.3.1 First-order energy minimization methods

As evident from the name, these are the methods that simply use first derivatives for the energy minimization of molecules. First-order energy minimization methods are fundamental techniques employed in CADD for energy minimization of molecular systems. These methods aim to find the local energy minimum, where the forces acting on the atoms are approximately zero. While not as computationally intensive as higher-order methods, they are efficient and valuable for exploring the potential energy surface of molecules. One of the most common first-order methods is the steepest descent or gradient descent algorithm. It works by iteratively adjusting atomic positions in the direction of the negative gradient of the potential energy until a minimum is reached. However, the steepest descent can suffer from slow convergence and can become trapped in shallow energy minima. Conjugate gradient methods are another class of first-order techniques used in CADD. They address the convergence issues of steepest descent by ensuring that the search directions are conjugate, leading to faster convergence. These first-order energy minimization methods are integral in refining molecular structures, optimizing binding geometries of drug candidates with target proteins, and predicting the stability of molecular complexes – all essential steps in drug discovery and development. They provide valuable insights into the energy landscape of molecules and aid in the rational design of more effective drugs [1].

8.3.1.1 Steepest descent method

This is one of the simplest and oldest methods for the minimization of energy that uses the first derivative for its calculation [15]. The steepest descent method for minimizing any nonlinear function was first introduced in 1947 by Cauchy [16]. In CADD, the steepest descent method finds its application in the optimization of the geometry of the molecules in 3D. The steepest descent method is also known as the name of the gradient descent method of energy minimization [17].

This is an iterative method for the energy minimization of the molecules. Steps involved in the steepest descent energy minimization include the following (Fig. 8.5):

- We initiate with an arbitrary point (x_1) and calculate the energy at that point of our molecule.
- From x_1, moving along the negative gradient of energy towards the point x_2, energy calculations are performed.
- This iteration and calculation is continued further by moving towards the negative gradient along the points x_3 and x_4.
- Finally, a point is reached where moving further (x_5 to x_6) no change in the gradient is obtained and this point is considered as the local minima for the energy.

To understand this, consider the above example again where we used energy equation (8.1). In this example, we directly concluded by employing calculus that at the

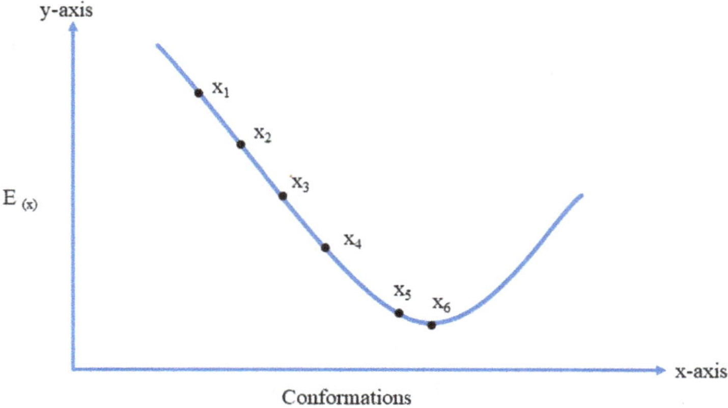

Fig. 8.5: Illustration of energy minimization using the steepest descent method.

value of $x = 2$ we have energy minima; however, the steepest descent method uses a slightly different methodology as discussed below:

- We start with an arbitrary value of x, suppose $x_1 = 3$, and consider it as the minima.
- But by plugging in $x_1 = 3$ in the first derivative of this energy function (equation (8.2); $dE/dx = 4x - 8$), we get the value of 4.
- It is evident that for minima, the value of $dE/dx = 0$, so this point cannot be our local minima, but it gives us an idea about the direction toward which it will lie.
- As the value of dE/dx is greater than 0 after plugging in the value of x_1 as 3, it indicates that we should search for our minima at a value of x_1 less than 3.
- The new point x_2 is calculated by the following equation:

$$x_2 = x_1 - \alpha \left(\frac{dE}{dx} \right) \tag{8.4}$$

  where α is the step size. Consider α as 0.2 in this case, so x_2 will be calculated to be 2.2 from equation (8.4).
- Now the new arbitrary point will be 2.2 and all the calculations will start from 2.2. Substituting its value in equation (8.1) (the equation of the first derivative), we get the value of 0.8, which is closer to 0 but not 0.
- Again we will calculate our new point x_3 as above by using the following equation:

$$x_3 = x_2 - \alpha \left(\frac{dE}{dx} \right) \tag{8.5}$$

  Keeping step size (α) as 0.2, the value of x_3 is calculated to be 2.04.
- These values will be used for finding the solution of equation (8.1) and the process will continue until we get the solution of equation (8.1) to be 0.

One obvious question that arises is that if we can get the solution of the above energy function directly by using the calculus method, then why we should use such a complex algorithm of gradient descent be employed. The explanation for this is that in the above example, the solution was calculated by using only one variable; however, in the real-life problem of energy minimization of molecules we have a combination of many variables, which are very complex to solve using simple calculus; hence, the steepest descent method finds its application for solving such complex equations.

The two important considerations while performing the steepest descent are the estimation of the step size at each time moving from one point to another and the calculation of the gradient along which one has to move. Both these aspects directly influence the accuracy in the overall minimization of the energy of the molecule. The major advantages of the steepest descent method are the ease of implementation of this method and the use of less storage on the computer while performing the same [18]. The biggest drawback of the steepest descent methodology is that it is a highly time-consuming process while performing the energy minimization calculations.

8.3.1.2 Conjugate gradient energy minimization (CGEM) method

Conjugate gradient energy minimization (CGEM) method is a vital technique in molecular modeling used to find the minimum energy configuration of a molecular system [5]. This method efficiently navigates the complex energy landscape of molecules, particularly in the context of protein–ligand interactions in drug discovery and structural biology. CGEM is an iterative optimization algorithm. It calculates the steepest descent direction and then adjusts it by considering past search directions, ensuring that it follows conjugate paths to the minimum energy state. This approach significantly accelerates convergence compared to simple gradient descent methods. In molecular modeling, this technique is employed to optimize molecular structures, refine protein–ligand complexes, and relax high-energy conformations. It helps predict stable conformations of molecules, which is crucial for understanding molecular interactions and designing drugs that can tightly bind to target proteins. Furthermore, the CGEM method is often integrated with other computational tools, like molecular mechanics or quantum mechanics, to provide a comprehensive and accurate description of molecular behavior. Overall, it plays a pivotal role in enhancing our understanding of molecular systems, ultimately contributing to advancements in drug discovery and structural biology [19, 20].

CGEM is similar to the steepest descent method but it uses a lesser number of steps to reach the minima and hence is a time-efficient approach. In the conjugate gradient method, the first step is the same as that of the steepest descent, that is, to move toward the next point along the direction of the gradient. After the first step, the difference in the algorithms of the two methods becomes evident. As suggested by the name, there occurs a “conjugation” of two things (direction of the gradient and history of the previous direction) for making the next move in search of the minima.

The difference in selecting the next point for the iteration in the CGEM method is represented in the following equations:

$$x_{\text{new}} = x_{\text{old}} - \left[\left(\frac{dE}{dx} \right) + \beta d_0 \right] \tag{8.6}$$

$$\beta = \left[\frac{(dE/dx)}{(dE/dx)_{\text{prev}}} \right]^2 \tag{8.7}$$

where d_0 represents the direction of traversal in the previous step and $(dE/dx)_{\text{prev}}$ is the gradient value at the previous step. There are several algorithms used in the conjugate gradient method for the calculations of d_0.

8.3.2 Second-order energy minimization methods

Second-order energy minimization methods are vital tools in CADD for achieving stable and accurate energy minimization of molecular structures. These methods aim to find the minimum energy conformation of a molecule by iteratively adjusting its atomic coordinates while considering both the first and second derivatives of the energy function. One prominent second-order method is the limited-memory Broyden-Fletcher-Goldfarb-Shanno (L-BFGS) algorithm. L-BFGS is particularly useful in CADD because it efficiently handles large molecular systems. It employs an approximation of the Hessian matrix, which describes the curvature of the energy surface, making it computationally feasible for extensive molecular structures. Second-order methods offer advantages over first-order methods like steepest descent due to their ability to converge more rapidly to the energy minimum. They are especially valuable when dealing with complex molecular systems or when high precision in energy minimization is required for docking studies, molecular dynamics simulations, or accurate predictions of binding energies in drug discovery efforts. Second-order energy minimization methods are indispensable tools in CADD, contributing to the efficient exploration of molecular conformations and enhancing the accuracy of drug discovery processes [19].

Here we will discuss the methods that use both the first- and second-order derivatives for energy minimization of the molecules. The first derivative of the energy provides us with the information regarding the gradient, whereas, the second derivative contains the information regarding the curvature of the energy function. The most common second-order energy minimization methods employed for these studies in CADD are discussed below.

8.3.2.1 Newton-Raphson method for energy minimization

The Newton-Raphson method is a powerful numerical technique employed in CADD for energy minimization, a critical step in understanding molecular structures and interactions. This iterative optimization algorithm aims to find the local minimum of a potential energy function, often described by a force field. In CADD, the Newton-Raphson method iteratively adjusts atomic coordinates to minimize the energy of a molecular system. It utilizes both the first and second derivatives of the energy function (gradient and Hessian matrix) to calculate the optimal step size for each iteration. This makes it a second-order optimization method, which converges more rapidly toward the energy minimum compared to first-order methods like gradient descent. The Newton-Raphson method is particularly useful in drug discovery because it allows for highly accurate energy minimization, enabling precise predictions of binding energies and molecular conformations. However, it does require the computation of the Hessian matrix, which can be computationally expensive for large molecular systems [21].

This is a very fast method of energy minimization but requires complex equations and hence limited for the energy minimization calculations of molecules that contain up to 100 atoms. Consider the following example to understand this algorithm of the energy minimization:

– Let us assume we have an energy function (E) with respect to one variable x which can be represented by the following equations:

$$E = \frac{x^4}{4 - x^2 - 5x} \tag{8.8}$$

$$E = \frac{x^4}{4} - x^2 - 5x \tag{8.9}$$

– We now require a certain value of x where we can have the minimum value of energy that we call local minima.
– To calculate the same we first calculate the first and second derivatives of the above equation by employing the following equations:

$$\frac{dE}{dx} = x^3 - 2x - 5 \tag{8.10}$$

$$\frac{d^2E}{dx^2} = 3x^2 - 2x \tag{8.11}$$

– From the previous discussion, we already know that for local minima the dE/dx = 0. This will be applied to find the solution of equation (8.10) to be 0 by using the Newton-Raphson method.
– The equation (8.10) of dE/dx will be solved starting from x = 0 until we get a consecutive solution of one negative value and one positive value as depicted below:

  At x = 0, dE/dx = −5 (negative value)

  At x = 1, dE/dx = −6 (negative value)

At $x = 2$, $dE/dx = -1$ (negative value)
At $x = 3$, $dE/dx = 16$ (positive value)

- At $x = 2$ and 3, we get consecutive negative and positive solutions of equation (8.10), hence our solution lies between these two values of x.
- To find the exact solution, we start from any arbitrary value of x between these two values and our minima will lie closer to that value, which has a solution closer to zero. From the above discussion, we can identify that our minima will lie close to 2.
- Assume that arbitrary point x_1 is 2.2; then the next point will be calculated using the Newton-Raphson formulae by the following equation:

$$x_{(n+1)} = x_n - \left[\frac{(dE/dx)}{(d^2E/dx^2)}\right] \tag{8.12}$$

where $x_{(n+1)}$ is the next variable point and x_n represents the previous point.

- The next point will be calculated by using the following equations:

$$x_2 = x_1 - \left[\frac{(dE/dx)}{(d^2E/dx^2)}\right] \tag{8.13}$$

$$x_2 = x_1 - \left[\frac{(x_1^3 - 2x_1 - 5)}{(3x_1^2 - 2x_1)}\right] \tag{8.14}$$

$$x_2 = 2.2 - \left[\frac{2.2^3 - (2 \times 2.2) - 5)}{(3 \times 2.2^3) - (2 \times 2.2)}\right]$$

$$x_2 = 2.100$$

- The same step is repeated for x_3 and calculated to be 2.095.
- The same step is repeated for x_4 and calculated to be 2.095.
- These steps are repeated until we get the same values of our variable in two consecutive calculations. As we get the same values of x_3 and x_4, it implies that the value of x variable is where our minima lies.
- Therefore it can be said that by using the Newton-Raphson method, at $x = 2.095$, we will get the minimum energy of the energy function ($E = x^4/4 - x^2 - 5x$).

8.3.2.2 Quasi-Newton method for energy minimization

The quasi-Newton method is a robust and efficient technique for energy minimization in CADD, particularly valuable for optimizing molecular structures in drug discovery. It is part of the family of second-order optimization algorithms. Unlike the steepest descent method, which uses only first-order derivatives (gradients) to guide the minimization process, the quasi-Newton method incorporates information about the curvature of the energy surface through an approximation of the Hessian matrix. This incorporation of second-order information allows for faster convergence towards the energy minimum. One of the most widely used quasi-Newton algorithms in CADD is the Broyden-Fletcher-Goldfarb-Shanno (BFGS) method. It iteratively updates an approximation of the inverse Hessian matrix to determine the optimal step size and direction for energy minimization. It strikes a balance between computational efficiency and convergence speed, making it suitable for optimizing the structures of drug candidates with reasonable accuracy. The quasi-Newton method enhances the accuracy and efficiency of energy minimization in CADD, facilitating the exploration of molecular conformations and aiding in the rational design of pharmaceutical compounds [22].

Quasi-Newton methods are very helpful for energy minimization where the above Newton-Raphson method is very complex and not easy for the calculations. This is the modification of Newton's method with certain approximations [23]. Let us try to understand the algorithm of these methods with the same example as explained for the Newton-Raphson method:

- We have the energy term, which is the function of one variable x and is given in equation (8.8).
- For the Newton-Raphson method, we used equation (8.12); however, in quasi-Newton methods, instead of calculating the second derivatives, d^2E/dx^2, we use certain approximations as for multivariable system it becomes a very tedious task to calculate the same.
- The approximation of d^2E/dx^2 is done by employing the following equation:

$$\frac{d^2E}{dx^2} = \frac{(dE/dx) - (dE/dx_{n-1})}{x - x_{n-1}} \tag{8.15}$$

  where dE/dx is the solution obtained by putting the value of x_n in the first derivative and dE/dx_{n-1} is the solution obtained by putting the value of x_{n-1} in the first derivative.
- To make it more clear, if we have to calculate the value of x_4 from the example mentioned in the Newton-Raphson method, we will use the following equation:

$$x_4 = x_3 - \frac{(dE/dx_3)}{d^2E/dx_3^2} \tag{8.16}$$

  The above equation is as per the Newton-Raphson formulae; however, in the quasi-Newton method d^2E/dx_3^2 will be replaced by the approximation value $[(dE/dx_3) - (dE/dx_2)]/(x_3 - x_2)$.
- All the other algorithms in the quasi-Newton methods remain the same as per the Newton-Raphson method.

8.4 Nonderivative energy minimization methods (NDEMM)

Nonderivative energy minimization methods (NDEMM) are essential techniques in CADD for optimizing molecular structures without explicitly calculating derivatives of the energy function. Unlike derivative-based methods, NDEM methods do not require gradient information, making them computationally efficient and suitable for complex molecular systems. One widely used NDEM method is the conjugate gradient method. It iteratively adjusts atomic coordinates along conjugate directions to minimize energy. The simplicity and efficiency of this approach make it valuable in molecular modeling tasks, such as protein–ligand docking and molecular dynamics simulations. Another notable NDEMM is the Monte Carlo method, which explores various conformational space by probabilistically accepting or rejecting coordinate changes based on energy differences. While Monte Carlo methods are often associated with sampling rather than energy minimization, they can be combined with energy-based criteria to achieve minimization-like behavior.

NDEM methods are particularly useful when dealing with systems where analytical derivatives of the energy function are challenging to compute, such as proteins with flexible loops or large ligands. Their versatility and ability to navigate complex energy landscapes make them valuable tools in the drug discovery process, aiding in the prediction of ligand binding and conformational changes. These energy minimization methods are also known as the zero-order derivative energy minimization and they do not use any derivatives for the energy minimization calculations. Various nonderivative methods employed for energy minimization are discussed below [1].

8.4.1 Downhill simplex method of energy minimization

The downhill simplex method is a nonderivative energy minimization technique frequently utilized in CADD for optimizing molecular structures without relying on gradient information. Unlike derivative-based methods, the downhill simplex method is particularly useful when analytical gradients are unavailable or expensive to compute [24]. This optimization algorithm operates by forming a simplex, a geometric figure with $n+1$ vertices in an n-dimensional space, where n represents the number of atomic coordinates to be optimized. It then iteratively adapts the simplex through a series of transformations, such as reflection, expansion, and contraction, to explore the energy landscape. The downhill simplex method does not require knowledge of the energy gradient, making it robust and adaptable for various molecular systems. However, its convergence to the global energy minimum may be slower than gradient-based methods, especially for high-dimensional spaces. In CADD, the downhill simplex method can be advantageous for exploring potential energy surfaces and identifying low-energy conformations of drug candidates. Its versatility and independence from gradient information make it a valuable tool when dealing with complex molecular systems in drug discovery [25].

The downhill simplex method is a multidimensional energy minimization approach that employs the use of simplex for the minimization purpose. Let us assume we use a one-variable then our simplex will have two vertexes and will be represented by a straight line, whereas, for a two-variable energy function it will be represented by the simplex with three vertexes, that is, by a triangle (Fig. 8.6). The main idea behind using $n + 1$ vertexes is that it helps in calculating the energy more efficiently. The simple idea behind finding the energy minima through the downhill simplex approach is that initially a simplex is defined and energy at each vertex of the simplex is calculated. From here onwards different operations on simplex are performed to find out the point that corresponds to the minima. Different actions performed on the simplex include:

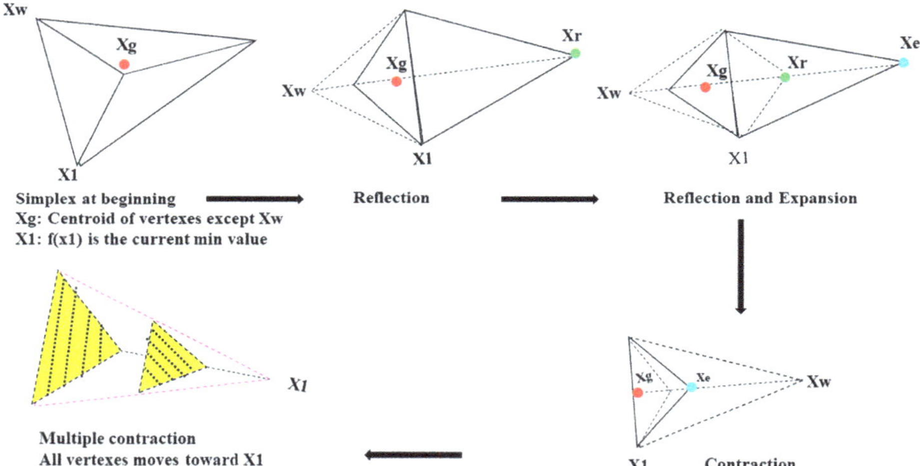

Fig. 8.6: Downhill simplex method of energy minimization illustrating the simplex, reflex, reflex and contraction, contraction, and multiple contractions.

- **Reflection**: This is the very first action that is performed on the simplex in the search for energy minima. In the beginning, the vertex having the highest energy in the simplex is located. Afterward, a reflection of the simplex is made along the opposite face of the highest energy vertex. The energy of the new vertex is calculated.
- **Expansion**: If the energy of the new vertex is found to be better than the energy of the previous vertexes then the simplex is expanded further along the line of that vertex which has the lowest energy.
- **Contraction**: On the other hand if the energy of the new vertex is not better than the previous point, then the contraction action is taken on the simplex. In contraction, the simplex is contracted in one dimension from the vertex having the

highest energy. There is one more condition here that arises. Suppose the new vertex is far worse than all the previous vertexes, then the action of multiple contraction is performed in which simplex is contracted in all the dimensions and towards the point that has the lowest energy in the simplex.

To understand this, consider an example where we have an imaginary energy function having a value given in the following equation:

$$E = 3x + 6y \tag{8.17}$$

where x and y are two variables that define the energy of the system.

Since the system has two variables, we will be requiring a simplex with three (2 + 1) vertexes, which is a triangle (Fig. 8.7). Consider any three points in the system from where we initiate our calculations, *a* (9,9), *b* (9,11), and *c* (11,9) points. The energy calculation will follow the following actions:

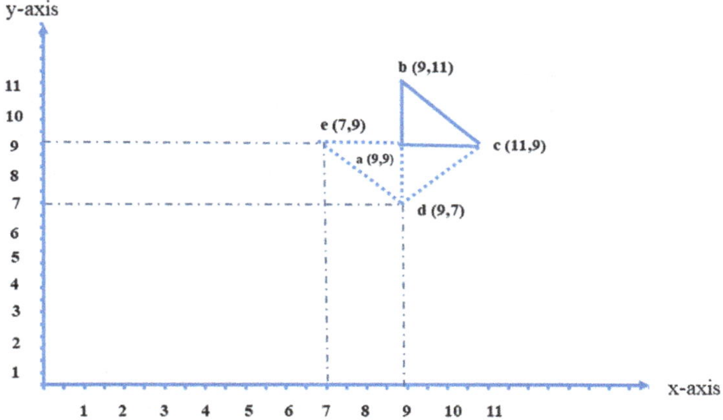

Fig. 8.7: Illustration of the downhill simplex method of energy minimization.

- At the initial stage of our simplex, three points, *a*, *b*, and *c*, will have values of energy function 81, 93, and 87, respectively, as per equation (8.15).
- Here, point *b* has the highest energy, therefore, from here we will take a reflection along the opposite face, that is, *ac*. The new point will be at *d* (9,7) and the energy function value at this point will be 69.
- Now, the new simplex will have *a*, *c*, and *d* vertexes. Upon calculations using equation (8.15) point *c* will now have the highest energy function.
- Therefore the next reflection will be followed from point *c* along the face ad. The new point will be *e* (7,9) and the energy function value corresponding to it will be 75, which is not the lowest among the three but better than the *c* point.

- The new vertex will now have vertexes *a*, *d*, and *e* and the highest energy point will be vertex *a*.
- This whole process of performing different actions on the simplex continues until we reach energy minima.
- The biggest drawback of the downhill simplex approach is that it is highly time-consuming and hence it is highly recommended to initially start with this method and after certain iterations switch to other fast methods for finding the energy minima.

8.4.2 The sequential univariate method for energy minimization

As discussed, the downhill simplex method is not recommended for the quantum mechanical calculations of energy minimization as it is a highly time-consuming method. Therefore, a more acceptable approach, the sequential univariate, ethod, is applied [26, 27]. The sequential univariate method is a nonderivative energy minimization method often employed in CADD to optimize the energy of molecular systems. Unlike derivative-based methods that use gradient information, this approach iteratively adjusts one atomic coordinate at a time, making it computationally less intensive but less efficient for large and complex molecules. In the sequential univariate method, the energy of the system is minimized by varying the coordinates of each atom sequentially while keeping the positions of the other atoms fixed. The method evaluates the energy of the system at each step and continues the process until a minimum energy configuration is reached or convergence criteria are met. While this approach is less sophisticated than second-order methods, it is advantageous for its simplicity and lower computational cost. Sequential univariate method can be suitable for initial energy minimization or as part of a larger conformational search strategy in CADD for drug discovery. However, it may not always provide the level of precision and convergence seen in second-order methods, especially for complex molecular systems. Researchers often choose the appropriate minimization method based on the specific needs and computational resources available in their drug discovery projects [28, 29].

Sequential univariate method is a highly efficient method for minimizing the energy of a multidimensional system. As the name suggests we reach energy minima step by step by taking a single variable at a time. To understand the methodology, assume that we have to minimize the energy of a system having three variables, x, y, and z. The steps to be followed are as follows:

- To initiate the energy minimization process, we start with any of the three variables (suppose we started with the variable x).
- Now, we first calculate the energy of the structure at different positions of the x, say x_1, x_2, and x_3.
- The graph between the energy and variable x will be a parabola (Fig. 8.8).

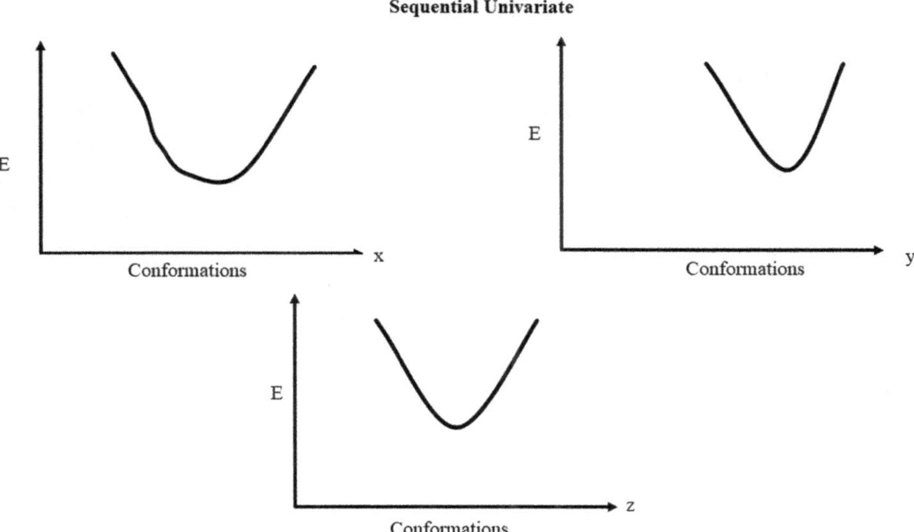

Fig. 8.8: A sequential univariate method for energy minimization. The graph is between the energy and position of the atom along with the different axis, that is, at the *x*-, *y*-, and *z*-axes.

– The equation of the parabola in this case can be represented as follows:

$$E = a(x - h)^2 + k^2 \tag{8.18}$$

where h represents the value of x where E will be the minimum, k is the minimum energy, and a represents the coefficient of the squared term.
– From the values of x and the value of energy at these points, a parabola will be plotted to obtain the value of x where the energy will be minimum.
– The same procedure is applied for finding out the values of variables y and z, where the structure will have the minimum value.
– Combining the results, we will find our local minima where the energy of our structure will be minimal with respect to all the variables.

8.5 Concept of global minima and bioactive conformer

As already discussed in the previous section of the chapter, the global minima of a molecule refers to the conformation of the drug molecule, which has the lowest energy among all possible conformations of that molecule. It is energetically the most stable conformer of that molecule. On the other hand, the bioactive conformation of a drug molecule is the conformation that produces the maximum pharmacological response when it binds with the receptor protein. In the drug designing process, it is the bioactive conformation that is of interest because ultimately we are in search of that molecule that can produce maximum pharmacological response against a particular disease. These two terms are interconnected but global minima of a drug molecule does not need to always act as a bioactive conformer against a particular receptor. Then a conceptual question that should arise is why we search for a global minimum if it can be different from the bioactive conformer. The answer to this is that although it is possible that the global minima and bioactive conformation can be different, they always lie close to each other. Therefore, once we find out the global minima of a drug molecule, we can search further from that point to find out the bioactive conformer. In other words, global minima provide us with a narrow search space in which we can search for the bioactive conformer of any drug molecule (Fig. 8.9).

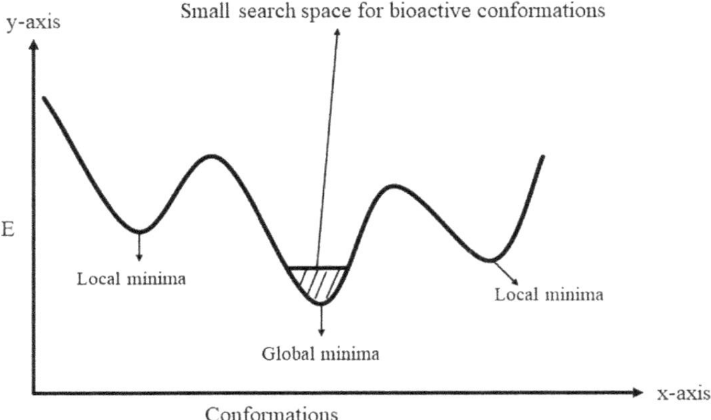

Fig. 8.9: Concept of bioactive conformer. The search space for the bioactive conformer always lies near the global minima.

The concept of global minima and bioactive conformers is fundamental in CADD, as it pertains to the determination of the most energetically favorable and biologically active molecular conformations. Understanding these concepts is crucial for identifying drug candidates with optimal binding affinities and pharmacological effects.

8.5.1 Global minima

In CADD, a molecular system can adopt numerous conformations, each associated with a specific energy level. The global minimum represents the most stable and energetically favorable conformation among all possible ones. It is the conformation that the molecule would ideally adopt under given environmental conditions, such as temperature and pressure. Locating the global minimum is essential in drug discovery as it provides insight into the most stable state of a molecule, which is often the biologically active form. In the context of ligand-receptor interactions, identifying the global minimum of a ligand's conformation can help predict its binding affinity to a target protein. Molecular mechanics and energy minimization methods are commonly used to search for global minima by iteratively adjusting atomic coordinates to reach the lowest possible energy state. However, finding the true global minimum can be computationally challenging, especially for large and flexible molecules, and it may require the use of advanced search algorithms.

8.5.2 Bioactive conformers

In the context of drug discovery, bioactive conformers refer to specific molecular conformations of a drug or ligand that exhibit a high affinity for a target protein, resulting in a pharmacologically relevant biological response. These conformations are crucial because they represent the geometric arrangements of atoms that facilitate optimal interactions with the target, such as binding to an enzyme's active site or receptor site on a cell surface. The concept of bioactive conformers underscores the importance of molecular flexibility. Molecules are not rigid structures; they can adopt a range of conformations due to the rotation of bonds and other degrees of freedom. To identify bioactive conformers, researchers use various computational techniques, including molecular docking and molecular dynamics simulations.

8.5.3 Molecular docking

Molecular docking involves the prediction of how a ligand interacts with a target protein. The goal is to identify the most favorable binding pose and orientation of the ligand within the protein's binding site. Docking programs sample different ligand conformations and orientations to find those that optimize binding interactions. The bioactive conformers are typically the ones that exhibit the strongest binding affinity.

8.5.4 Molecular dynamics simulations

Molecular dynamics simulations track the dynamic behavior of molecules over time. They are used to explore the flexibility and motion of both ligands and target proteins. Through these simulations, researchers can identify bioactive conformers by observing which ligand conformations maintain stable interactions with the protein throughout the simulation period. These conformations represent states with high biological activity.

In summary, the concepts of global minima and bioactive conformers are integral to CADD in drug discovery. Global minima represent the most stable molecular conformations, while bioactive conformers are specific arrangements of atoms within a molecule that exhibit high binding affinity and biological activity. Identifying these conformations helps researchers design and optimize drug candidates with enhanced pharmacological properties, ultimately contributing to the development of safer and more effective drugs [30].

References

[1] Gautam B. Energy minimization. Homology Molecular Modeling-Perspectives and Applications, 2020.

[2] Karamertzanis PG, Price SL. Energy minimization of crystal structures containing flexible molecules. Journal of Chemical Theory and Computation, 2006, 2(4), 1184–99.

[3] Tripathi MK, Ahmad S, Tyagi R, Dahiya V, Yadav MK. Fundamentals of molecular modeling in drug design. In: Computer Aided Drug Design (CADD): From Ligand-Based Methods to Structure-Based Approaches. Elsevier, 2022, 125–55.

[4] Vinter JG, Davis A, Saunders MR. Strategic approaches to drug design. I. An integrated software framework for molecular modelling. Journal of Computer-Aided Molecular Design, 1987, 1, 31–51.

[5] Fletcher R, Reeves CM. Function minimization by conjugate gradients. The Computer Journal, 1964, 7(2), 149–54.

[6] Saputro DR, Widyaningsih P. Limited memory Broyden-Fletcher-Goldfarb-Shanno (L-BFGS) method for the parameter estimation on geographically weighted ordinal logistic regression model (GWOLR). In AIP Conference Proceedings, AIP Publishing, 2017, 1868(1).

[7] Singh DB, Pathak RK. Computational approaches in drug designing and their applications. Experimental Protocols in Biotechnology, 2020, 95–117.

[8] Pissurlenkar RR, Shaikh MS, Iyer RP, Coutinho EC. Molecular mechanics force fields and their applications in drug design. Anti-Infective Agents in Medicinal Chemistry (Formerly Current Medicinal Chemistry-Anti-Infective Agents), 2009, 8(2), 128–50.

[9] Sheppard D, Terrell R, Henkelman G. Optimization methods for finding minimum energy paths. The Journal of Chemical Physics, 2008, 128(13), 134106.

[10] Leach AR. Energy minimisation and related methods for exploring the energy surface. Molecular Modelling: Principles and Applications, 2001, 253–302.

[11] Floudas CA, Pardalos PM, (ed). Optimization in Computational Chemistry and Molecular Biology: Local and Global Approaches. Springer Science & Business Media, 2000.

[12] Neumaier A. Molecular modeling of proteins and mathematical prediction of protein structure. SIAM Review, 1997, 39(3), 407–60.

[13] Holtje HD, Sippl W, Rognan D, Folkers G. Molecular Modeling. Wiley-VCH, Weinheim, Germany, 2003.

[14] Wiberg KB. A scheme for strain energy minimization. Application to the cycloalkanes1. Journal of the American Chemical Society, 1965, 87(5), 1070–8.

[15] Sun W, Yuan YX. Optimization Theory and Methods: Nonlinear Programming. Springer Science & Business Media, 2006.

[16] Cauchy A. Méthode générale pour la résolution des systemes d'équations simultanées. Comptes rendus de l'Académie des Sciences, 1847, 25(1847), 536–8.

[17] Elber R, Meller J, Olender R. Stochastic path approach to compute atomically detailed trajectories: Application to the folding of C peptide. The Journal of Physical Chemistry B, 1999, 103(6), 899–911.

[18] Hsieh WT, Kuo MK, Yau HF, Chang CC. A simple arbitrary phase-step digital holographic reconstruction approach without blurring using two holograms. Optical Review, 2009, 16, 466–71.

[19] Spang, III HA. A review of minimization techniques for nonlinear functions. SIAM Review, 1962, 4(4), 343–65.

[20] Godard R. Finding the roots of a non-linear equation: history and reliability. In: Research in History and Philosophy of Mathematics: The CSHPM 2015 Annual Meeting. Springer International Publishing, Washington, DC, 2016, 57–68.

[21] Hilderbrandt RL. Application of Newton-Raphson optimization techniques in molecular mechanics calculations. Computers & Chemistry, 1977, 1(3), 179–86.

[22] Schaefer B, Alireza Ghasemi S, Roy S, Goedecker S. Stabilized quasi-Newton optimization of noisy potential energy surfaces. The Journal of Chemical Physics, 2015, 142(3), 034112.

[23] Schlick T, Overton M. A powerful truncated Newton method for potential energy minimization. Journal of Computational Chemistry, 1987, 8(7), 1025–39.

[24] Disser Y, Skutella M. The simplex algorithm is NP-mighty. ACM Transactions on Algorithms (TALG), 2018, 15(1), 1–9.

[25] Black A, De Loera J, Kafer S, Sanità L. On the simplex method for 0/1 polytopes. arXiv preprint arXiv:2111.14050. 2021.

[26] Seeger R, Pople JA. Self-consistent molecular orbital methods. XVI. Numerically stable direct energy minimization procedures for solution of Hartree–Fock equations. The Journal of Chemical Physics, 1976, 65(1), 265–71.

[27] Zheng H, Cai C, Tan XM. Optimization of partial constrained layer damping treatment for vibrational energy minimization of vibrating beams. Computers & Structures, 2004, 82(29–30), 2493–507.

[28] Ohanomah MO, Thompson DW. Computation of multicomponent phase equilibria – Part I. Vapour-liquid equilibria. Computers & Chemical Engineering, 1984, 8(3–4), 147–56.

[29] Wang Y, Senatore G. Design of adaptive structures through energy minimization: Extension to tensegrity. Structural and Multidisciplinary Optimization, 2021, 64(3), 1079–110.

[30] Hollingsworth SA, Dror RO. Molecular dynamics simulation for all. Neuron, 2018, 99(6), 1129–43.

Chapter 9
Molecular docking and drug–receptor interactions

Molecular docking is like a key fitting into a lock, unlocking the secrets of drug-receptor interactions.
– Dr. Arthur J. Olson

9.1 Introduction

Molecular docking in computer-aided drug designing (CADD) is a pivotal computational technique employed to predict the interactions between small drug-like molecules and biological macromolecules, primarily proteins [1, 2]. This approach plays a crucial role in the early stages of drug discovery and development by assisting researchers in identifying potential drug candidates with high binding affinities and specificities for target proteins [3]. By simulating the binding process, molecular docking aids in the rational design of new drugs, optimization of existing ones, and the exploration of possible binding modes and conformations [4]. Molecular docking in CADD is a crucial computational technique that aids in the identification and optimization of potential drug candidates by simulating and predicting their interactions with target proteins. It enhances the efficiency and success rates of drug discovery programs by facilitating the rational design of novel therapeutic agents [5, 6].

Molecular docking operates on the principles of molecular physics and chemistry, utilizing algorithms and scoring functions (SFs) to evaluate the binding energy and affinity between ligands (drug molecules) and receptors (target proteins) [7, 8]. The main objectives of molecular docking in CADD include identifying lead compounds, understanding the binding mechanism, and estimating binding affinities to prioritize molecules for further experimental validation. This computational approach requires detailed knowledge of the three-dimensional structures of the target proteins and the ligands of interest [9]. Often, these structures are obtained through experimental techniques like X-ray crystallography or NMR spectroscopy. Alternatively, homology modeling can be employed to predict protein structures when experimental data is unavailable [10]. The success of molecular docking relies heavily on the accuracy of SFs, which estimate the free binding energy between the ligand and the receptor. Various SFs and algorithms, such as Autodock, Vina, and Glide, are employed to evaluate binding affinities and rank potential drug candidates [11]. Molecular docking in CADD is a versatile tool applicable to a wide range of therapeutic areas, including oncology, infectious diseases, neurology, and cardiovascular diseases. It accelerates the drug discovery process by reducing the time and cost required for experimental screening and by providing valuable insights into the interactions between molecules and their targets [12].

https://doi.org/10.1515/9783111434858-009

From its emergence in the early 1980s till date, there has been lots of improvement in the molecular docking algorithm and it has given us many successful marketed drugs against various diseases. Molecular docking has gained so much popularity in the research fraternity that whenever a researcher performs any CADD approach, it is the molecular docking that he performs first in most cases. As discussed already in the previous chapters, the CADD approach is classified into two types, structure-based drug designing (SBDD) and ligand-based drug designing (LBDD). Molecular docking is an SBDD approach in which we try to predict how efficiently the proposed drug molecules could bind to the desired receptor whose 3D structure is known to us [13, 14]. The basic principle behind any molecular docking approach is that it works on the hypothesis of lock and key, where the receptor acts as the lock and the ligand is the key. Like how every lock is opened with a specific key, every receptor shows its biological activity only when it binds with its specific ligand (Fig. 9.1)

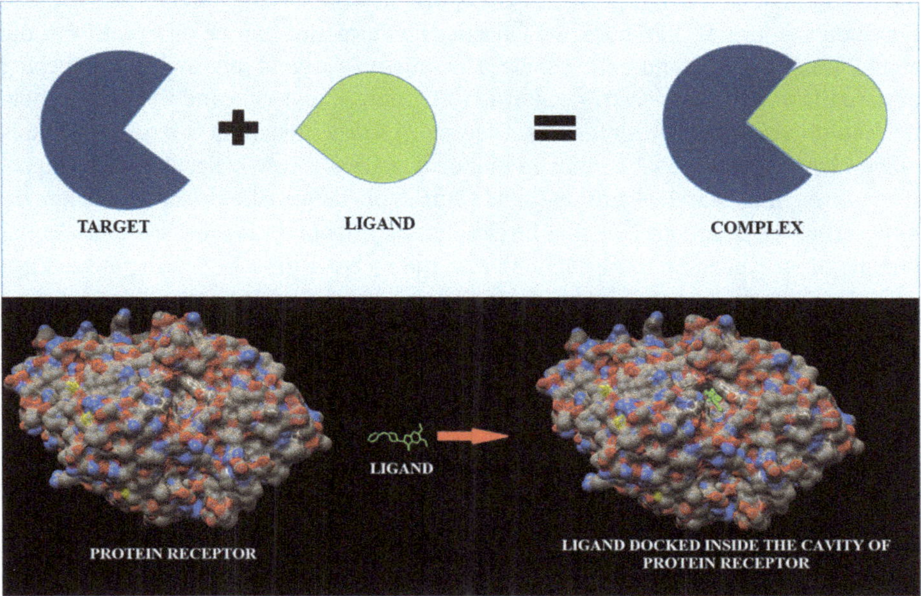

Fig. 9.1: The concept of molecular docking resembles the lock (receptor) and key (ligand) where every ligand (lock) has a specific key (receptor).

9.2 Theory of molecular docking

At present, there are numerous molecular docking softwares available; some of them are commercial while some are available for open access. Whichever software we use for the study, all have a similar algorithm that performs two functions [15]:

- 2.1 Conformational search (search algorithm)
- 2.2 Scoring of the ligand–receptor binding

Both of these functions are of utmost importance for performing successful molecular docking and assessing whether the proposed ligand will bind with the receptor or not [16]. Molecular docking software differ from each other in the approach they use for performing this conformational search and scoring of the ligand–receptor binding. To understand the concept of molecular docking, it becomes indispensable to understand the different algorithms used for performing these two basic functions of any molecular docking program [17, 18].

In the previous chapters, the concepts of bioactive conformer and global minima of a drug molecule have been explained in detail. A bioactive conformer is the conformation of a drug molecule that has the highest probability of binding with the receptor molecule to generate a biological response. Likewise, global minimum is considered as that conformation of the drug molecule that has the lowest energy among all the possible conformations of that molecule. Bioactive conformation can be different from the global minima or it could be the same. The only objective of any molecular docking program is to find the bioactive conformer of the drug molecule and then assess how this conformation binds with the target receptor protein. Generally, a bioactive conformation lies near the global minima of that drug molecule. Any molecular docking program, through its search algorithm, finds a bunch of conformations (commonly known as poses) of the drug molecule that lie close to the global minima and then checks the binding of each of these conformations with the receptor protein. Scoring of bindings with the receptor protein is given to each conformation and the conformation that shows the best binding with the receptor, in terms of the score, is considered the bioactive conformation.

9.2.1 Conformational search (search algorithms of the molecular docking)

The bioactive conformer of any drug is of primary interest to any computational scientist. It is quite possible that the energy of the bioactive conformer can vary slightly from the global minima. The very first purpose of any molecular docking software is to search for ligand conformations that can produce the desired pharmacological response. Search algorithms are defined as the set of certain rules and parameters that are used for finding those conformations of a ligand that can act as the probable bioactive conformation, which mostly falls near the global minima. Hence, through any search algorithm, at first, the global minimum is located, and then certain conformations are selected near the global minima, out of which the bioactive conformation is to be selected through the SF [19]. There are various search algorithms applied by molecular docking programs in search of the conformations of the ligand that can act as potential bioactive conformers (Fig. 9.2).

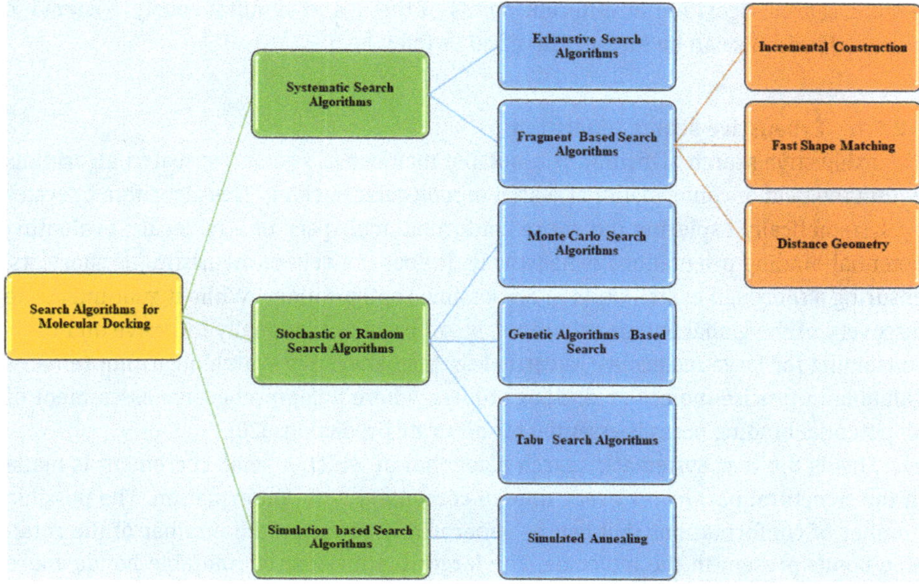

Fig. 9.2: Different search algorithms used by various molecular docking programs in search of the bioactive conformer.

9.2.1.1 Systematic search algorithms

Systematic search algorithms are fundamental in conformational search during molecular docking. They explore the vast conformational space of ligands to identify energetically favorable binding poses [20]. These algorithms systematically sample different conformations by varying dihedral angles, torsion angles, and bond rotations. Methods like Monte Carlo and genetic algorithms perform exhaustive searches while minimizing energy and maximizing ligand–receptor complementarity. By rigorously examining multiple conformations, systematic search algorithms enhance the accuracy of molecular docking predictions, enabling the identification of optimal ligand poses for drug–target interactions. Their systematic approach significantly contributes to the success of computational drug discovery and the design of effective therapeutic agents [21].

These are the most exhaustive search algorithms in which all the possible conformations of the ligand are explored in the search for the possible conformations for a bioactive conformer. As the degree of freedom is increased in any drug molecule, the possible conformations due to the various combinations also increase exponentially. This problem can be overcome by employing an incremental construction approach for the search for possible bioactive conformations. One another major drawback of the systematic search algorithm is that it can result in finding out the local minima instead of the global minima. This drawback can be overcome by initiating the systematic search algorithm at different points of the ligand simultaneously. Systematic search algorithms can be further classified as follows.

9.2.1.1.1 Exhaustive search algorithm

The exhaustive search algorithm is a notable member of systematic search algorithms within the field of conformational search in molecular docking. This algorithm operates by systematically exploring the entire conformational space of a molecule, evaluating potential binding orientations exhaustively. It does not rely on heuristics or shortcuts, ensuring a thorough examination of all possible configurations. While it guarantees the discovery of the global energy minimum, it can be computationally intensive and time-consuming for large molecules. Nevertheless, the exhaustive search algorithm remains valuable in precise molecular docking studies, where a comprehensive assessment of all potential binding poses is essential for accurate predictions [20].

This is the true systematic search algorithm in which a small increment is made in the structural parameter every time to compute a new conformation. The possible number of conformations that can be generated depends on the number of the rotatable bonds present in the molecule. The larger the number of rotatable bonds, more will be the conformations that can be generated. In other words, the complexity of calculating all the possible conformations will be increased in the molecules in which the number of rotatable bonds is higher. Hence, the exhaustive search algorithm is only applicable to the less flexible molecules, and for more flexible molecules, other algorithms of systematic search are preferred.

9.2.1.1.2 Fragment-based search algorithms

Fragment-based search algorithms are an integral part of conformational search techniques in molecular docking. These algorithms systematically explore the vast conformational space of molecules by dividing them into smaller, more manageable fragments. They then search for energetically favorable combinations of these fragments within the binding site of a target protein. Fragment-based algorithms can be classified into systematic search methods, which thoroughly explore conformational space by exhaustively evaluating various fragment combinations. This classification aids in the precise prediction of molecular interactions and contributes significantly to drug discovery by identifying potential ligand–protein binding modes efficiently. These systematic search algorithms can cater to the problem of high flexibility in calculating the different conformations. There are various fragment-based search algorithms which are described below.

Incremental construction approach

As the name suggests, this algorithm initially breaks the drug molecule into small fragments. These small fragments consist of a rigid portion, also known as the anchor, and flexible fragments. Initially, the anchor of the molecule is docked inside the binding site of the receptor and then one by one, flexible portions of the molecule are docked by taking into consideration their flexibility [22]. The one major advantage of this method is that the flexibility or number of rotatable bonds is considered separately for each fragment and their most active conformation is calculated separately. All the fragments are then combined to obtain the probable most active conformations of the drug molecule [23]. The most common molecular docking software that employs this algorithm includes DOCK, FLEXX, FLOG, and Surflex.

Fast shape matching approach

The basic principle behind this approach is that the shape of the receptor's cavity receptor is considered first and the conformations of the ligand that match best inside the shape of that cavity are then considered for finding the bioactive conformation (Fig. 9.3). There are various approaches to matching the shape of the cavity and the ligand's conformations. The shape of the cavity can be mapped either through shape complimentary, desolvation, or electrostatic parameters. ZDOCK, DOCK, EUDOK, SYSDOCK, e-HiTS, and MS-DOCK are the common molecular docking software that employs this approach for the search algorithm [24].

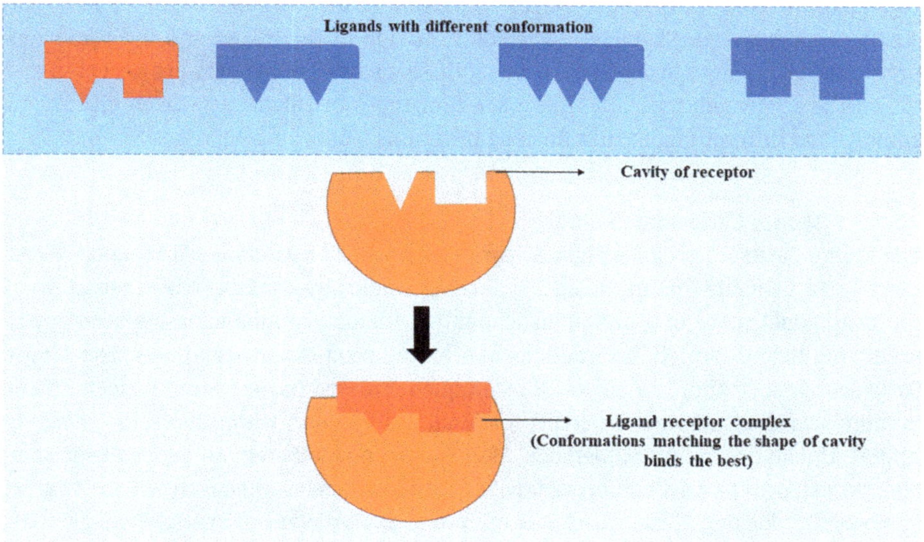

Fig. 9.3: Fast shape matching approach for finding the bioactive conformer of the ligand that fits best inside the cavity of the receptor.

Distance geometry approach

The distance geometry approach is the application of mathematical geometric algebra to computational chemistry. The basic principle of this approach is that at first, a GRID is created based on the cavity of the receptor and then different conformations of the ligand are placed inside this GRID. Based on the distance-based calculations, conformations are accepted or rejected for a probable bioactive conformer. FLOG is the molecular docking software that employs this search algorithm [25].

9.2.1.2 Stochastic or random search algorithms

Stochastic or random search algorithms are essential tools in the field of conformational search within molecular docking. These methods explore the vast conformational space of molecules by generating random conformations or perturbations, making them invaluable for finding low-energy configurations. Examples include Monte Carlo and genetic algorithms. Unlike deterministic methods, stochastic approaches do not rely on predefined search paths, allowing for a more comprehensive exploration of potential binding poses. While inherently probabilistic, these algorithms often incorporate energy evaluations to guide the search toward energetically favorable conformations, enhancing their effectiveness in predicting ligand–protein binding modes in the context of molecular docking. In this type of algorithm, the use of probability concept is employed for the search of the probable bioactive conformers. A random change is made in the conformation of the ligand and the conformations that have a high probability of being the bioactive conformers are accepted and the rest are rejected. Basically, there are four types of stochastic search algorithms employed in different molecular docking programs [26].

9.2.1.2.1 Monte Carlo search algorithm

The Monte Carlo search algorithm, as part of stochastic or random search approaches, serves as a valuable tool in molecular docking studies by enabling the exploration of conformational space in a probabilistic manner, ultimately enhancing the accuracy of predicting ligand–protein interactions and aiding drug discovery efforts. The Monte Carlo search algorithm is a powerful technique employed in the realm of stochastic or random search algorithms for conformational search tasks within molecular docking studies. Unlike deterministic methods, Monte Carlo methods rely on probabilistic sampling to explore the vast conformational space of molecules effectively. In the context of molecular docking, this algorithm is utilized to explore various spatial arrangements and orientations of a ligand (drug-like molecule) within the binding site of a target protein. It achieves this by randomly generating molecular conformations and evaluating their energies using SFs. The key idea is to accept or reject these conformations, based on their energy differences and temperature-dependent probabilities, allowing the algorithm to explore both high-energy and low-energy regions of the conformational landscape. Monte Carlo search excels in its ability to escape local energy minima, which can be challenging for deterministic methods. By employing random movements and energy-based acceptance criteria, it can efficiently sample diverse conformations and identify potential binding modes that might be missed by other search strategies.

The Monte Carlo simulation was initially introduced as the energy minimization procedure for molecular dynamics and was later accepted for the search algorithms in molecular docking programs. In this method, initially, a ligand is placed within the cavity of the receptor having a random conformation, and certain parameters, such as energy, are calculated. Afterward, a random change is made in the conformation of the ligand and all the previously calculated parameters are recalculated. Whether to accept or reject the new conformation is decided based on the probability score of the new conformation that is calculated, based on the following Boltzmann probability function.

$$P = Ae^{-E1+E0/KT} \tag{9.1}$$

where A is a constant, $E1$ is the energy of the new conformation, $E0$ is the energy of the previous conformation, K is the Boltzmann constant, and T is the temperature at which the calculations are performed.

A certain threshold value of P is decided and the conformations having a value below this threshold value are rejected and the rest are accepted. MCDOCK, ICM, DockVision, and Prodock are molecular docking software that use this type of search algorithm [27].

9.2.1.2.2 Genetic algorithm (GA)-based search

Genetic algorithm (GA), as stochastic search algorithms, are a valuable tool in the conformational search process of molecular docking. They leverage principles of evolution to efficiently explore complex binding landscapes, aiding in the identification of biologically relevant and energetically favorable binding conformations for drug discovery and molecular interaction studies. GAs are a class of stochastic or random search algorithms widely applied in the field of conformational search within molecular docking studies. The conformational search involves exploring the three-dimensional spatial arrangements of molecules to identify energetically favorable binding conformations, which is crucial for understanding molecular interactions and drug discovery. GAs simulate the process of natural selection and evolution by maintaining a population of potential solutions, or individuals, each representing a molecular conformation. These individuals are subjected to genetic operations such as mutation and crossover, mimicking genetic recombination and mutation in biology. Through successive generations, GAs iteratively improve the population's fitness, by evaluating how well each conformation fits the binding site, and its energy score. The stochastic nature of GAs allows them to explore a vast conformational space efficiently, even in the presence of complex energy landscapes. They are particularly useful when dealing with flexible molecules and large search spaces, as they have the potential to discover diverse and energetically favorable binding poses.

These algorithms take inspiration from the theory of evolution and natural selection of the biological system. It was introduced for the first time by John Holland in 1975. In this algorithm, the conformation of a ligand is defined by the combination of certain set parameters. These parameters are known as "genes" in the genetic algorithm. These genes are then combined to form a chromosome, and a population of chromosomes is generated by the different combinations of genes. From this population, the most favorable chromosomes (having the least energy) are selected for mimicking another biological process known as "crossover". By the crossover process, new population of the chromosomes (offspring) is generated that possess better properties than the parent chromosomes. Another phenomenon from the biological system, "mutation", does occur in these offspring, which ultimately leads the conformation close to the global minima. A flow diagram of the steps involved in the GA-based search algorithm is given in Fig. 9.4. The two most widely accepted molecular docking software worldwide, GOLD and AUTODOCK, employ this type of search algorithm. Other molecular docking software based on this algorithm are MolDock, FLIPDock, and EADock [28–30].

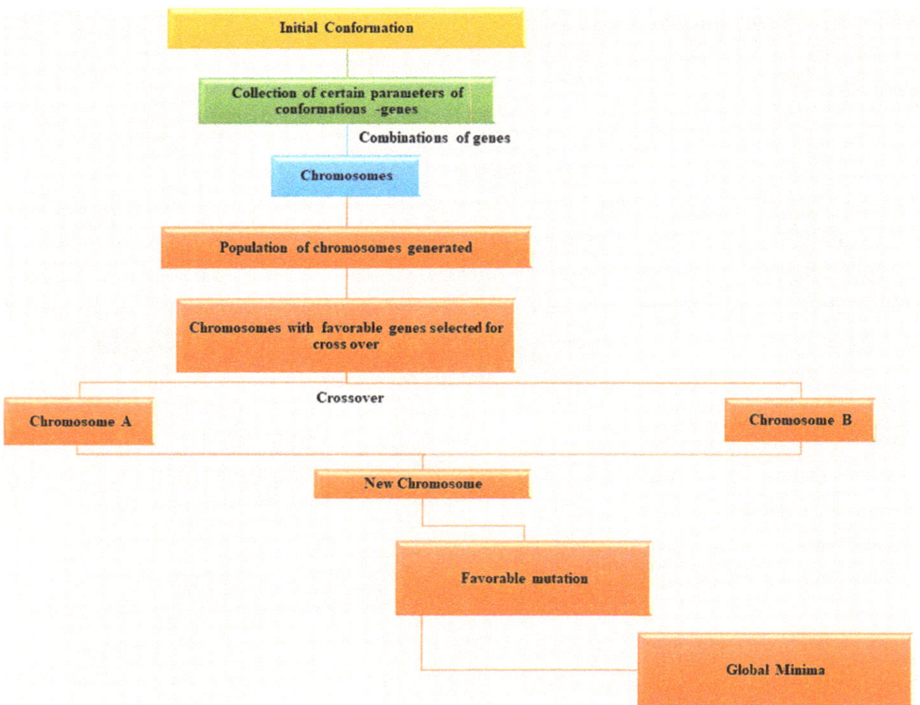

Fig. 9.4: Overview of the process of the genetic algorithm of the conformational search of the molecular docking.

9.2.1.2.3 Tabu search (TS) algorithm

Tabu search (TS) algorithm is a prominent stochastic or random search algorithm utilized in the field of molecular docking for conformational search. In this context, TS serves as a powerful optimization technique to explore the vast conformational space of molecules and efficiently identify energetically favorable binding configurations. TS, inspired by the principles of local search and optimization, employs a memory-based strategy to navigate through the solution space. It maintains a short-term memory, known as the "Tabu list," to prevent revisiting previously explored solutions, and facilitate diversification during the search process. This memory mechanism encourages the algorithm to escape local minima and explore different regions of the conformational landscape. In molecular docking, TS plays a crucial role in predicting the binding conformations of ligands with the target proteins. It aims to discover the most energetically favorable binding pose by iteratively perturbing and refining the legend's position and orientation within the protein's binding site. By efficiently sampling diverse conformational possibilities and avoiding redundant searches, TS contributes to the accurate estimation of binding affinities and enhances the success rate of drug discovery efforts.

TS is an iterative algorithm developed by Glover and designed initially for the optimization of complex systems. This program is based on the principle that a certain amount of change in the conformation of an initial ligand is made, and its RMSD, with all the conformations available in the Tabu list, is compared. If its RMSD comes more than the cutoff value with any of the conformation of the Tabu list, then this conformation is rejected; otherwise, it is accepted. Pro-LEADS and PSI-DOCK are two molecular docking software that implement this search algorithm [31–33].

9.2.1.3 Simulation method-based search algorithms

Simulation method-based search algorithms are integral in the conformational search process within molecular docking studies. These algorithms play a pivotal role in exploring the vast conformational space of molecules, particularly when predicting the binding modes of ligands to target proteins. Rather than exhaustively sampling all possible conformations, simulation-based approaches efficiently navigate through this space to identify energetically favorable conformations. One prominent technique is molecular dynamics simulation, which employs Newtonian mechanics to simulate the motion of atoms and molecules over time. By iteratively adjusting atomic positions, based on forces and energies, molecular dynamics explores various conformations, allowing researchers to capture dynamic interactions between ligands and receptors. Incorporating simulation methods into conformational searches enhances the accuracy and reliability of molecular docking results, providing valuable insights into the binding mechanisms and energetically favorable configurations of drug-like molecules in complex biological systems.

These are the most recent search algorithms employed for the search of the probable bioactive conformation of the ligand. The most widely employed technique for this purpose is the simulated annealing (SA) approach [34].

9.2.1.3.1 Simulated annealing (SA)

SA-based search algorithm is suitable for those ligand molecules that contain multiple local minima. A major advantage of this type of algorithm is that it can search for the minima in both uphill and downhill directions. This approach follows a temperature-dependent molecular dynamics methodology. Initially, a random conformer is placed inside the cavity of the receptor and allowed to move inside the cavity as per Newton's second law of motion. In the next cycle, the conformer is allowed to move to the lower temperature. These cycles are repeated until we move to 0 K temperature. As the movement of the molecules is temperature-dependent and it is considered at 0 K, there are zero velocities in the system. Hence, the conformation obtained at 0 K is considered as the minima.

Although SA is a very accurate algorithm, it is highly time-consuming if the number of conformations that go under simulations is large. To overcome this problem, the SA approach is often coupled with other approaches, such as Monte Carlo, to decrease the number of conformations, and ultimately reduce the total time. AutoDock employs this kind of hybrid Monte Carlo-SA search algorithm [35, 36].

9.2.2 Scoring of the ligand–receptor binding

Scoring of ligand–receptor binding in molecular docking for CADD is a critical process that evaluates and quantifies the strength of binding interactions between ligands (potential drug molecules) and their target receptors (typically proteins). This scoring is essential for ranking and selecting the most promising ligand–receptor complexes from a pool of candidate poses generated during the docking process. SFs aim to estimate the binding affinity or free energy of a given ligand–receptor complex. They consider various intermolecular forces and interactions, such as van der Waals forces, electrostatic interactions, hydrogen bonding, and solvation effects. The ultimate goal is to predict which ligand poses have the highest likelihood of forming stable and biologically relevant complexes.

There are several types of SFs, including empirical, knowledge-based, and physics-based approaches. Empirical SFs are based on statistical analysis of known ligand–receptor complexes and utilize parameters derived from experimental data. Knowledge-based scoring relies on databases of known protein–ligand interactions to make predictions. Physics-based SFs employ molecular mechanics and quantum mechanics principles to calculate binding energies from first principles. While SFs are crucial for molecular docking, it is important to acknowledge that accurately predicting binding affinities remains a complex challenge due to the inherent limitations of current computational models. However, these SFs play a pivotal role in virtual screening and lead optimization, helping researchers identify potential drug candidates and prioritize them for further experimental validation. Refining and improving SFs continue to be a focus of research in the field of CADD to enhance the reliability of molecular docking results.

A set of conformations obtained through the search algorithms of the molecular docking program are put in for the scoring, based on their binding with the target protein receptor. Based on this scoring, the best binding conformation is considered as the bioactive conformation and analyzed further for the type of interactions. Similar to the search algorithms, the scoring of the ligand–receptor binding is done by the molecular docking programs by using different algorithms. Scoring of the ligand–receptor binding is done by calculating the change in Gibb's Free energy (ΔG) using the following equation:

$$\Delta G = \Delta H - T\Delta S \tag{9.2}$$

where ΔH is the change in the enthalpy, T is the temperature in kelvin, and ΔS represents the change in the enthalpy.

This ΔG can also be related to the ligand–receptor binding constant, as given in the following equation:

$$\Delta G = -RT \ \ln K \tag{9.3}$$

where R is the ideal gas constant, T is the temperature in kelvin, and K is the binding constant.

Each scoring algorithm differs from another in the calculation of ΔG value. In other words, it can be said that the SF tells us about the best binding mode of any drug candidate with a receptor's cavity.

Previously, SFs of any molecular docking software were classified into force-field-based, empirical-based, and knowledge-based SFs [37]. However, with the emergence of more sophisticated methodologies, SFs are nowadays classified into the following types:

- Physics-based SFs
- Empirical-based SFs
- Knowledge-based SFs
- Machine-learning-based SFs

All these methods have their algorithms for scoring the protein–ligand interactions in molecular docking.

9.2.2.1 Physics-based scoring functions

The scoring of the protein–ligand binding was started by the force-field-based scoring, which is also the first physics-based SF. Initial credit for the development of various force fields goes to Martin Karplus, who with his coworkers, has done tremendous work in this field in the 1970s. The Van der Waals interactions and electrostatic interactions were the initial noncovalent interactions that were taken into account for the initial scoring of the protein–ligand binding by the force fields, and hence the initial scoring was based on the following equation:

$$\Delta G = \Delta E_{\text{vdw}} + \Delta E_{\text{elec}} \tag{9.4}$$

where ΔG is the change in Gibb's free energy, ΔE_{vdw} is the change in the Van der Waals interaction energy, and ΔE_{elec} represents the change in energy due to the electrostatic interactions.

With the increased understanding of the protein–ligand interaction, it was observed that hydrogen bonding also plays an important role in such interactions; therefore, equation (9.4) is further modified to the following equation:

$$\Delta G = \Delta E_{\text{vdw}} + \Delta E_{\text{elec}} + \Delta E_{\text{H-bonding}} \tag{9.5}$$

where $\Delta E_{\text{H-bonding}}$ is a change in the energy due to hydrogen bonding.

DOCK and AUTODOCK software in their initial phases used this kind of scoring, based on the AMBER force field.

The one major drawback of this kind of scoring is that it does not account for the effect of the solvent on the protein–ligand docking. This is overcome by adding solvation energy terms in the SF to form the following equation:

$$\Delta G = \Delta E_{\text{vdw}} + \Delta E_{\text{elec}} + \Delta E_{\text{H-bonding}} + \Delta G_{\text{desolvation}} \tag{9.6}$$

where $\Delta G_{\text{desolvation}}$ is the change in Gibb's free energy after removing the effect of solvent from the system.

Till now, the scoring of the protein–ligand binding is based on the noncovalent interactions and the effect of the solvent. Incorporating covalent interactions, polarization, and charge transfer parameters in the SF is not possible by merely using molecular mechanics (MM) methods. These new parameters can only be studied by using quantum mechanics (QM); however, QM methodology, despite its high accuracy, is a costly and time-consuming process. To make the balance in scoring the protein–ligand binding in terms of accuracy, time, and cost, the scientists have introduced a hybrid QM/MM approach, given by the following equation:

$$\Delta G = \Delta E_{\text{QM/MM}} + \Delta G_{\text{desolvation}} \tag{9.7}$$

where $\Delta E_{\text{QM/MM}}$ is the energy change, calculated by using the hybrid method of QM and MM [38–41].

9.2.2.2 Empirical-based scoring function

Empirical-based scoring is another approach for calculating the binding between the protein and ligand in molecular docking. In this approach, the total binding affinity of the protein ligand is considered the empirical sum of the individual energy terms. A generalized equation for calculating the binding energy using empirical scoring is given in the following equation:

$$\Delta G = \Sigma w \Delta G_i \tag{9.8}$$

where ΔG is the binding affinity of the protein–ligand, ΔG_i is the individual energy terms parameters, such as Van der Waals, electrostatic, and H-bonding, and w is the coefficient that is the individual weight of that energy term calculated through least-fit square or multiple linear regression analysis.

PLP, ChemScore, GlideScore, X-score, LudiScore, and LigScore are some of the popular SFs that use empirical scoring in calculating the binding affinity of the protein and the ligand.

Let us understand the empirical SFs with arguably one of the most sophisticated SF, the GlideScore:

$$\text{GScore} = 0.05(\text{vdw}) + 0.15(\text{Coul}) + \text{Lipo} + \text{H bond} + \text{Metal} + \text{Rewards} + \text{RotB} + \text{Site} \tag{9.9}$$

where vdW is the Van der Waals energy term, Coul represents the Coulomb's electrostatic energy terms, lipo is the lipophilic energy term that assigns certain value for favorable hydrophobic interactions, H bond represents the hydrogen-bonding terms, metal is the metal binding term that comes into play when there is any metal present in the receptor protein, Rewards represents other values that are not included above, RotB is the penalty imposed for freezing the rotatable bonds, and Site is the value given to the polar interactions inside the active site of the target protein [42–45].

9.2.2.3 Knowledge-based scoring function

The biggest advantage of the knowledge-based SF is its high accuracy and less time consumption, when compared to the previously described methods. In this type of SF, binding affinity is calculated in terms of energy potential, also known as statistical potential. The name of this SF comes from the fact that the knowledge for the calculation of binding affinity is based on the already reported protein–ligand complexes in the protein data bank. In other words, this database of protein–ligand complexes obtained from the protein data bank is used for the calculation of the pairwise potentials between the atom pairs using the inverse Boltzmann equation. Suppose there is a pair of atoms x–y, then their distance-dependent potential, based on the inverse Boltzmann equation, will be given by the following equation:

$$w_{xy}(r) = -k_b T \ \ln\left[g_{xy}(r)\right] \tag{9.10}$$

The value of $g_{xy}(r)$ can be calculated from the following equation.

$$g_{xy}(r) = \frac{\rho(r)}{\rho} \times (r) \tag{9.11}$$

where $w_{xy}(r)$ is the distance potential at distance r, k_b is the Boltzmann constant, T is the absolute temperature, $\rho(r)$ is the numeric density of the atom pair at distance r, and $\rho^*(r)$ is the numeric density for the same at reference state where inter atomic interactions are zero. The overall binding affinity (W) can be calculated as the sum of all of these calculated pairwise potentials using the following equation:

$$W = \Sigma w_{xy}(r) \tag{9.12}$$

IT-Score, DrugScore, BLEEP, PMF, and KScore are a few of the knowledge-based SFs [46, 47].

9.2.2.4 Machine learning scoring functions

This is the most recent type of SF used extensively, in which QSAR knowledge is used for assessing the protein–ligand binding. These functions utilize data-driven algorithms, often based on neural networks or random forests, to predict binding affinities between ligands and receptors. By leveraging extensive training datasets of experimentally measured binding data, they capture complex, nonlinear relationships between molecular features and binding energies. Machine learning SFs enhance the accuracy of ligand–receptor binding predictions, enabling the prioritization of potential drug candidates. Their ability to account for intricate interactions and diverse molecular structures makes them valuable in rational drug design, expediting the discovery of novel therapeutics. As discussed in previous chapters, QSAR is a tool in which biological activity is statistically correlated with certain properties of ligands. Likewise, in machine learning SFs, protein–ligand interactions are first measured in some numerical terms and are then statistically correlated with the binding affinity. These entire machine learning SFs differ from each other in the use of different methods for the development of these QSAR models for assessing the protein–ligand binding. The most common machine-learning approaches for this purpose include:
- Support vector machine
- Random forest
- Artificial neural network
- Deep learning

A workflow for the machine learning-based SFs is given in Fig. 9.5. RI-Score, RF-Score, DLSCORE, NNScore, SFScore, and so on are the popular machine learning-based SFs for molecular docking [48–51].

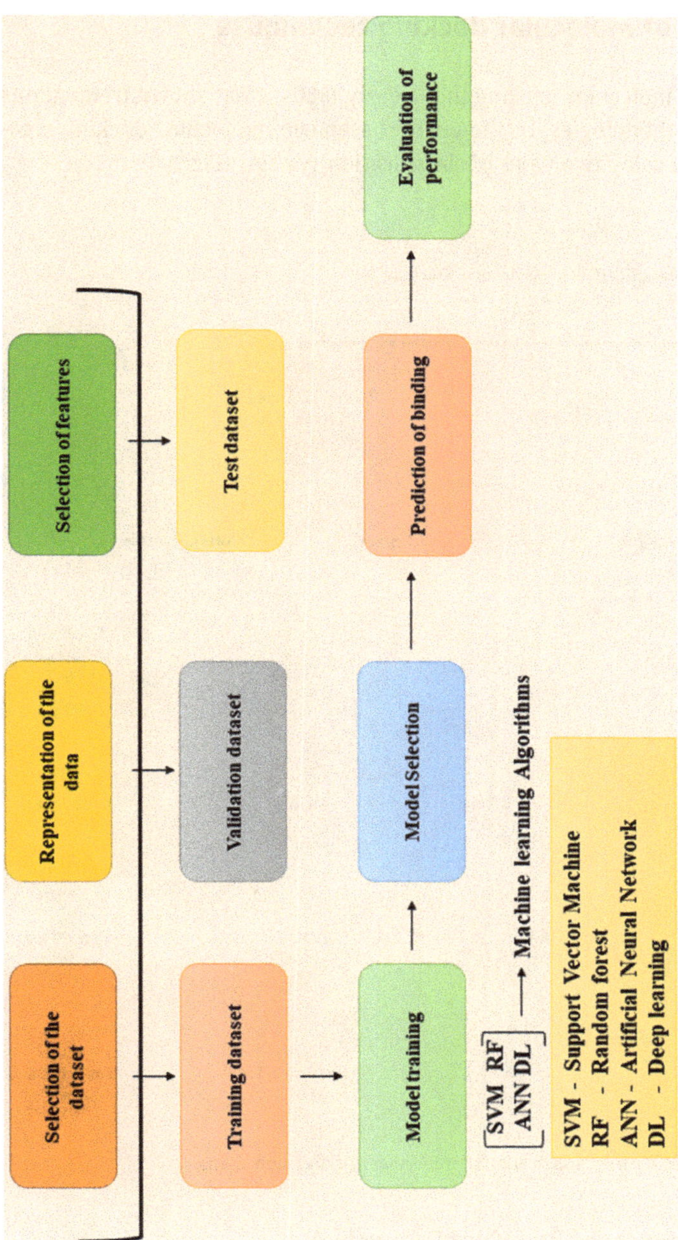

Fig. 9.5: A complete overview of the machine learning (ML) algorithms for calculating the scoring functions of the ligand–receptor docking.

9.3 Classification of molecular docking techniques

With the emergence of molecular docking in the early 1980s, it has shown tremendous development in its methodologies. In the current scenario, molecular docking algorithms are classified by using two types of classification systems (Fig. 9.6):

Fig. 9.6: Different approaches for the classification of molecular docking algorithms.

- Classification is based on the flexibility of the molecules
- Classification based on types of molecules used for molecular docking

9.3.1 Classification based on the flexibility of the molecules

This is the initial system of classification for all the molecular docking programs. As the name suggests, in this type of classification, all the molecular dockings are classified based on the amount of flexibility of the molecules that are allowed in the particular molecular docking. Molecular docking, based on the flexibility of molecules, can be classified into three main categories, rigid docking, semiflexible docking, and flexible docking. The choice of docking approach depends on the specific research objectives and the degree of accuracy required, with flexible docking being most suitable for capturing complex interactions in drug discovery and molecular modeling [37].

9.3.1.1 Rigid docking

Rigid docking is a classification of molecular docking that assumes both the ligand (typically a small molecule) and the receptor (usually a protein) to be completely inflexible during the docking process. This simplification is based on the assumption that the binding site and the ligand maintain fixed conformations, disregarding any conformational changes that may occur upon binding. In rigid docking, the ligand is treated as a static structure, and the receptor is held in a fixed conformation, obtained from experimental data or molecular modeling. While this approach is computationally less demanding than flexible docking methods, it has limitations. Rigid docking may not accurately capture the full range of interactions that occur in real biological systems, as it cannot account for induced-fit effects where binding induces conformational changes in the receptor or ligand. Rigid docking is often used in cases where structural data is limited, and researchers want a quick estimate of the binding affinity or pose. However, it may not be suitable for situations where flexibility plays a critical role in the binding process, such as in many protein–ligand interactions. In such cases, flexible docking methods that allow for conformational changes are preferred to provide a more realistic representation of the molecular interactions. FRODOCK, RDOCK, ZDOCK, MEGADOCK, SOFTDOCK, and MS-DOCK are popular software that perform rigid molecular docking [52].

9.3.1.1.1 Advantages of rigid docking

- **Speed:** Rigid docking is computationally efficient because it simplifies the conformational search space, making it faster for screening large compound libraries.
- **Simplicity:** Rigid docking is straightforward to implement, and requires fewer computational resources compared to flexible docking methods.
- **Predictive power:** In cases where the ligand or receptor flexibility is minimal, rigid docking can provide accurate predictions of binding modes.

- **Initial screening:** Rigid docking is useful for initial virtual screening to identify potential hits, helping prioritize compounds for more detailed studies.
- **Insights into rigid interactions:** It can highlight the importance of rigid interactions, which are often the dominant forces in binding.

9.3.1.1.2 Disadvantages of rigid docking

- **Inaccurate for flexible molecules:** Rigid docking may fail to account for important conformational changes in either the ligand or the receptor, leading to inaccurate binding predictions.
- **Missing key interactions:** It cannot capture critical interactions that require flexibility, such as induced fit or conformational adaptation, upon binding.
- **Limited applicability:** Rigid docking is less suitable for targets with significant flexibility, such as proteins with flexible binding sites or multi-domain proteins.
- **False positives/negatives:** It may generate false positives (binding predictions that do not occur in reality) or false negatives (missed true binding interactions).
- **Reduced drug discovery success:** Rigid docking may overlook potential drug candidates that could have been identified using more flexible docking methods, potentially delaying drug discovery efforts.

9.3.1.2 Semiflexible docking

Semiflexible docking is a crucial classification within molecular docking that strikes a balance between computational feasibility and the representation of molecular flexibility. In this approach, one or both molecules involved in the binding event are allowed a degree of flexibility, while keeping the other(s) relatively rigid. This captures essential conformational changes that may occur upon binding, while managing computational complexity. Typically, the receptor (often a protein) is treated as rigid, representing its core structure, while the ligand (the drug candidate) is allowed to adopt multiple conformations. These flexible ligand conformations are systematically explored within the binding site to identify energetically favorable binding modes. Various algorithms and techniques, such as Monte Carlo or molecular dynamics simulations, are employed to perform semiflexible docking. Semiflexible docking is particularly valuable in drug discovery as it considers the adaptability of ligands to fit the binding site, accommodating induced-fit and induced-conformational changes. This approach enhances the accuracy of binding affinity predictions and can lead to the discovery of more effective drug candidates. Overall, semiflexible docking is a versatile and widely used classification in molecular docking, offering a compromise between computational efficiency and the realistic representation of molecular flexibility, making it an essential tool in structure-based drug design and molecular modeling [3].

9.3.1.2.1 Advantages of semiflexible docking

- **Increased realism:** Semiflexible docking allows for limited flexibility in ligands or receptors, making it more biologically relevant by accommodating minor conformational changes during binding.
- **Improved accuracy:** This approach often provides more accurate binding predictions compared to rigid docking because it considers some level of molecular flexibility, which is critical in capturing important interactions.
- **Identification of flexible binding sites:** Semiflexible docking can reveal potential binding sites that might be missed in rigid docking, providing insights into allosteric or induced-fit binding mechanisms.
- **Reduced computational demands:** Semiflexible docking is computationally less demanding than fully flexible docking, making it a practical choice for screening large compound libraries.
- **Balance of speed and accuracy:** It strikes a balance between the speed of rigid docking and the accuracy of fully flexible docking, making it suitable for various stages of drug discovery.

9.3.1.2.2 Disadvantages of semiflexible docking

- **Limited flexibility:** Despite offering some flexibility, semiflexible docking may not capture extensive conformational changes that are crucial for certain binding events, potentially leading to inaccuracies.
- **Parameterization challenges:** It requires careful parameterization to define the extent of flexibility, which can be a challenging and subjective task, affecting the reliability of results.
- **Increased computational costs:** Compared to rigid docking, semiflexible methods are computationally intensive, consuming additional time and resources.
- **Risk of overfitting:** The inclusion of flexibility may lead to overfitting, if not properly controlled, resulting in unrealistic binding poses.
- **Complexity of analysis:** Interpreting the results can be more complex than rigid docking, as researchers must consider the implications of the partially flexible molecular structures, in terms of binding energy and affinity.

9.3.1.3 Flexible docking

Flexible docking is a pivotal classification within molecular docking, emphasizing the consideration of molecular flexibility during the simulation of ligand–receptor interactions. In flexible docking, both the ligand (drug molecule) and the receptor (typically a protein) are allowed to undergo conformational changes to explore various binding poses. This approach recognizes that molecules in a biological environment are rarely rigid; they adapt to achieve optimal interactions. Flexible docking employs advanced algorithms and force fields to model these conformational changes accurately. This accounts for side-chain and backbone flexibility in proteins and torsional rotations in ligands. Consequently, it provides a more realistic representation of the binding process and yields insights into dynamic interactions. Flexible docking is particularly valuable in drug discovery as it captures the nuances of binding, allowing the identification of compounds with higher binding affinities. However, it is computationally intensive and demands substantial computational resources. Researchers often choose between rigid, semiflexible, and flexible docking approaches, based on their specific research goals, available resources, and the extent of molecular flexibility required for an accurate representation of the biological system. Software such as AutoDock Vina, Gold, Glide, and FleXX employ flexible types of molecular docking [53–55].

9.3.1.3.1 Advantages of flexible docking

- **Improved accuracy:** Flexible docking allows for the modeling of conformational changes in both ligands and receptors, resulting in more accurate predictions of binding interactions.
- **Enhanced predictive power:** It can identify binding modes that rigid or semiflexible docking methods might miss, making it a valuable tool for exploring diverse binding conformations.
- **Realistic biological insights:** Flexible docking better reflects the dynamic nature of biomolecular interactions, providing insights into the energetically favorable conformations and binding mechanisms.
- **Exploration of binding site flexibility:** It enables the study of how binding sites adapt to different ligand shapes and sizes, aiding in the design of ligands that accommodate flexible target sites.
- **Lead optimization**: Flexible docking can be crucial in lead optimization efforts, allowing for the refinement of ligand structures to improve binding affinity and selectivity.

9.3.1.3.2 Disadvantages of flexible docking

- **Computational intensity:** The increased flexibility significantly raises the computational cost and time required for docking simulations, limiting its applicability for high-throughput screening.
- **Complex parameterization:** Setting up parameters for flexible docking, including defining rotatable bonds and energy functions, can be challenging and may require extensive tuning.
- **Increased risk of overfitting:** Flexible docking runs the risk of overfitting the ligand to the receptor, potentially yielding unrealistic binding poses.
- **Higher resource requirements:** It demands access to substantial computational resources, such as high-performance computing clusters, making it less accessible for smaller research groups.

- **Potential for convergence issues:** Flexible docking simulations may suffer from convergence issues, leading to inconsistent results or difficulties in identifying the most energetically favorable binding mode.

9.3.2 Classification based on types of molecules used for molecular docking

In this type of classification, molecular docking methods are classified based on the types of molecules docked inside the target receptor. It can be further classified into the following types.

9.3.2.1 Protein–ligand docking

Protein–ligand docking is the most common type of molecular docking, which accounts for approximately 90% of the docking studies being done worldwide. In this approach, small molecules, often representing potential drug candidates (small molecules that can be either of synthetic origin or from natural origin), are docked into the binding sites of target proteins, such as enzymes or receptors. The primary goal is to predict the binding geometry, affinity, and interactions between the ligand and the protein. This information is crucial for identifying promising drug candidates and understanding their mechanisms of action. Protein–ligand docking employs computational algorithms to search through a vast conformational space, evaluating various orientations and conformations of the ligand within the protein's binding site. By ranking and scoring these poses, researchers can prioritize compounds for experimental testing, significantly expediting the drug development process, and reducing costs. Most of the molecular docking software we use in the current scenario are based on protein–ligand docking [4, 5].

9.3.2.2 Protein–protein docking

Protein–protein docking focuses on understanding the interactions between two protein molecules and predicting the formation of protein complexes. This computational approach is essential for uncovering the mechanisms behind various biological processes, including signal transduction, immune responses, and cell cycle regulation. By simulating the docking of two proteins, researchers can predict the binding interfaces, orientations, and affinities of the interacting molecules. This information aids in deciphering protein–protein interaction networks and provides insights into the functions and dysregulations associated with specific diseases. Protein–protein docking algorithms take into account the structural flexibility of the proteins, allowing for conformational changes that occur upon complex formation. This dynamic view of protein–protein interactions is vital for designing therapeutic strategies and understanding intricate cellular pathways. ClusPro, PatchDock, HAD-

DOCK, RosettaDock, and SwarmDock are the most commonly employed web-based servers for performing protein–protein docking [54].

9.3.2.3 Protein–nucleic acid docking

Protein–nucleic acid docking involves the study of interactions between proteins (e.g., transcription factors or enzymes) and nucleic acids (such as DNA or RNA). This computational technique sheds light on the molecular mechanisms governing processes like gene regulation, replication, and DNA repair. By simulating the docking of proteins with nucleic acids, researchers can predict the binding modes, affinity, and structural changes associated with these interactions. Understanding these interactions is crucial for unraveling the genetic and epigenetic processes that underlie various diseases and for designing targeted therapies. Protein–nucleic acid docking methods consider the flexibility of both the protein and the nucleic acid, accounting for conformational changes during binding events. This dynamic perspective enhances the accuracy of predictions and allows for a more comprehensive understanding of molecular recognition in nucleic acid-related processes [56, 57].

9.3.2.4 Ligand–ligand docking

Ligand–ligand docking explores the interactions between two or more small molecules or ligands. This approach is valuable in studying drug combinations, identifying synergistic effects, and understanding chemical reactions in complex mixtures. It provides insights into how different molecules interact with one another, which is essential for optimizing drug formulations, designing combination therapies, or investigating chemical reactions in biochemical systems. Ligand–ligand docking algorithms consider the spatial arrangement and intermolecular forces between the ligands, predicting their binding modes and energetics. This information aids researchers in selecting the most effective combinations of molecules for specific applications, whether in drug development, material science, or chemical engineering [58, 59].

9.4 Steps involved in the molecular docking (protein–ligand docking)

The basic steps involved in performing molecular docking through any docking software are similar and the basic workflow is given in Fig. 9.7. The results of any molecular docking are highly dependent on how accurately each step of the molecular docking is performed. The basic steps involved in any type of molecular docking algorithm include the following [5, 7].

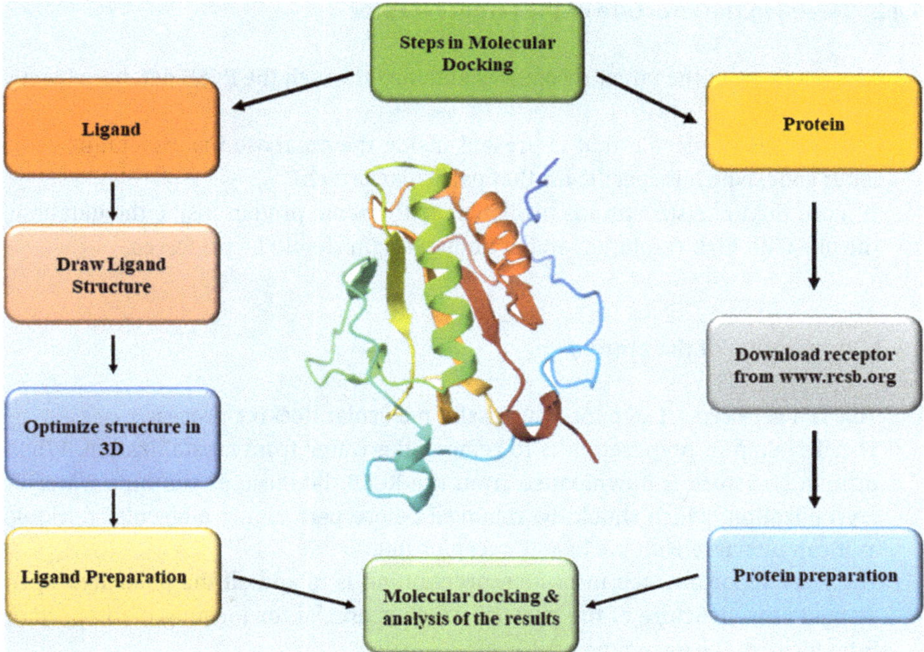

Fig. 9.7: An overview of the steps involved in performing the molecular docking simulations.

9.4.1 Obtaining and optimizing the ligand structure

- This is the very first step for performing molecular docking studies. Molecules that are to be docked inside a certain target protein are either drawn through any molecular modeling software or directly obtained through certain databases.
- The most popular molecular modeling software packages for drawing the chemical structure of the proposed ligands are MarvinSketch, ChemSkecth, and ChemDraw.
- Ligands can also be obtained directly, if available in the drug databases such as ZINC, ChEMBL, DrugBank, and PubChem.
- Once the structure of the desired ligand is obtained, the next step is to optimize the structure of that ligand. This step is the step of energy minimization of the ligand and is performed to obtain the global minima.
- After the energy minimization of the ligand structure, the next step is the addition of charges to the atoms inside the ligand, based on the polarity. These charges are added to accurately predict the type of interactions of the ligand inside the cavity of the protein.

9.4.2 Obtaining the structure of the target receptor

- The structure of the target receptor is obtained through the RCSB database (www.rcsb.org).
- Each protein molecule that is present inside the database has its unique four-letter code, which is specific for that particular protein.
- If more than one structure is available for the same protein inside the database, the one with high resolution and of human origin should be preferred.

9.4.3 Preparation of the protein

- This is a very crucial step for a successful molecular docking process.
- The first step in preparation is to remove the water from crystallization. When protein structure is downloaded from the RCSB database, it contains water of crystallization, which should be removed before performing molecular docking, as it can interfere with the ligand–receptor interaction.
- The next important step in protein preparation is to add all the polar hydrogen atoms to the structure of the protein, which is crucial for forming the important H-bonds with the ligand molecule.
- The addition of charges to the receptor molecule is the next step so that proper interaction with the ligand molecule can be assessed.
- A search space inside the protein receptor is defined around the cavity (binding site) of the protein. However, if the cavity of the receptor is not known, then the blind docking is performed without defining the specific search space.

9.4.4 Performing the molecular docking

- After ligand and receptor preparation, the next step is to perform molecular docking, which is the core of this whole process.
- As discussed already, any molecular docking program performs two basic functions. The first is to generate certain conformations of the ligand from its minimized conformations and the second is to rank these conformations based on their binding affinity with the protein receptor.
- The conformation that has the best binding is considered close to the bioactive conformer and is used for further evaluation.
- The steps to be performed for molecular docking vary with the molecular docking software.

9.4.5 Analysis of the molecular docking results

- The best binding pose of the ligand is then analyzed to evaluate its binding with the target protein.
- Evaluation is done to assess the docking score, number of hydrogen bonding interactions, hydrophobic interactions, pie–pie interactions, and bond length.
- These results are visualized with many open-access software such as PyMol, Discovery Studio, AutoDock Tools, and Schrodinger Maestro.

9.5 Some examples of molecular docking

To get a better understanding of molecular docking, consider the following examples of molecular docking studies on some of the established drugs against their target proteins.

9.5.1 Agents acting on dihydrofolate reductase (DHFR) enzyme

DHFR is an important key enzyme that catalyzes the conversion of dihydrofolic acid (DHFA) to tetrahydrofolic acid (THFA), both in mammalian and microbial cells. THFA plays a key role in DNA synthesis, in both microbial and mammalian cells, and hence, inhibiting DHFR is of pharmacological importance. This enzyme is a prime target of antibacterial and antimalarial agents. In humans, the same enzyme is targeted for controlling bacterial infection and for controlling the growth of cancer cells. The most common agents against DHFR are given in Tab. 9.1.

The interaction of these agents with DHFR can be evaluated through molecular docking simulations. The steps for performing the molecular docking are similar, as discussed in the common steps earlier:

- The very first step for performing molecular docking is to obtain the structure of the ligands. In this case, the structures of three ligands were drawn with any of the molecular modeling software.
- The next step is the energy minimization of these ligands, done using molecular modeling software. After the energy minimization, the charges are added to the polar atoms.
- These three receptors of DHFR, 2HM9, 2BL9, and 1RG7, complexed with trimethoprim, pyrimethamine, and methotrexate, respectively, are downloaded from the protein data bank (www.rcsb.org) in the form of PDBs.
- The next step is to remove the water of crystallization from all of these PDBs to rule out any false interaction with the ligands.
- In the next step, polar hydrogens and charges are added to the PDBs of these three receptors.

Tab. 9.1: List of molecules tested against DHFR.

S. no.	Name and structure of the ligand (drug)	Class	Name and code of the protein (receptor)
1	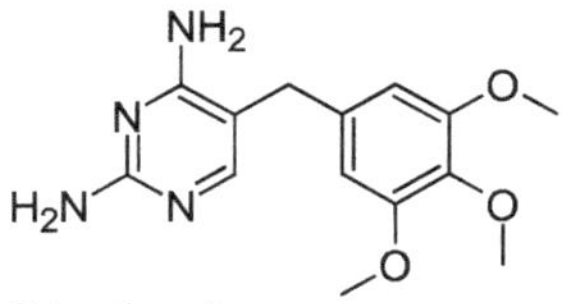 Trimethoprim	Antibacterial	X-ray structure of *Escherichia coli* dihydrofolate reductase (6XG5)
2	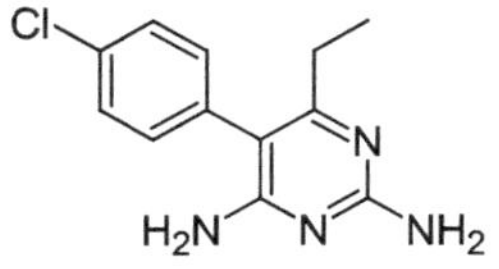 Pyrimethamine	Antimalarial	X-ray crystal structure of *Plasmodium vivax* dihydrofolate reductase (2BL9)
3	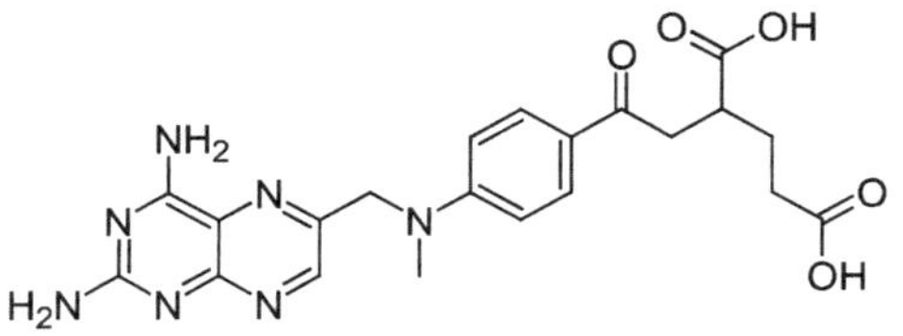 Methotrexate	Anticancer	X-ray crystal structure of dihydrofolate reductase (1RG7)

- This is followed by the defining of the search space where these three ligands will bind inside the cavity of the respective receptors.
- Finally, molecular docking is performed on DHFR receptors with these ligands with the help of any of the molecular docking software.
- Finally, visualization of docking results is performed with an available docking result visualizer.

The results of the molecular docking of all these three ligands are shown in Tab. 9.2.

From these results, it can be observed that all of the three ligands are binding with different binding energies. The highest binding to the DHFR is demonstrated by methotrexate (−8.2 kcal/mol), followed by trimethoprim (−7.6 kcal/mol) and pyrimethamine (−6.8 kcal/mol). The interaction diagram of all three is shown in Fig. 9.8. Considering hydrogen bonding interactions and hydrophobic interactions, it is also clear that all three ligands bind differently with the DHFR enzyme.

Tab. 9.2: Results of the molecular docking study.

S. no.	Ligand	Receptor	Dock score (kcal/mol)	No. of H-bonds	Amino acids involved in H-bonding	Amino acids involved in hydrophobic interactions
1	Trimethoprim	6XG5	−7.6	04	GLY-97 TYR-100 ALA-07 THR-46	ILE-50 ILE-14 ALA-06
2	Pyrimethamine	2BL9	−6.8	02	SER-117 SER-120	ILE-13 ALA-15 PHE-57 CYS-14
3	Methotrexate	1RG7	−8.2	NIL	NIL	LEU-28 LYS-32 ILE-50 PHE-31

9.5.2 Agents acting on the HMG-CoA reductase

HMG-CoA (3-hydroxy-3-methylglutaryl coenzyme A reductase) is an enzyme responsible for catalyzing the synthesis of mevalonate in the body. Mevalonate is further required for the synthesis of sterols, including cholesterol, in the body. By inhibiting HMG-CoA, the level of cholesterol is lowered as a pharmacological response.

Statins are the class of drugs that are used to inhibit this HMG-CoA reductase for lowering the blood cholesterol level and are clinically prescribed for disorders like diabetes mellitus, cardiovascular complications, and obesity. Common statins employed for this example are given in Tab. 9.3.

To understand the molecular interactions of the above statins against the HMG-CoA, molecular docking was performed. The structure of the HMG-CoA was obtained from the protein data bank (PDB ID: 1HWK) and the structures of the above-mentioned statins were drawn through the molecular modeling software. All other steps for performing molecular docking are similar as mentioned above. The results of our molecular docking are shown in Tab. 9.4.

From the results depicted in Tab. 9.4, it is clearly understood that all of the three mentioned statins show similar types of binding energy with the HMG-CoA. They differ in their interaction with the HMG-CoA in the number and type of H-bonding and hydrophobic interactions. Amino acid residue, ASN-755, shows hydrogen-bonding interactions with all of the three statins. LYS-692 and SER-684 are common interacting amino acids between lovastatin and rosuvastatin. Amino acid, ARG-590, is common between rosuvaststin and atorvastatin. The molecular interaction images of all three statins are given in Fig. 9.9.

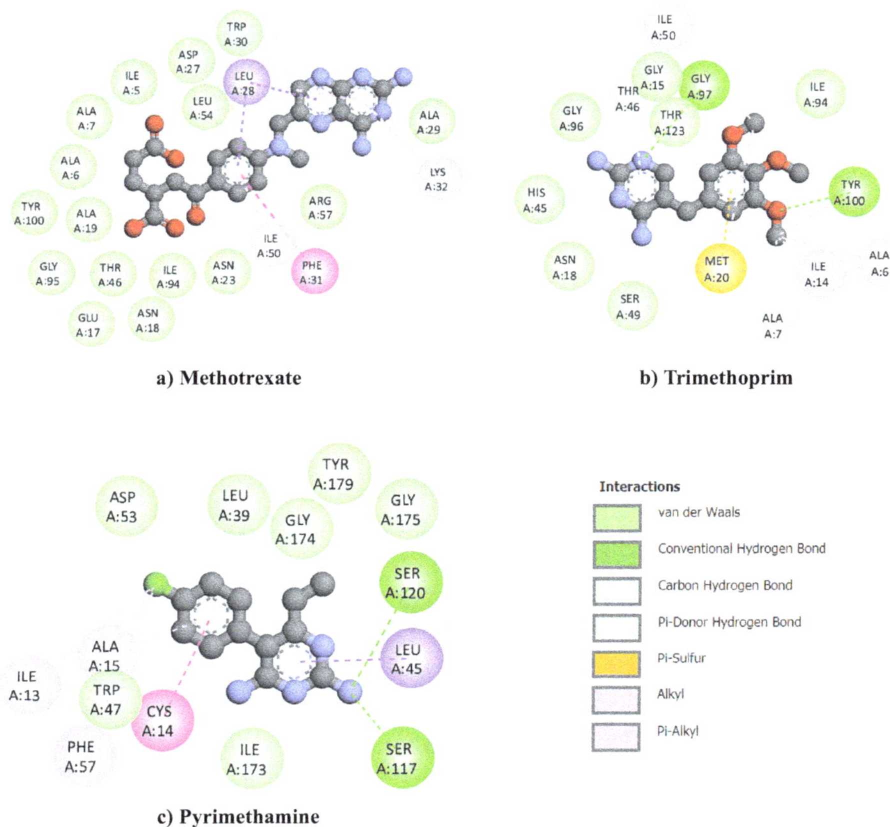

Fig. 9.8: The molecular docking interaction diagram of the binding of trimethoprim, pyrimethamine, and methotrexate with the dihydrofolate reductase (DHFR) enzyme.

Tab. 9.3: List of molecules tested against HMG–CoA.

S. no.	Name and structure of ligand (drug)
1	Lovastatin

Tab. 9.3 (continued)

S. no.	Name and structure of ligand (drug)
2	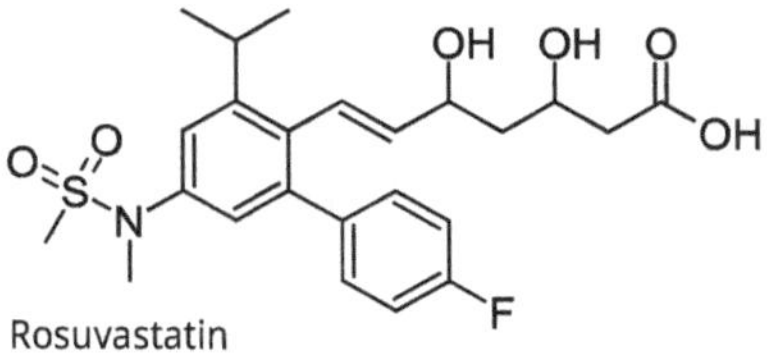 Rosuvastatin
3	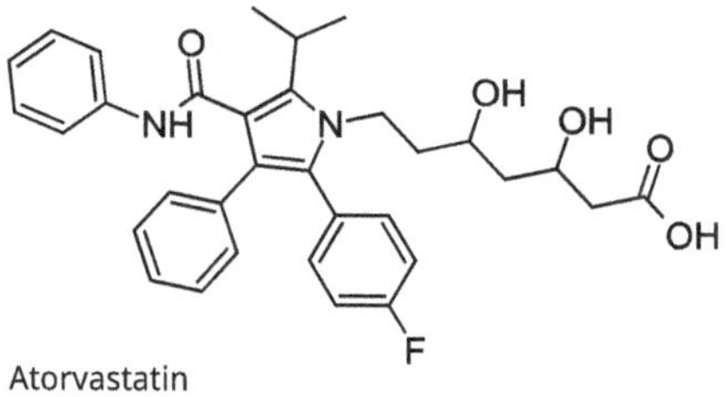 Atorvastatin

Tab. 9.4: Results of the molecular docking study.

S. no.	Ligand	Dock score (kcal/mol)	No. of H-bonds	Amino acids involved in H-bonding	Amino acids involved in hydrophobic interactions
1	Lovastatin	−7.4	04	HIS-752 ASN-755 SER-684 LYS-692	NIL
2	Rosuvastatin	−7.4	06	ARG-590 SER-684 LYS-692 LYS-691 ASN-755 LYS-735	ALA-856 LEU-853
3	Atorvastatin	−7.5	07	GLU-559 ASP-690 (02) ASN-755 ARG-590 (02) SER-661	LEU-853 VAL-683 ALA-856 ALA-564

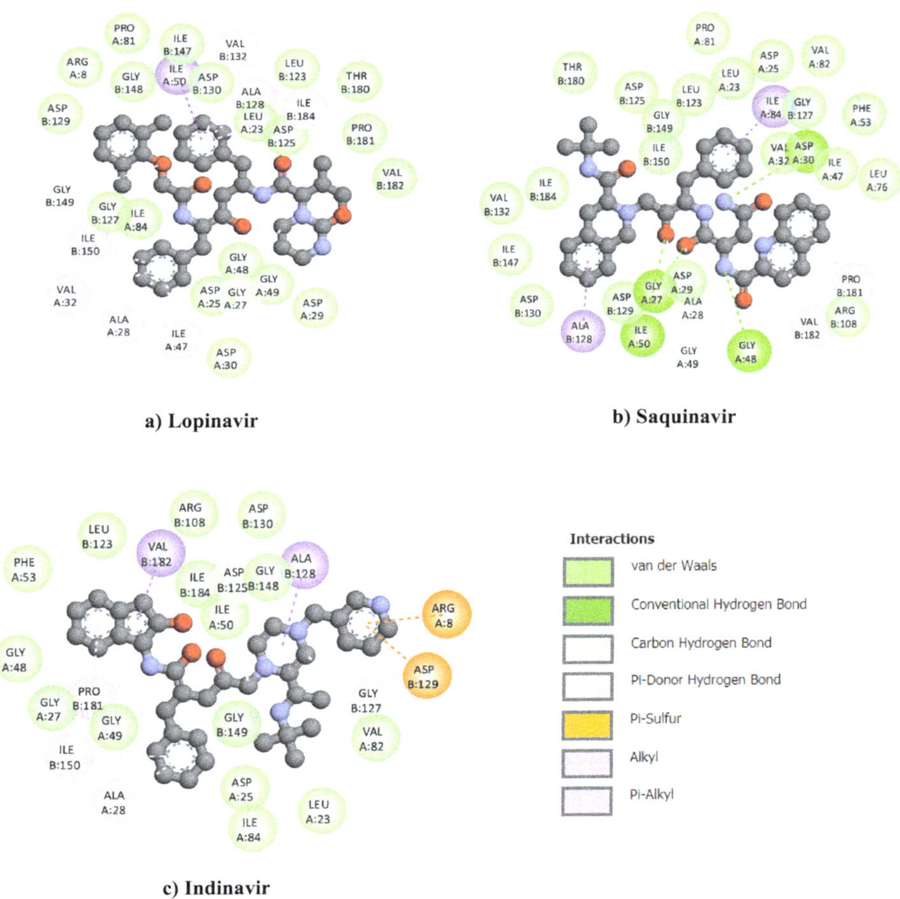

Fig. 9.9: The molecular docking interaction diagram of the binding of lovastatin, rosuvastatin, and atorvastatin with the HMG-CoA (3-hydroxy-3-methylglutaryl coenzyme A) reductase enzyme.

9.5.3 Agents acting on HIV-protease

HIV protease is a key enzyme required by the human immunodeficiency virus (HIV) for its replication. Blocking these protease enzymes results in the inhibition of viral replication inside the human body and hence, reducing the virus load to the desired value. Therefore, use of HIV protease inhibitors is an important strategy in anti-HIV therapy. Some of the important drugs marketed under this category are given in Tab. 9.5. Interestingly, the discovery of these protease inhibitors is credited to molecular docking studies.

Tab. 9.5: List of molecules tested against HIV.

S. no.	Name of drug
1	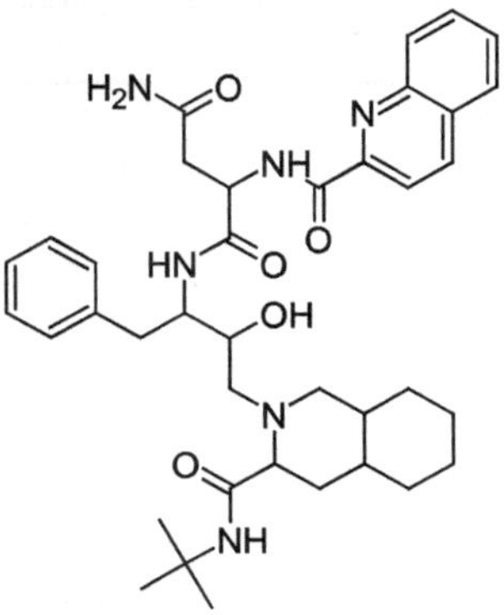 Saquinavir
2	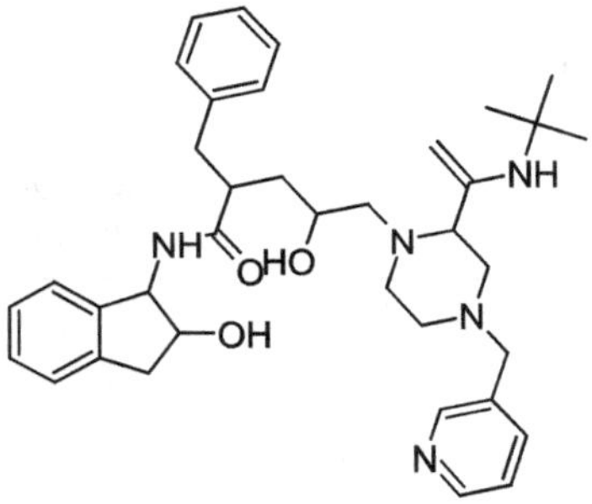 Indinavir
3	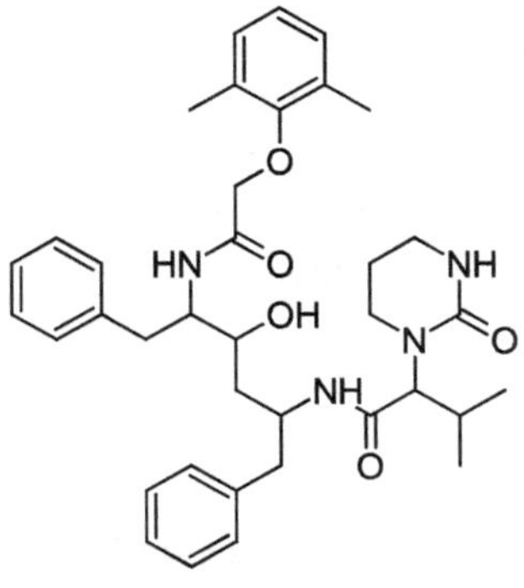 Lopinavir

Tab. 9.6: Results of the molecular docking study.

S. no.	Ligand	Dock score (kcal/mol)	No. of H-bonds	Amino acids involved in H-bonding	Amino acids involved in hydrophobic interactions
1	Saquinavir	−11.5	04	ASP-30 GLY-48 GLY-27 ILE-50	ALA-128 ILE-84 PRO-181 VAL-182
2	Indinavir	−11.0	02	ASP-125 GLY-127	PRO-181 VAL-182 ALA-128 ALA-28 ILE-150
3	Lopinavir	−11.3	02	GLY-48 GLY-149	ILE-150 VAL-32 ALA-28 ILE-47 ILE-184 ILE-50 ALA-128 VAL-132

The molecular docking of these protease inhibitors was performed using the HIV protease enzyme (PDB ID: 3OXC). By using the steps mentioned above for the molecular docking, the results are depicted in Tab. 9.6.

From the above results, it is evident that all three protease inhibitors are highly potent as they show excellent binding affinity with the receptors. Saquinavir demonstrated the highest binding in terms of binding energy (−11.5 kcal/mol) as well as the number of hydrogen bond interactions (four). Among the amino acid residues, GLY-48 is observed to be the common amino acid showing hydrogen bonding interaction with Saquinavir and Lopinavir. The molecular interaction images of these drugs docked with the protease enzyme are shown in Fig. 9.10.

9.5.4 Agents acting on acetylcholinesterase (AChE)

Acetylcholine (ACh) is a neurotransmitter that is broken down by the AChE enzyme into acetate and choline part. During Alzheimer's disorder, there is a degeneration of cholinergic neurons resulting in impaired cholinergic signaling. Therefore, inhibiting AChE results in the increased concentration of ACh in synapses and consequently improved cholinergic signaling during Alzheimer's disorder. Some of the clinically used drugs for inhibiting this AChE enzyme are given in Tab. 9.7.

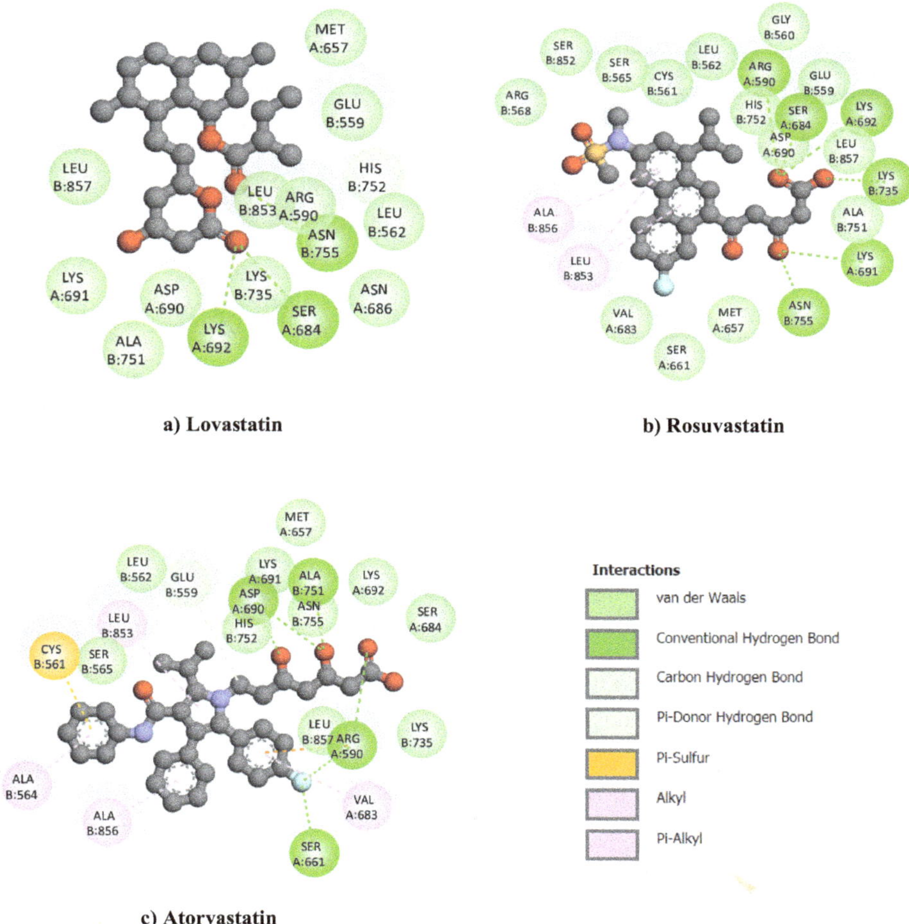

Fig. 9.10: The molecular docking interaction diagram of the binding of saquinavir, indinavir, and lopinavir with the HIV-protease enzyme.

Tab. 9.7: List of molecules tested against AChE.

S. no.	Name of drug
1	Rivastigmine

Tab. 9.7 (continued)

S. no.	Name of drug
2	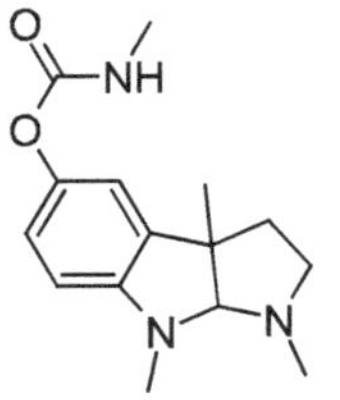 Physostigmine
3	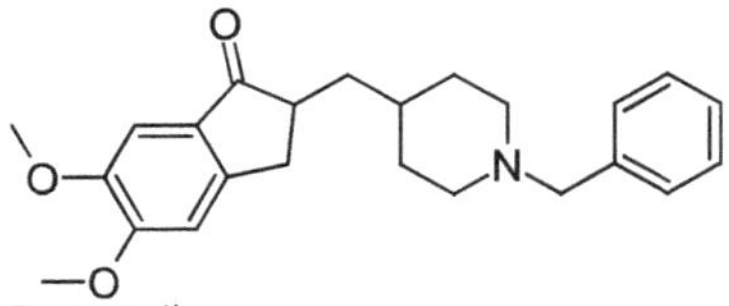 Donepezil

The molecular docking studies of the above AChE inhibitors were performed on the AChE enzyme (PDB ID: 1GQR). The results of the molecular docking simulations are summarized in Tab. 9.8.

Tab. 9.8: Results of the molecular docking study.

S. no.	Ligand	Dock score (kcal/mol)	No. of H-bonds	Amino acids involved in H-bonding	Amino acids involved in hydrophobic interactions
1.	Rivastigmine	−7.8	02	TYR-124 SER-203	TRP-286 TYR-337 PHE-338 TYR-341
2.	Physostigmine	−8.8	02	HIS-447 ARG-296	TYR-337 PHE-338 TYR-341
3.	Donepezil	−11.8	03	SER-125 (02) GLY-120	PHE-338 TYR-337 TYR-341 LEU130 TRP-86 TRP-286

From the results depicted in Tab. 9.8, it is evident that donepezil shows the best binding affinity (−11.8 kcal/mol) with AChE, followed by physostigmine (−8.8 kcal/mol), and rivastigmine (−7.8 kcal/mol) (Fig. 9.11).

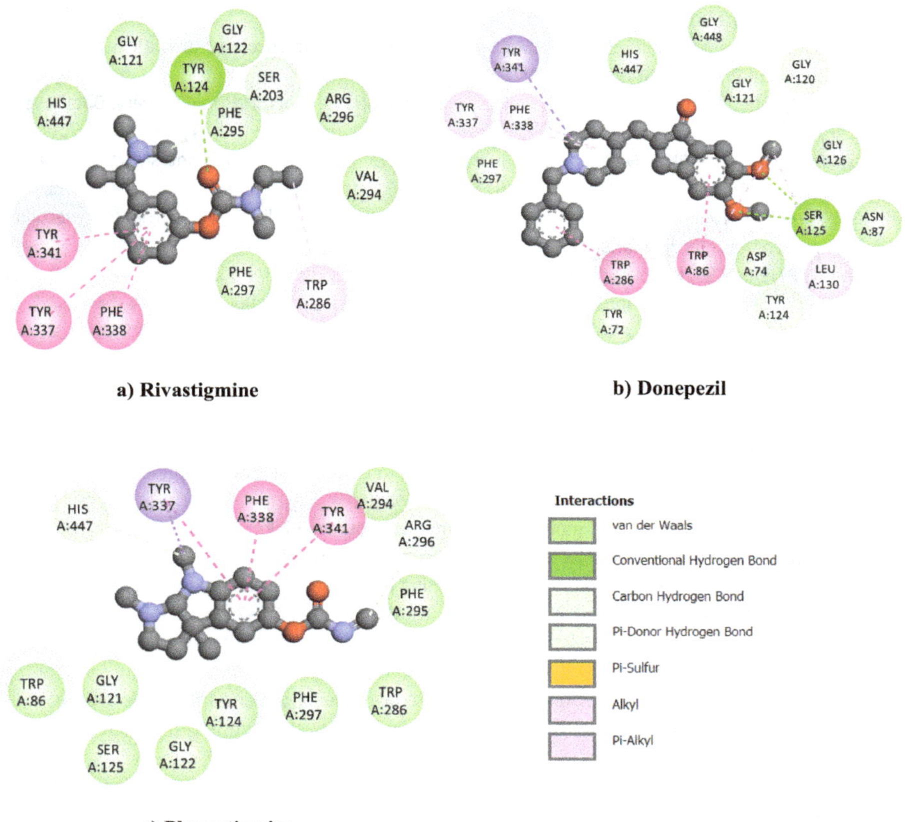

Fig. 9.11: The molecular docking interaction diagram of the binding of the rivastigmine, physostigmine, and donepezil with the acetylcholinesterase (AChE) enzyme.

References

[1] Ferreira LG, Dos Santos RN, Oliva G, Andricopulo AD. Molecular docking and structure-based drug design strategies. Molecules, 2015, 20(7), 13384–421.

[2] Torres PH, Sodero AC, Jofily P, Silva-Jr FP. Key topics in molecular docking for drug design. International Journal of Molecular Sciences, 2019, 20(18), 4574.

[3] Jakhar R, Dangi M, Khichi A, Chhillar AK. Relevance of molecular docking studies in drug designing. Current Bioinformatics, 2020, 15(4), 270–8.

[4] De Ruyck J, Brysbaert G, Blossey R, Lensink MF. Molecular docking as a popular tool in drug design, an in silico travel. Advances and Applications in Bioinformatics and Chemistry, 2016, 1–11.
[5] Tripathi A, Misra K. Molecular docking: A structure-based drug designing approach. JSM Chemistry, 2017, 5(2), 1042–7.
[6] Saikia S, Bordoloi M. Molecular docking: Challenges, advances and its use in drug discovery perspective. Current drug targets, 2019, 20(5), 501–21.
[7] Naqvi AA, Hassan MI. Methods for docking and drug designing. InOncology: Breakthroughs in Research and Practice. IGI Global, 2017, 876–90.
[8] Chaudhary KK, Mishra N. A review on molecular docking: Novel tool for drug discovery. Databases, 2016, 3(4), 1029.
[9] Agnihotry S, Pathak RK, Srivastav A, Shukla PK, Gautam B. Molecular docking and structure-based drug design. Computer-Aided Drug Design, 2020, 115–31.
[10] Gschwend DA, Good AC, Kuntz ID. Molecular docking towards drug discovery. Journal of Molecular Recognition: An Interdisciplinary Journal, 1996, 9(2), 175–86.
[11] Bhagat RT, Butle SR, Khobragade DS, Wankhede SB, Prasad CC, Mahure DS, Armarkar AV. Molecular docking in drug discovery. Journal of Pharmaceutical Research International, 2021, 46–58.
[12] Sethi A, Joshi K, Sasikala K, Alvala M. Molecular docking in modern drug discovery: Principles and recent applications. Drug Discovery and Development-New Advances, 2019, 1–21.
[13] Muhammed MT, Aki-Yalcin E. Molecular docking: Principles, advances, and its applications in drug discovery. Letters in Drug Design & Discovery, 2024, 21(3), 480–95.
[14] Sivakumar KC, Haixiao J, Naman CB, Sajeevan TP. Prospects of multitarget drug designing strategies by linking molecular docking and molecular dynamics to explore the protein–ligand recognition process. Drug Development Research, 2020, 81(6), 685–99.
[15] Fan J, Fu A, Zhang L. Progress in molecular docking. Quantitative Biology, 2019, 7, 83–9.
[16] Morris GM, Lim-Wilby M. Molecular docking. Molecular Modeling of Proteins, 2008, 365–82.
[17] Ghasemi JB, Abdolmaleki A, Shiri F. Molecular docking challenges and limitations. InPharmaceutical sciences: Breakthroughs in research and practice. IGI Global, 2017, 770–794.
[18] Prieto-Martínez FD, Arciniega M, Medina-Franco JL. Molecular docking: Current advances and challenges. Revista Especializada En Ciencias Químico-Biológicas, 2018, 21.
[19] Weill N, Therrien E, Campagna-Slater V, Moitessier N. Methods for docking small molecules to macromolecules: A user's perspective. 1. The theory. Current Pharmaceutical Design, 2014, 20(20), 3338–59.
[20] Ewing TJ, Kuntz ID. Critical evaluation of search algorithms for automated molecular docking and database screening. Journal of Computational Chemistry, 1997, 18(9), 1175–89.
[21] Dias R, de Azevedo J, Walter F. Molecular docking algorithms. Current Drug Targets, 2008, 9(12), 1040–7.
[22] Rarey M, Kramer B, Lengauer T, Klebe G. A fast flexible docking method using an incremental construction algorithm. Journal of Molecular Biology, 1996, 261(3), 470–89.
[23] Kramer B, Rarey M, Lengauer T. Evaluation of the FLEXX incremental construction algorithm for protein–ligand docking. Proteins: Structure, Function, and Bioinformatics, 1999, 37(2), 228–41.
[24] Kuntz ID, Blaney JM, Oatley SJ, Langridge R, Ferrin TE. A geometric approach to macromolecule-ligand interactions. Journal of Molecular Biology, 1982, 161(2), 269–88.
[25] Moré JJ, Wu Z. Distance geometry optimization for protein structures. Journal of Global Optimization, 1999, 15, 219–34.
[26] Westhead DR, Clark DE, Murray CW. A comparison of heuristic search algorithms for molecular docking. Journal of Computer-Aided Molecular Design, 1997, 11, 209–28.
[27] Liu M, Wang S. MCDOCK: A Monte Carlo simulation approach to the molecular docking problem. Journal of Computer-Aided Molecular Design, 1999, 13, 435–51.

[28] Morris GM, Goodsell DS, Halliday RS, Huey R, Hart WE, Belew RK, Olson AJ. Automated docking using a Lamarckian genetic algorithm and an empirical binding free energy function. Journal of Computational Chemistry, 1998, 19(14), 1639–62.

[29] Jones G, Willett P, Glen RC, Leach AR, Taylor R. Development and validation of a genetic algorithm for flexible docking. Journal of Molecular Biology, 1997, 267(3), 727–48.

[30] Brooijmans N, Kuntz ID. Molecular recognition and docking algorithms. Annual Review of Biophysics and Biomolecular Structure, 2003, 32(1), 335–73.

[31] Reeves CR, (ed). Modern Heuristic Techniques for Combinatorial Problems. John Wiley & Sons, Inc., 1993.

[32] Glover F. Future paths for integer programming and links to artificial intelligence. Computers & Operations Research, 1986, 13(5), 533–49.

[33] Betzi S, Suhre K, Chétrit B, Guerlesquin F, Morelli X. GFscore: A general nonlinear consensus scoring function for high-throughput docking. Journal of Chemical Information and Modeling, 2006, 46(4), 1704–12.

[34] Becker OM. Conformational Analysis. Computational Biochemistry and Biophysics. CRC Press, 2001, 81–102.

[35] Kirkpatrick S, Gelatt Jr CD, Vecchi MP. Optimization by simulated annealing. Science, 1983, 220(4598), 671–80.

[36] Goodsell DS, Olson AJ. Automated docking of substrates to proteins by simulated annealing. Proteins: Structure, Function, and Bioinformatics, 1990, 8(3), 195–202.

[37] Li J, Fu A, Zhang L. An overview of scoring functions used for protein–ligand interactions in molecular docking. Interdisciplinary Sciences: Computational Life Sciences, 2019, 11, 320–8.

[38] Meng EC, Shoichet BK, Kuntz ID. Automated docking with grid-based energy evaluation. Journal of Computational Chemistry, 1992, 13(4), 505–24.

[39] Jorgensen WL, Chandrasekhar J, Madura JD, Impey RW, Klein ML. Comparison of simple potential functions for simulating liquid water. The Journal of Chemical Physics, 1983, 79(2), 926–35.

[40] Raha K, Peters MB, Wang B, Yu N, Wollacott AM, Westerhoff LM, Merz Jr KM. The role of quantum mechanics in structure-based drug design. Drug Discovery Today, 2007, 12(17–18), 725–31.

[41] Senn HM, Thiel W. QM/MM methods for biomolecular systems. Angewandte Chemie International Edition, 2009, 48(7), 1198–229.

[42] Eldridge MD, Murray CW, Auton TR, Paolini GV, Mee RP. Empirical scoring functions: I. The development of a fast empirical scoring function to estimate the binding affinity of ligands in receptor complexes. Journal of Computer-Aided Molecular Design, 1997, 11, 425–45.

[43] Murray CW, Auton TR, Eldridge MD. Empirical scoring functions. II. The testing of an empirical scoring function for the prediction of ligand-receptor binding affinities and the use of Bayesian regression to improve the quality of the model. Journal of Computer-Aided Molecular Design, 1998, 12, 503–19.

[44] Friesner RA, Murphy RB, Repasky MP, et al. Extra precision glide: Docking and scoring incorporating a model of hydrophobic enclosure for protein– ligand complexes. Journal of Medicinal Chemistry, 2006, 49(21), 6177–96.

[45] Zheng Z, Merz Jr KM. Ligand identification scoring algorithm (LISA). Journal of Chemical Information and Modeling, 2011, 51(6), 1296–306.

[46] Muegge I, Martin YC. A general and fast scoring function for protein– ligand interactions: A simplified potential approach. Journal of Medicinal Chemistry, 1999, 42(5), 791–804.

[47] Gohlke H, Hendlich M, Klebe G. Knowledge-based scoring function to predict protein-ligand interactions. Journal of Molecular Biology, 2000, 295(2), 337–56.

[48] Ma DL, Chan DS, Leung CH. Drug repositioning by structure-based virtual screening. Chemical Society Reviews, 2013, 42(5), 2130–41.

[49] Cheng T, Li Q, Zhou Z, Wang Y, Bryant SH. Structure-based virtual screening for drug discovery: A problem-centric review. The AAPS Journal, 2012, 14, 133–41.
[50] Zhang L, Qiao M, Gao H, Hu B, Tan H, Zhou X, Li CM. Investigation of mechanism of bone regeneration in a porous biodegradable calcium phosphate (CaP) scaffold by a combination of a multi-scale agent-based model and experimental optimization/validation. Nanoscale, 2016, 8(31), 14877–87.
[51] Zhang L, Zhang S. Using game theory to investigate the epigenetic control mechanisms of embryo development: Comment on: "Epigenetic game theory: How to compute the epigenetic control of maternal-to-zygotic transition" by Qian Wang et al. Physics of Life Reviews, 2017, 20, 140–2.
[52] Fan J, Fu A, Zhang L. Progress in molecular docking. Quantitative Biology, 2019, 7, 83–9.
[53] Rosenfeld R, Vajda S, DeLisi C. Flexible docking and design. Annual Review of Biophysics and Biomolecular Structure, 1995, 24(1), 677–700.
[54] Bonvin AM. Flexible protein–protein docking. Current Opinion in Structural Biology, 2006, 16(2), 194–200.
[55] Rarey M, Kramer B, Lengauer T, Klebe G. A fast flexible docking method using an incremental construction algorithm. Journal of Molecular Biology, 1996, 261(3), 470–89.
[56] Tuszynska I, Magnus M, Jonak K, Dawson W, Bujnicki JM. NPDock: A web server for protein–nucleic acid docking. Nucleic Acids Research, 2015, 43(W1), W425–30.
[57] He J, Wang J, Tao H, Xiao Y, Huang SY. HNADOCK: A nucleic acid docking server for modeling RNA/DNA–RNA/DNA 3D complex structures. Nucleic Acids Research, 2019, 47(W1), W35–42.
[58] Carboni S, Gennari C, Pignataro L, Piarulli U. Supramolecular ligand–ligand and ligand–substrate interactions for highly selective transition metal catalysis. Dalton Transactions, 2011, 40(17), 4355–73.
[59] Kim SH, Martin RB. Noncovalent ligand-metal and ligand-ligand interactions in tridentate (dipeptide) palladium (II) complexes. Journal of the American Chemical Society, 1984, 106(6), 1707–12.

Chapter 10
ADMET analysis in CADD

ADMET properties can make or break a drug candidate, so integrating ADMET analysis early in drug design is crucial for success in drug discovery. – Dr. A. N. Jain

10.1 Introduction

Absorption, distribution, metabolism, excretion, and toxicity (ADMET) analysis is a pivotal component of CADD, playing a crucial role in the drug discovery and development process [1]. This multidisciplinary approach combines computational methods with experimental data to assess the pharmacokinetic and pharmacodynamic properties of potential drug candidates. ADMET analysis begins with the assessment of a compound's absorption, determining its ability to enter the bloodstream through various routes such as oral ingestion or injection. Distribution focuses on how a drug spreads throughout the body, affecting different tissues and organs [2, 3]. Metabolism evaluates the compound's susceptibility to enzymatic transformation, which can impact its bioavailability and activity. Excretion assesses how efficiently the body eliminates the drug, influencing its duration of action. Finally, toxicity analysis aims to predict any adverse effects a compound may have on the body. CADD utilizes a range of computational tools and databases to predict these ADMET properties, helping researchers filter out potential drug candidates with undesirable characteristics early in the drug development pipeline. This accelerates the discovery process, reduces costs, and minimizes the likelihood of adverse effects in clinical trials. Overall, ADMET analysis in CADD is a crucial step in optimizing the safety and efficacy of new drug candidates, ultimately benefiting patients and the pharmaceutical industry [4–6].

It is indispensable for any drug molecule that its sufficient concentration should reach its site of action for executing the desired pharmacological action. In achieving this, absorption, distribution, metabolism, and excretion (ADME) properties play a very crucial role whenever a new drug molecule is designed for any target disease. This study of the ADME properties of the drug molecules is often referred to as the pharmacokinetics of the drug molecule. The importance of studying the ADME properties can be understood by the fact that approximately 40% of the drugs demonstrating good pharmacological activity fail because of their poor ADME properties. This failure rate can be reduced if the ADME properties of any new potential drug moiety are studied and optimized simultaneously with its pharmacological profile. The overall representation of the journey of a drug molecule inside the body is given in Fig. 10.1. Therefore, to design new drug molecules, one should have a clear understanding of the ADME parameters, which should be considered at the time of the drug designing process [7].

https://doi.org/10.1515/9783111434858-010

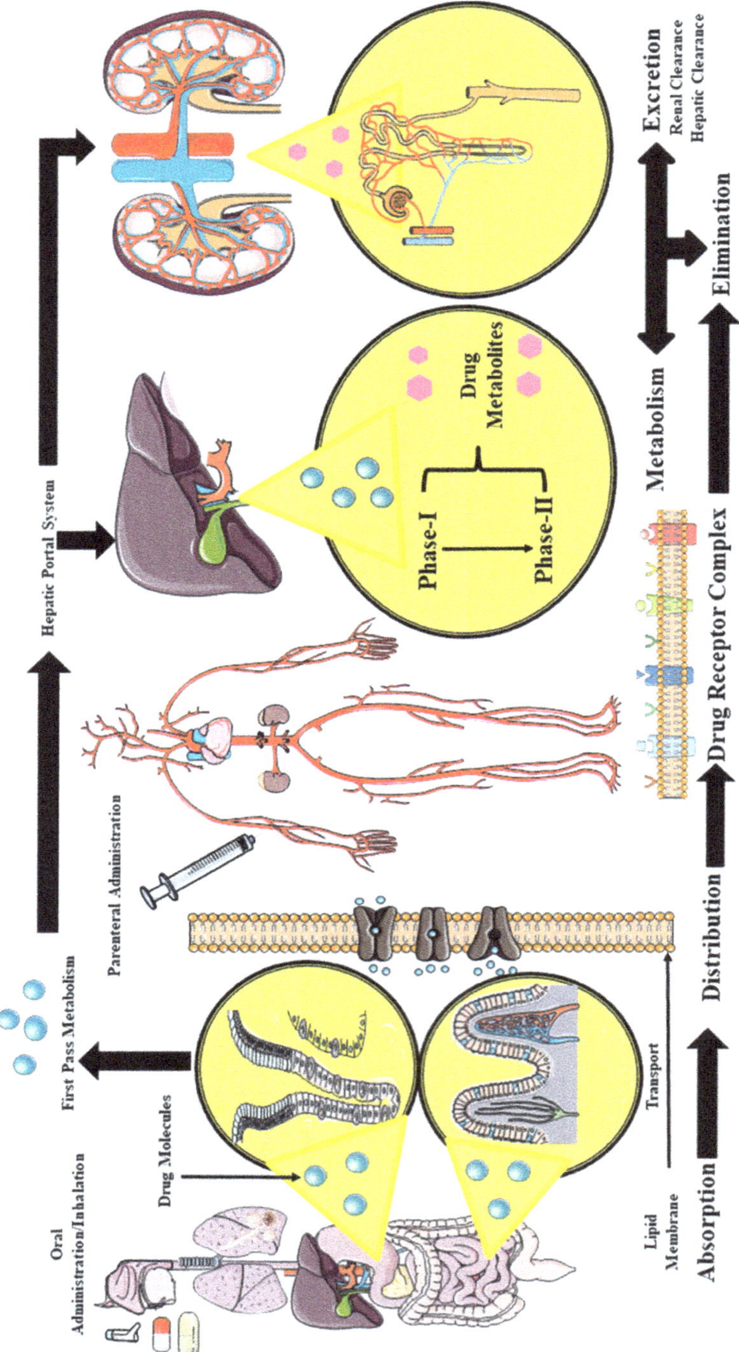

Fig. 10.1: A complete overview of the processes of absorption, distribution, metabolism, and excretion of a drug inside the body.

ADME analysis in CADD is crucial for selecting promising drug candidates, reducing toxicity risks, speeding up drug development, improving drug efficacy, and ensuring compliance with regulatory standards. This integrated approach harnesses the power of computational methods to revolutionize the way pharmaceuticals are discovered and developed, ultimately leading to safer and more effective medicines for patients worldwide [8, 9].

10.2 Importance of ADME studies in CADD

- **Optimizing drug candidates:** ADME analysis helps identify molecules with favorable properties, ensuring that drugs are effectively absorbed into the bloodstream, distributed to target tissues, metabolized efficiently, and excreted appropriately. This streamlines the drug development process by focusing efforts on compounds with the greatest potential for success.
- **Reducing toxicity and side effects:** Poorly metabolized or excreted drugs can lead to toxic effects or unwanted side effects. By predicting ADME properties early in the drug discovery process, researchers can eliminate or modify compounds with unfavorable profiles, minimizing risks to patients.
- **Cost and time efficiency:** Traditional drug development is time-consuming and expensive, with many potential candidates failing due to ADME issues late in the process. CADD, with ADME analysis, allows researchers to screen and prioritize compounds in silico, significantly reducing the cost and time associated with experimental testing.
- **Enhancing drug efficacy:** ADME analysis is not just about avoiding problems; it is also about improving drug performance. By optimizing a drug's pharmacokinetic properties, it is possible to enhance its efficacy, ensuring that it reaches its target in the right concentration and for the right duration.
- **Meeting regulatory requirements:** Regulatory agencies like the FDA require comprehensive ADME data during the drug approval processes. CADD can provide a wealth of predictive ADME information, helping companies meet these requirements more effectively.

10.3 Absorption parameters of the drug molecule

Entry of the drug inside the systemic circulation from the site of administration is known as the process of absorption. Absorption is a critical phase in the pharmacokinetic profile of a drug molecule, influencing how efficiently it enters the bloodstream and reaches its target site within the body [10, 11]. Several key parameters govern this process, which include lipophilicity, molecular size, ionization, etc. Lipophilicity, often measured as LogP, represents the drug's affinity for lipids or fats. A higher LogP suggests greater fat solubility, which can enhance absorption as many biological membranes consist of lipid bilayers. Likewise, smaller molecules can permeate biological barriers more easily than larger ones, facilitating absorption. The charge state of a drug molecule is another determinant. Drugs with a balanced charge are typically better absorbed due to their ability to traverse aqueous and lipid environments. Ionization, which is dependent on pH, impacts absorption. A drug's ionization state can change along the gastrointestinal tract, affecting its ability to cross biological membranes. Collectively, these absorption parameters guide drug design, ensuring that potential candidates possess the characteristics necessary for effective absorption into the bloodstream, a crucial step in the development of pharmaceuticals [12].

The majority of the clinically available drugs are administered through the oral route and, therefore, are absorbed through the gastrointestinal tract (GIT). GIT is considered as the major site of the drug absorption, although it can also take place through skin, lungs, etc., which are considered as minor sites of drug absorption. Regardless of the site of absorption, any drug that is administered inside the body has to cross the bilayered cell membrane to reach the systemic circulation. The cell membrane consists of a bilayer of phospholipids, which contains a hydrophilic head and lipophilic tail part (Fig. 10.2). Absorption of the drugs mostly occurs by passive diffusion through this cell membrane. This passive diffusion of the small drug molecules is totally dependent on the concentration gradient along the cell membrane and the partition coefficient value of the drug molecule (lipid/water partition coefficient). Thus for designing any drug molecule, it is important to consider its partition coefficient value and it is the very first parameter to be considered in the drug-designing process.

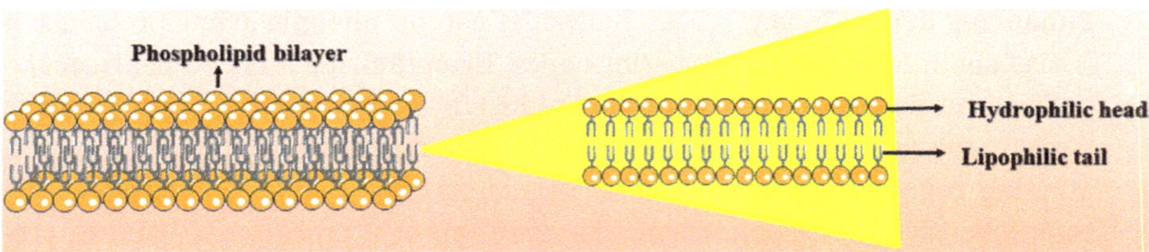

Fig. 10.2: Bilayered phospholipid cell membrane of the gastrointestinal tract (GIT) through which absorption of the drug takes place inside the body. It consists of a hydrophilic head and a lipophilic tail.

10.3.1 Partition coefficient/log *P*

Partition coefficient (log *P* or log *D*) is a fundamental parameter in ADME studies, particularly in the field of pharmaceutical research. It quantifies the distribution of a drug molecule between two immiscible phases, typically a lipid-like organic solvent and water [13]. This parameter offers valuable insights into a compound's lipophilicity, which plays a pivotal role in its pharmacokinetics. A higher LogP value indicates greater lipophilicity, suggesting that the molecule has an affinity for lipid-rich environments, such as biological membranes. This property can enhance a drug's absorption and distribution within the body, enabling it to traverse cellular barriers more effectively. Conversely, a lower LogP value implies higher hydrophilicity, indicating that the drug is more soluble in water. While hydrophilicity can be advantageous for certain aspects of drug development, such as solubility, an optimal balance is often sought to ensure efficient absorption and distribution while maintaining a suitable aqueous solubility. Therefore, the partition coefficient is a vital parameter in ADME analysis, aiding researchers in the selection and optimization of drug candidates. It guides the design of molecules with the appropriate balance of lipophilicity and hydrophilicity, ultimately contributing to the development of safe and effective pharmaceuticals [14].

Mathematically, the partition coefficient can be defined as in the following equation:

$$\text{Partition coefficient } (P) = \frac{\text{The concentration of a drug in the lipid phase}}{\text{The concentration of a drug in aqueous phase}} \tag{10.1}$$

The logarithmic form, log P, is commonly used because it simplifies the expression of the data and is more convenient for comparisons. The LogP value can be positive, negative, or zero, indicating whether a compound is more soluble in the organic phase, aqueous phase, or equally distributed between the two, respectively. Generally, an optimum value of partition coefficient is required for a drug molecule so that it can cross the cell membrane and reach systemic circulation. A too-low partition coefficient value will result in a highly aqueous soluble drug that will not cross the cell membrane. Likewise, too high a partition coefficient value results in a drug molecule that has high lipid solubility, making it to be entrapped in the lipid layer of the membrane and, therefore, will not reach the aqueous systemic circulation.

There are numerous tools available through which this parameter of the proposed drug molecule can be calculated through in silico approaches. Any drug molecule having a logP value less than 0 is considered aqueous soluble with poor lipid solubility, whereas the molecule having a logP value more than 0 is considered lipid-soluble. As per Lipinski's rule, an ideal drug molecule should have a LogP value of less than 5 [15, 16].

10.3.2 Acid–base dissociation constant (pK_a)

Acid–base dissociation constant, often referred to as pK_a, is a fundamental parameter in drug development that plays a crucial role in understanding and manipulating a drug's pharmacological properties. This constant quantifies the degree of ionization of a compound in a solution, specifically indicating the equilibrium between its protonated (acidic) and deprotonated (basic) forms [15]. The solubility of a drug in physiological fluids is strongly influenced by its pK_a. Compounds with pK_a values close to the physiological pH (around 7.4) are typically more soluble, ensuring their effective dissolution and absorption within the body. The ionization state of a drug significantly impacts its ability to traverse biological barriers. A compound's charge (resulting from its pK_a) affects its transport across cell membranes and its distribution within tissues. Likewise, the pK_a of a drug can influence its binding to target receptors or enzymes. Variations in pH at different physiological sites can modulate a drug's affinity for its target, impacting its therapeutic effectiveness. Moreover, the enzymatic processes in the body often depend on the ionization state of a drug. Knowledge of a drug's pK_a can aid in predicting its metabolic fate and designing prodrugs that become active through enzymatic transformations. Therefore, the acid–base dissociation constant (pK_a) is a critical parameter in drug development, influencing drug solubility, transport, receptor interactions, metabolism, and formulation. Understanding and manipulating the pK_a of a drug candidate is essential for designing pharmaceuticals with the desired pharmacokinetic and pharmacodynamic properties, ultimately leading to safer and more effective medications.

It is estimated that overall 75% of the drugs are weak bases, 20% are weak acids, and the remaining are either neutral substances or ampholytic in nature. This type of chemical behavior of the drugs also affects their absorption inside the body. It is a universal fact that a unionized drug is lipophilic in nature and is absorbed easily through the cell membrane, whereas, a drug in the ionic form is considered polar and is not absorbed through the cell membrane. As already discussed, most of the drugs available are weak electrolytes. Their dissociation into ions is dependent on the pH of the system. Hence, the pK_a value of any drug molecule gives a fair idea about the pH-dependent dissociation of the drug inside the body, and ultimately about its absorption. pK_a is defined as the negative logarithm of the dissociation constant (K_a) of the drug molecule [17, 18]. Consider an example of an acidic drug HA whose dissociation will be given by the following equation:

$$\mathrm{pH} = \mathrm{p}K_\mathrm{a} + \log\frac{[\mathrm{A}^-]}{[\mathrm{HA}]} \tag{10.2}$$

where $[A^-]$ is the concentration of the ionic (non-protonated) state and [HA] is the concentration of the acid in the unionized (protonated) state.

If at any stage, $[A^-] = [HA]$, then log $[A^-]/[HA]$ will become 0 and pH will be equal to pKa ($pH = pK_a$). From here, it can be determined that the pK_a is equal to the pH value when half of the drug is ionized. From equation (10.2) it becomes evident that if the value of pH is increased by 1, the concentration of ionized or non-protonated state ($[A^-]$) will be increased by 10 folds. Similarly, for the decrease in the pH value by 1, there will be a 10-fold increase in the concentration of unionized or protonated state ([HA]). Since the unionized form of the drug is absorbed readily from the cellular membrane, acidic drugs are absorbed more readily in the acidic medium (stomach) and basic drugs are absorbed more readily in the basic environment (intestine) inside the body

Considering a similar example for the basic drug B, the dissociation of the basic drug will be given by the following equation:

$$pH = pK_a + \log\frac{[B]}{[HB^+]} \tag{10.3}$$

where [B] is the concentration of the unionized basic drug and $[HB^+]$ is the concentration of the basic drug in its protonated or ionized form.

The value of pK_a will be equal to the pH when half of this basic drug is ionized, that is, $[B] = [HB^+]$. However, in the case of the basic drugs, the situation will be totally opposite as compared to the acidic drug. If there is an increase in the pH value by 1, the concentration of the unionized drug will increase 10-fold. Likewise, for a decrease in the pH by a value of 1, the concentration of ionized form will increase by 10 folds. Therefore, it can be said that basic drugs that increase in pH will increase the concentration of the unionized form of the drug and therefore will enhance their absorption in the basic medium. This dependence of absorption on the pK_a of the drug can be best understood with the example given in Fig. 10.3. Log *P* and pK_a values of some of the commonly prescribed drugs are given in Tab. 10.1, which was calculated through the Marvin Sketch of Chemaxon [19].

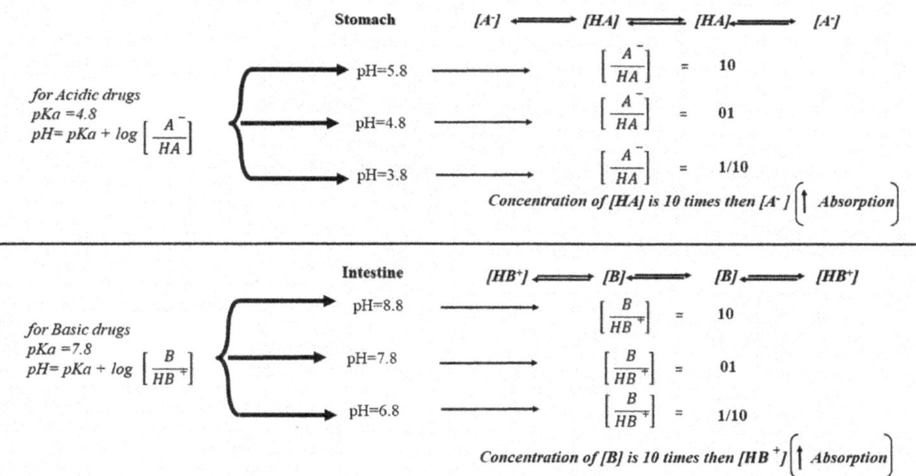

Fig. 10.3: The dependence of the drug absorption on its pK_a value, which determines whether the drug will be absorbed in the stomach or the intestine.

10.3.3 Bioavailability (*F*)

Bioavailability plays a pivotal role in the field of CADD when analyzing the ADME properties of potential drug candidates. Understanding bioavailability is essential in predicting a

Tab. 10.1: LogP and pK_a values of some of the commonly prescribed drugs calculated through the Marvin Sketch of Chemaxon.

S. no.	Name of drug	Calculated log *P* value	Calculated pK_a value
1	Paracetamol	0.91	9.46
2	Clonazepam	3.15	11.63
3	Captopril	0.73	8.71
4	Rosuvastatin	1.92	4.00
5	Cetirizine	3.58	7.97
6	Pantoprazole	2.18	9.15
7	Glimepride	3.12	14.06
8	Donepezil	4.21	9.12
9	Quinine	2.51	13.89
10	Lidocaine	2.84	7.75

drug's effectiveness and safety in vivo. In the context of CADD, bioavailability refers to the fraction of an orally administered drug that reaches the bloodstream and is available to interact with the target receptor. It serves as a critical parameter for assessing drug candidates' viability and optimizing their pharmacokinetic profiles. CADD models take into account a compound's physicochemical properties, such as lipophilicity and solubility, to estimate its likelihood of being absorbed through biological membranes. This prediction helps identify compounds with favorable bioavailability profiles. Bioavailability impacts the distribution of drugs within the body. By considering factors like protein binding and tissue permeability, CADD helps assess a drug's ability to reach its target site effectively. Bioavailability also affects a drug's susceptibility to metabolic processes in the liver and other tissues. CADD tools predict potential metabolites and their impact on bioavailability, aiding in drug design to minimize unwanted transformations. Predicting a drug's elimination rate and potential excretion routes, such as renal or hepatic clearance, is also crucial in optimizing bioavailability and overall pharmacokinetics. Therefore, bioavailability serves as a critical parameter in CADD's ADME analysis, guiding drug design efforts to develop compounds that maximize their therapeutic potential while minimizing adverse effects. By leveraging computational techniques, researchers can efficiently screen and prioritize potential drug candidates with improved bioavailability, ultimately accelerating the drug discovery process [20–23]. Table 10.2 depicts the values of bioavailability of the commonly marketed drugs calculated through the SwissADME server [24].

Tab. 10.2: Values of bioavailability of the commonly marketed drugs calculated through the SwissADME server.

S. no.	Name of drug	Calculated *F* value
1	Paracetamol	0.55
2	Clonazepam	0.55

Tab. 10.2 (continued)

S. no.	Name of drug	Calculated *F* value
3	Captopril	0.11
4	Rosuvastatin	0.56
5	Cetirizine	0.55
6	Pantoprazole	0.55
7	Glimepride	0.55
8	Donepezil	0.55
9	Quinine	0.55
10	Lidocaine	0.55

10.4 Distribution parameters of the drug molecule

After the absorption of the drug inside the systemic circulation, its distribution inside the body comes into effect. Distribution, the second component of ADME is a pivotal phase in a drug's journey through the body. It involves the movement of a drug from the bloodstream to various tissues and compartments, impacting its efficacy and safety. Distribution is influenced by several factors, including a drug's physicochemical properties, such as size, lipophilicity, and charge. Lipophilic drugs tend to penetrate cell membranes more easily, while charged molecules may face barriers. Protein binding also plays a significant role; drugs bound to plasma proteins have a limited distribution, while those unbound are more readily available for action. Tissues vary in blood flow and permeability, affecting drug distribution. Organs with high blood flow, like the heart and liver, receive drugs more rapidly, while those with low blood flow, such as fat tissue, may accumulate lipophilic drugs. Distribution can impact a drug's therapeutic window and potential side effects. Understanding these factors is crucial in drug development to optimize dosing regimens and ensure the drug reaches its target site effectively while minimizing unwanted distribution to other tissues [25–28]. There are certain parameters that determine the distribution of the drug inside the body and can be calculated through in silico approaches.

10.4.1 Apparent volume of distribution (V_d)

Apparent volume of distribution (V_d) is a pharmacokinetic parameter that reflects the extent of a drug's distribution within the body relative to its concentration in the bloodstream. It helps assess a drug's distribution characteristics and is a key factor in dosing and drug design. In silico prediction of Vd involves computational models that use physicochemical and pharmacological properties to estimate this parameter. These models leverage data on a drug's lipophilicity, molecular weight, protein binding, and tissue-specific factors to simulate its distribution behavior. By predicting Vd, researchers can anticipate a drug's distribution profile before conducting costly and time-consuming in vivo experiments, aiding in drug development and optimizing dosing strategies for therapeutic efficacy. Apparent volume of distribution is an important parameter that is defined as the volume required to dissolve any drug inside the body to reach a steady state concentration in the blood plasma at equilibrium. Mathematically, it can be expressed as the following equation:

$$V_d = \frac{\text{The dose of the drug administered}}{\text{Steady-state concentration of the drug in plasma}} \tag{10.4}$$

A drug will have a high value of V_d if a larger fraction of it remains inside the tissues as compared to plasma due to certain factors such as high lipophilicity. Therefore, a larger dose of such drugs is required to attain a steady-state plasma concentration. Similarly, if the major fraction of the drug remains in the blood plasma due to high plasma protein binding or high molecular weight, then the drug will have a low value of V_d [29–32].

10.4.2 Plasma protein binding

Plasma protein binding is a crucial pharmacokinetic parameter that influences a drug's distribution, efficacy, and safety within the body. It refers to the degree to which a drug molecule binds to proteins found in the bloodstream, primarily albumin and alpha-1 acid glycoprotein. Only the unbound, or free fraction of a drug is pharmacologically active and able to distribute to target tissues. The bound fraction acts as a reservoir, gradually releasing the drug over time. In silico prediction methods are essential tools in drug discovery and development for estimating plasma protein binding. These computational approaches leverage the drug's physicochemical properties, such as molecular weight, lipophilicity, hydrogen bonding capacity, and 3D molecular structure, to predict its binding affinity for specific plasma proteins. Several methods, including molecular docking, QSAR modeling, and machine learning algorithms, are employed for this purpose. Accurate prediction of plasma protein binding is invaluable for several reasons. It helps researchers understand a drug's distribution characteristics, assists in optimizing dosage regimens, and can flag potential drug–drug interactions. Moreover, it aids in prioritizing drug candidates early in the development process, ensuring that compounds with appropriate binding profiles progress through the pipeline. Therefore, in silico prediction of plasma protein binding is a vital step in drug discovery and development, providing valuable insights into a drug's pharmacokinetic behavior and supporting informed decision-making throughout the drug development process.

Drugs that show high affinity to the plasma proteins tend to have a low apparent volume of distribution. The fraction of the drug that binds with the plasma protein cannot cross the lipid membrane and hence does not reach the tissues to show their desirable pharmacological action. This plasma protein binding is a reversible phenomenon and it is important to mention that the presence of multiple drugs in the systemic circulation can lead to altered plasma protein binding of drugs as some drugs do displace other plasma protein-bound drugs. Albumin is one of the most common proteins to which most of the drugs bind in the blood plasma [33–36]. Table 10.3 depicts the volume of distribution (V_d) and plasma protein binding values of the commonly marketed drugs calculated from the ADMETlab 2.0 online server [37].

Tab. 10.3: Values of bioavailability of the commonly marketed drugs calculated through the SwissADME server.

S. no.	Name of drug	Calculated V_d value (liters)	Plasma protein binding (%)
1	Paracetamol	0.923	14.863
2	Clonazepam	1.799	95.727
3	Captopril	0.351	16.486
4	Rosuvastatin	0.20	98.792
5	Cetirizine	0.963	91.207
6	Pantoprazole	0.276	98.095
7	Glimepride	0.286	95.395
8	Donepezil	1.589	87.743
9	Quinine	2.192	86.377
10	Lidocaine	1.46	68.787

10.4.3 CNS MPO score for the distribution of the drugs in the brain

CNS-active drugs for the management of disorders like Alzheimer's disease, Parkinson's disorder, depression, anxiety, etc. can show their pharmacological action only when they cross the blood–brain barrier (BBB) and reach the brain. Therefore, the first prerequisite for any drug to show its activity in the brain is that it should cross the BBB. For assessing this ability of any drug molecule, the Central Nervous System Multiparameter Optimization (CNS MPO) score was developed by Wager et al. The CNS MPO score for any drug can range from 0 to 6 and for any drug to exert its CNS pharmacological activity, it is desired that its CNS MPO score should be greater or equal to 4.

The CNS MPO score is a valuable tool in drug development, specifically designed to predict and optimize the distribution of drugs within the brain. As the brain is a highly specialized and protected organ with a complex BBB, assessing a drug's ability to reach its target site within the CNS is critical for neuropharmacology. The CNS MPO score incorporates multiple parameters, such as molecular size, lipophilicity, polarity, and the presence of specific functional groups, to evaluate a drug candidate's potential for effective brain penetration. Here is how the CNS MPO score aids in the distribution of drugs in the brain:

- **Size and lipophilicity:** Small, lipophilic molecules tend to cross the BBB more efficiently. The CNS MPO score assesses these properties to determine a drug's likelihood of reaching its intended site of action.
- **Polarity:** Polar molecules generally have more difficulty crossing the BBB. The score considers the balance between lipophilicity and polarity to optimize brain distribution.
- **Functional groups:** Certain functional groups can enhance or inhibit brain penetration. The CNS MPO score evaluates the presence of these groups to predict a drug's BBB permeability.
- **Optimization:** Researchers can use the CNS MPO score to fine-tune drug candidates, modifying their chemical structures to improve brain distribution while maintaining their pharmacological activity.

By employing the CNS MPO score, pharmaceutical scientists can make informed decisions during drug design and development, leading to more effective CNS drugs with improved brain distribution, potentially increasing therapeutic efficacy, and minimizing side effects in neurological disorders and conditions [38–40].

10.4.4 Blood–brain barrier (BBB) score

Blood–brain barrier (BBB) score is a critical parameter in drug development and distribution within the brain. The BBB is a specialized barrier that separates the circulating blood from the brain's extracellular fluid, ensuring the brain's microenvironment remains stable and protected. However, it also poses a significant challenge for drug delivery to the central nervous system. The BBB score is a computational tool used in silico to predict a drug's ability to penetrate the BBB. It takes into account various physicochemical properties of a compound, such as molecular size, lipophilicity, charge, and hydrogen bonding capacity, to estimate its likelihood of crossing the BBB. A high BBB score indicates a drug with a better chance of penetrating this barrier and reaching its target within the brain. Accurate BBB score prediction is crucial in drug development for neurological and psychiatric conditions, as it helps identify promising drug candidates with optimal brain distribution. Drugs designed to treat diseases of the CNS must possess the ability to cross the BBB effectively to exert their therapeutic effects without causing unwanted side effects. Therefore, the BBB score is a valuable computational tool that aids in the early screening and prioritization of drug candidates for CNS-related conditions. It plays a pivotal role in optimizing drug design and dosage regimens, ultimately contributing to the development of safer and more effective treatments for brain-related disorders.

This parameter was also developed with the same aim to identify the drugs that can cross the BBB and act on the CNS. BBB score was calculated by involving 5 parameters of the drug molecule; its value lies between 0 and 6. Any drug having a value less than 4 is considered as non-CNS drugs and those having a score equal to or greater than 4 are considered as CNS drugs [41]. The CNS MPO and BBB scores can be calculated by in silico methods using many tools. One such tool is Marvin Sketch of the ChemAxon [24]. Table 10.4 illustrates the CNS MPO and BBB scores of certain drugs calculated through this tool.

Tab. 10.4: CNS MPO and BBB score of certain drugs calculated through the Marvin Sketch of ChemAxon.

S. no.	Name of drug	Calculated CNS MPO score value	Calculated BBB score value
1	Paracetamol	5.5	4.49
2	Clonazepam	5.1	4.64
3	Captopril	5.5	3.55
4	Rosuvastatin	3.38	1.67
5	Cetirizine	5.54	4.84
6	Pantoprazole	5.5	3.84
7	Glimepride	3.14	2.58
8	Donepezil	4.39	5.28
9	Quinine	5.48	5.16
10	Lidocaine	5.20	5.31

10.5 Metabolism parameters of the drug molecule

After the distribution of the drug inside the body, it undergoes one or more of the following changes inside the body:

- Excreted, unchanged
- Remains unchanged inside the body
- Chemical modification occurs to the drug moiety
- Enzymatic biotransformation

Drugs that are highly polar in nature are excreted unchanged through the urine as they are not absorbed through the lipid membrane and not distributed much to the tissues. On the contrary, highly polar drugs are retained inside the tissues of the body through the lipid membranes. Certain drugs also undergo chemical modification such as hydrolysis inside the body. These chemical modifications are done to impart polarity to the drug molecules so that they can be excreted from the body. It is the enzymatic metabolism of the drugs that predominates over all these processes and occurs mostly inside the liver. Besides the liver, this enzymatic metabolism also occurs to a smaller extent inside the kidney, lungs, skin, and gastrointestinal mucosa. For enzymatic metabolism, the cytochrome-P450 enzyme superfamily plays a very important role as most of the enzymatic metabolism is performed by these enzymes inside the body. The most important enzymes of the cytochrome-P450 family that are involved in drug metabolism are CYP1A, CYP2C, CYP2D, and CYP3A subfamilies.

Metabolism is a critical parameter in ADME studies, particularly in the context of in silico drug design. In silico methods harness computational techniques to predict and understand a drug candidate's metabolic fate within the body, providing essential insights to streamline drug development. In silico metabolism and prediction involves modeling the enzymatic transformations that drugs undergo, primarily in the liver. Cytochrome P450 enzymes (CYPs) play a central role in drug metabolism, and in silico tools predict how these enzymes may modify a drug's chemical structure. Understanding the metabolic pathways a drug may follow helps identify potential metabolites and assess their pharmacological activity, toxicity, and impact on bioavailability. Moreover, in silico metabolism analysis aids in optimizing a drug's chemical structure to minimize susceptibility to metabolic processes that could render it ineffective or harmful. Researchers can explore structural modifications to enhance a drug's metabolic stability and bioavailability. In the early stages of drug design, in silico metabolism predictions allow for the prioritization of lead compounds that are less likely to be extensively metabolized or exhibit toxic metabolites [42–45]. This reduces the cost and time associated with experimental studies. Table 10.5 depicts the details of the cytochrome-P450 enzymes inhibited by some of the marketed drugs, assessed through the SwissADME online server [24].

Tab. 10.5: Details of the cytochrome-P450 enzymes inhibited by some of the marketed drugs, assessed through the SwissADME online server.

S. no.	Name of drug	CYP1A2 inhibitor	CYP2C19 inhibitor	CYP2C9 inhibitor	CYP2D6 inhibitor	CYP3A4 inhibitor
1	Paracetamol	No	No	No	No	No
2	Clonazepam	Yes	Yes	No	No	Yes
3	Captopril	No	No	No	No	No
4	Rosuvastatin	No	No	No	No	No
5	Cetirizine	No	No	No	Yes	No
6	Pantoprazole	Yes	Yes	Yes	Yes	Yes
7	Glimepride	No	No	Yes	No	Yes
8	Donepezil	No	No	No	Yes	Yes
9	Quinine	No	No	No	Yes	No
10	Lidocaine	No	No	No	Yes	No

10.6 Excretion parameters of the drug molecule

Excretion is the last step in the fate of any drug molecule that is present in the body. It is simply defined as the process of removing the drug that is absorbed systemically. The two most common routes of excretion of the drugs from the body are renal excretion and biliary excretion. Renal excretion is the most common type of excretion of the drug molecules from the body. If a drug is highly polar, it is readily excreted from the urine. For a lipophilic drug, it is first metabolized to a polar form and then excreted through the urine. Biliary excretion is the most preferable route for the excretion of large lipophilic drugs, such as steroids, which have high protein binding. Some other routes of excretion of less importance to the drug molecules also include respiration, saliva, and sweat.

In the realm of in silico drug design, the consideration of excretion is a pivotal aspect of ADME analysis. Understanding and predicting a drug's excretion profile is essential for optimizing its pharmacokinetics and ensuring safety. In silico methods are instrumental in assessing excretion parameters. Computational models take into account various factors, including a drug's molecular weight, lipophilicity, polarity, and metabolic stability, to estimate its propensity for renal or hepatic clearance. These models can predict the likelihood of a drug being excreted unchanged or as metabolites, providing critical insights into its elimination pathways. Accurate in silico predictions of excretion parameters aid in rational drug design by allowing researchers to:

- **Optimize dosage regimens:** Understanding a drug's excretion rate helps determine the appropriate dosing frequency and amount to maintain therapeutic levels in the body.
- **Minimize toxicity:** Predicting metabolites and their toxicity profiles helps identify and design drugs with safer excretion profiles.
- **Identify potential drug–drug interactions:** By assessing excretion pathways, in silico methods can flag potential interactions that might affect a drug's clearance, leading to adverse effects or reduced efficacy.

Therefore, the in silico prediction of excretion parameters in the context of ADME is a crucial step in drug design and development. It enables the creation of compounds with optimal pharmacokinetic profiles, enhancing the chances of success in clinical trials and ultimately improving patient outcomes [46–48]. The two most important parameters that are to be considered for designing any drug molecule include:

10.6.1 Clearance (CL) of the drug

Clearance (CL) of the drug is defined as the volume of blood cleared of a drug in a given period. Drug CL refers to the pharmacokinetic process by which the body eliminates a drug or its metabolites from the bloodstream. It represents the rate at which the drug is removed and is typically measured in units of volume per unit of time (e.g., milliliters per minute). CL is a critical parameter that influences a drug's concentration in the body over time, which in turn affects its efficacy and safety. CL can occur through various mechanisms, including renal CL (excretion of drugs or metabolites via the kidneys into urine), hepatic CL (metabolism and elimination of drugs by the liver), and other nonrenal and non-hepatic routes, such as metabolism in other tissues or excretion in feces, sweat, or saliva.

The concept of CL is central to pharmacokinetics, as it helps determine the appropriate dosing regimen for a drug. The CL rate, in combination with the volume of distribution (V_d), allows pharmacokinetic equations to estimate the drug's half-life, which is the time it takes for half of the drug concentration to be eliminated from the body. By understanding drug CL, healthcare professionals and researchers can tailor drug dosing to achieve desired therapeutic levels while minimizing the risk of toxicity or inadequate efficacy. Mathematically, CL can be defined through the following equation:

$$\mathrm{CL} = \frac{\text{Rate of elimination}}{\text{Plasma concentration of a drug}} \tag{10.5}$$

The value of CL greater than 15 is considered high, the value of CL between 5 and 15 is considered moderate, and the drug molecules possessing CL value less than 5 are considered to have low clearance inside the body [49, 50].

10.6.2 Plasma half-life of the drug ($t_{1/2}$)

Plasma half-life ($t_{1/2}$) is also a very important parameter for studying the excretion of the drug from the body. Plasma half-life is defined as the time taken by a drug to reduce to its half of initial concentration in the blood plasma. Mathematically, it is defined by the following equation:

$$t_{1/2} = \frac{0.693}{k} \tag{10.6}$$

where k is the elimination rate constant of the drug.

In the context of in silico drug design, understanding the plasma half-life of a drug is a crucial aspect of ADME analysis. This parameter is a key determinant of a drug's duration of action and dosing frequency. In silico methods are instrumental in predicting and optimizing the plasma half-life of potential drug candidates. Computational models take into account various factors such as the drug's chemical structure, physicochemical properties, metabolic stability, and clearance mechanisms to estimate its half-life. Machine learning algorithms and quantitative structure–activity relationship (QSAR) modeling are commonly employed for these predictions. A longer plasma half-life can be desirable for drugs intended for chronic conditions, as it reduces the frequency of administration and enhances patient compliance. Conversely, for drugs with potential safety concerns or narrow therapeutic windows, a shorter half-life may be preferred to allow for rapid elimination and better control of drug levels. By accurately predicting and manipulating the plasma half-life during the drug design phase, in silico methods help researchers make informed decisions about lead optimization, dosing regimens, and overall drug development strategies. This contributes to the creation of more efficacious and safer pharmaceuticals, speeding up the drug discovery process and reducing costs associated with experimental trials.

For any drug molecule to exert its desired pharmacological action, it must have the desired concentration inside the plasma, below which it cannot show its action. Hence, the $t_{1/2}$ value of any drug gives us an idea about the dosing duration of that drug moiety [51, 52]. Table 10.6 depicts the clearance value and $t_{1/2}$ values of some of the marketed drugs, calculated from the ADMETlabs 2.0 server [37].

Tab. 10.6: Clearance (CL) value and $t_{1/2}$ values of some of the marketed drugs calculated from the ADMETlabs 2.0 server.

S. no.	Name of drug	Calculated CL value	Calculated $t_{1/2}$ value
1	Paracetamol	8.547	0.907
2	Clonazepam	0.423	0.097
3	Captopril	12.500	0.871
4	Rosuvastatin	11.482	0.405
5	Cetirizine	3.595	0.239
6	Pantoprazole	3.395	0.900
7	Glimepride	0.174	0.076
8	Donepezil	10.635	0.164
9	Quinine	1.817	0.574
10	Lidocaine	16.174	0.888

10.7 Toxicity aspects of the drug molecule

Besides the pharmacological and pharmacokinetic profile of the drug, its toxicity profile should also be assessed before designing any drug molecule. The drug molecules that are highly potent and have a good pharmacokinetic profile may not be suitable drug candidates if they show high toxicity. Toxicity is a critical parameter within the ADMET framework in the context of in silico drug designing. In silico methods play a crucial role in predicting and evaluating drug toxicity, offering a cost-effective and time-efficient way to identify potential safety concerns early in the drug development process. In silico drug, design employs various computational techniques to assess toxicity, including QSAR modeling, molecular docking, and machine learning algorithms. These methods utilize information about a drug candidate's chemical structure, physicochemical properties, and interactions with biological targets to make predictions about its potential toxic effects.

Toxicity prediction encompasses the following aspects:

- **Predicting specific toxicity:** In silico tools can estimate the likelihood of a drug candidate causing specific types of toxicity, such as hepatotoxicity (liver damage), cardiotoxicity (heart toxicity), or nephrotoxicity (kidney toxicity).
- **Dose-response modeling:** Computational models can provide insights into the dose-response relationship, helping researchers determine safe dosage levels for a drug candidate.

- **Structural alerts:** Virtual screening can identify structural features or functional groups associated with toxicity, allowing for structural modifications to mitigate adverse effects.
- **Prediction of metabolite toxicity:** In silico methods can also predict the toxicity of metabolites produced during drug metabolism, aiding in the selection of candidates with safer metabolic profiles.

By integrating toxicity predictions into the drug design process, researchers can prioritize compounds with lower toxicity risks, optimize molecular structures to reduce toxic potential, and ultimately accelerate the development of safer and more effective drugs. This proactive approach to toxicity assessment is invaluable in reducing the attrition rate of drug candidates in later stages of development and improving patient safety [53–56].

There are numerous parameters available that should be checked for evaluating the toxicity of any new drug molecule. Table 10.7 depicts the important toxicity parameters obtained for some commonly marketed drugs calculated through the ADMETlabs 2.0 online server [37].

Tab. 10.7: Important toxicity parameters obtained for some commonly marketed drugs calculated through the ADMETlabs 2.0 online server.

S. no.	Name of drug	hERG blocker	Human hepatotoxicity	Skin sensitization	Carcinogenicity	Respiratory toxicity
1	Paracetamol	– (May not interfere)	– (Nontoxic)	+ (Little irritant to the skin)	+ (Little probability of being carcinogenic)	– (Nontoxic)
2	Clonazepam	– (Very little probability of being a blocker)	– (Very little probability of being toxic)	+ (Little irritant to the skin)	– (Very little probability of being carcinogenic)	– (Nontoxic)
3	Captopril	– (May not interfere)	+++ (Probably hepatotoxic)	++ (Little high probability of being an irritant to the skin)	– (Not a probable carcinogenic agent)	– (Little probability of being respiratory toxic)

Tab. 10.7 (continued)

S. no.	Name of drug	hERG blocker	Human hepatotoxicity	Skin sensitization	Carcinogenicity	Respiratory toxicity
4	Rosuvastatin	– (May not interfere)	++ (Probably hepatotoxic)	– (Not a skin irritant)	++ (Probably carcinogenic)	++ (Probably respiratory toxic)
5	Cetirizine	– (May not interfere)	++ (Probably hepatotoxic)	– (Not a skin irritant)	– (Not a probable carcinogenic)	++ (probably respiratory toxic)
6	Pantoprazole	– (May not interfere)	+++ (Probably hepatotoxic)	– (Not a skin irritant)	+++ (Probably carcinogenic)	+++ (Probably respiratory toxic)
7	Glimepiride	– (May not interfere)	+ (Little probability of hepatotoxic)	– (Not a skin irritant)	– (Not a probable carcinogenic)	– (Nontoxic)
8	Donepezil	+++ (Probably hERG blocker)	– (Very little probability of being toxic)	– (Not a skin irritant)	– (Not a probable carcinogenic)	+++ (probably respiratory toxic)
9	Quinine	– (Very little probability of being a blocker)	+++ (Probably hepatotoxic)	– (Not a skin irritant)	++ (Probably carcinogenic)	+++ (Probably respiratory toxic)
10	Lidocaine	– (May not interfere)	– (Nontoxic)	+ (Little irritant to the skin)	– (Not a probable carcinogenic)	+++ (Probably respiratory toxic)

Table 10.7 is self-explanatory for assessing the different parameters for the toxicity of any new drug molecule. It gives us a better idea about where our proposed drug molecule can be toxic and how exposure to that area can be avoided.

10.8 Drug-likeness profile of the newly proposed drug molecule

Besides the ADMET profile of any drug molecule, there is another popular concept of "drug-likeness" among the research fraternity for designing any new drug molecule. Drug-likeness is defined as how close our drug molecule is with respect to certain properties when compared with the already existing drug molecules. The drug-likeness profile assesses whether a compound possesses the characteristics and properties that are typically associated with successful drugs. A favorable drug-likeness profile increases the likelihood of a compound progressing through the drug development pipeline. Several key factors contribute to a drug's likeness profile in CADD and are given below.

- **Physicochemical properties:** CADD analyzes properties like molecular weight, lipophilicity, hydrogen bonding capacity, and polar surface area. A drug molecule should typically fall within specific ranges for these properties to be considered drug-like.
- **Structural features:** The compound's chemical structure is evaluated for specific motifs or functional groups associated with drug-likeness. Toxicophores or structural alerts that indicate potential toxicity are also screened.
- **Pharmacokinetics and ADME:** Computational tools assess how the molecule is likely to behave in terms of absorption, distribution, metabolism, and excretion (ADME). Favorable ADME properties are essential for a drug to be effective and safe.
- **Biological activity:** CADD predicts a compound's affinity for its target protein, ensuring that it is likely to bind effectively and exert the desired pharmacological effects.
- **Safety profile:** Potential toxicities and side effects are predicted, including risks to major organs like the liver and heart.
- **Comparative analysis:** The drug-likeness profile is often compared to existing drugs in the same therapeutic class or category to gauge its similarity and potential competitive advantage.

By rigorously assessing these factors, CADD helps researchers make informed decisions about whether a newly proposed drug molecule is worth pursuing, streamlining the drug discovery process, and reducing the costs and risks associated with bringing new medications to market. Different researchers have given different rules for the drug-likeness behavior of any newly proposed drug candidate and the most popularly employed among them in drug designing are:

10.8.1 Lipinski's rule of five

Lipinski's rule of five, a cornerstone of computer-aided drug design (CADD), offers a set of guidelines to assess the drug-likeness of a newly proposed molecule. These rules, established by Dr. Christopher Lipinski in 1997, are used to predict a compound's oral bioavailability and its potential as a viable drug candidate [57].

The rule stipulates that a drug molecule should meet the following criteria:

- **Molecular weight (MW) ≤ 500 Da:** Compounds exceeding this limit may have difficulty crossing biological membranes.
- **LogP (Octanol–water partition coefficient) ≤ 5:** This parameter measures a molecule's lipophilicity; excessively lipophilic compounds may face absorption issues.
- **Hydrogen bond donors ≤ 5:** Compounds with too many hydrogen bond donors may struggle to permeate cell membranes.
- **Hydrogen bond acceptors ≤ 10:** A high number of hydrogen bond acceptors can hinder oral bioavailability.

Adhering to Lipinski's rule of five is not a strict requirement but serves as a practical guideline. Compounds that meet these criteria often have better chances of becoming successful drugs because they tend to exhibit improved pharmacokinetic properties and oral absorption. CADD tools and software can rapidly analyze a newly proposed molecule against these rules, aiding in the early-stage selection of drug candidates with favorable drug-like properties, thus saving time and resources in the drug development process.

The above conditions were proposed by Lipinski for orally active drugs and two more conditions to this are added by the Veber and are given below.

- No. of rotatable bonds in the molecule should not be greater than 10.
- The total polar surface area (TPSA) of the compound should be less than 140.

Table 10.8 depicts the values of the above conditions for some of the commonly marketed drugs, calculated from the SwissADME server [24].

10.8.2 Muegge (Bayer) filter of drug-likeness

Muegge (Bayer) filter of drug-likeness, developed by Dr. Irmgard Muegge at Bayer Pharmaceuticals in 2001, is an essential tool in CADD for assessing the drug-likeness of newly proposed drug molecules [58, 59]. It serves as a valuable filter to screen and prioritize compounds with a higher likelihood of success in the drug discovery process. This filter is designed to evaluate the chemical and physicochemical properties of a molecule to determine if it meets the criteria of a "drug-like" compound. The key features considered by the Muegge filter include molecular weight, lipophilicity, polar

Tab. 10.8: Values of the above conditions for some of the commonly marketed drugs calculated from the SwissADME server.

S. no.	Name of drug	Molecular weight	No. of H-bond donors	No. of H-bond acceptors	Log *P* value	No. of rotatable bonds	TPSA	No. of violations
1	Paracetamol	151	2	2	0.91	2	49.33	0
2	Clonazepam	315.71	1	4	1.42	2	87.28	0
3	Captopril	432	2	6	0.65	11	165.82	2
4	Rosuvastatin	481.54	3	9	0.94	10	149.30	1
5	Cetirizine	388.89	1	5	2.35	8	53.01	0
6	Pantoprazole	383.87	1	8	0.38	7	105.54	0
7	Glimepride	490.62	3	5	2.28	11	133.06	1
8	Donepezil	379.49	0	4	3.06	6	38.77	0
9	Quinine	324.42	1	4	2.23	4	45.59	0
10	Lidocaine	234.34	1	2	2.28	6	32.34	0

surface area, and the number of hydrogen bond donors and acceptors. These parameters are critical because they influence a compound's ADME properties and its overall pharmacokinetic profile. The Muegge filter helps medicinal chemists and drug designers to quickly eliminate molecules that are likely to exhibit poor bioavailability or have a high risk of toxicity, thus saving time and resources in the drug development process. Compounds that pass this filter are more likely to progress to further stages of development. In essence, the Muegge (Bayer) filter is a valuable component of CADD that aids in the rational selection and optimization of drug candidates by focusing on drug-likeness criteria, ultimately increasing the chances of identifying successful therapeutic agents with improved chances of reaching the market.

According to Muegge, a compound must meet the following conditions to be considered as a drug-like molecule:

- $200 \leq$ molecular weight ≤ 600
- $-2 \leq X \log p \leq 6$
- Total polar surface area (TPSA) ≤ 150
- The total number of rings in the molecule should be ≤ 7
- The number of hydrogen bond donors should not be greater than 5
- The number of hydrogen bond acceptors should not be greater than 10
- Total number of carbon should be greater than 4
- The number of heteroatoms should be greater than 1
- The number of rotatable bonds in the molecule should not be greater than 15

Table 10.9 depicts the values of these parameters for some of the marketed drugs as per Muegge's filter, calculated from the SwissADME online server.

Tab. 10.9: Values of various parameters for some of the marketed drugs as per Muegge's filter, calculated from the SwissADME online server.

S. No.	Name of drug	Molecular weight	No. of H-bond donors	No. of H-bond acceptors	Log *P* value	No. of rotatable bonds	TPSA	No. of rings	No. of heteroatoms	No. of carbon atoms	No. of violations
1	Paracetamol	151	2	2	0.91	2	49.33	01	03	8	0
2	Clonazepam	315.71	1	4	1.42	2	87.28	03	07	15	0
3	Captopril	432	2	6	0.65	11	165.82	02	10	18	1
4	Rosuvastatin	481.54	3	9	0.94	10	149.30	02	11	22	0
5	Cetirizine	388.89	1	5	2.35	8	53.01	03	06	21	0
6	Pantoprazole	383.87	1	8	0.38	7	105.54	03	10	16	0
7	Glimepride	490.62	3	5	2.28	11	133.06	03	10	24	0
8	Donepezil	379.49	0	4	3.06	6	38.77	04	04	24	0
9	Quinine	324.42	1	4	2.23	4	45.59	03	04	24	0
10	Lidocaine	234.34	1	2	2.28	6	32.34	01	03	14	0

Along with the above mentioned most common drug-likeness rules, there are certain additional rules that have been defined by researchers for drug-likeness of any new proposed drug candidate. These rules are summarized in Tab. 10.10.

Tab. 10.10: Additional rules for drug-likeness of any new proposed drug candidate.

S. no.	Name of the researcher	Drug-likeness rules proposed
1	Ghose	– $160 \leq$ molecular weight ≤ 480 – $-0.4 \leq X \log p \leq 5.6$ – $40 \leq$ molar refractivity ≤ 130 – $20 \leq$ number of atoms ≤ 70
2	Egan	– $W \log P \leq 5.88$ – TPSA ≤ 131.6
3	Rule of three for fragments by Congreve et al.	– The molecular weight of the fragment < 300 – $C \log P \leq 3$ – Number of H-bond donors ≤ 3 – Number of H-bond acceptors ≤ 3

References

[1] Van De Waterbeemd H, Gifford E. ADMET in silico modelling: Towards prediction paradise? Nature Reviews Drug Discovery, 2003, 2(3), 192–204.

[2] Yamashita F, Hashida M. In silico approaches for predicting ADME properties of drugs. Drug metabolism and Pharmacokinetics, 2004, 19(5), 327–38.

[3] Darvas F, Keseru G, Papp A, Dorman G, Urge L, Krajcsi P. In silico and ex silico ADME approaches for drug discovery. Current Topics in Medicinal Chemistry, 2002, 2(12), 1287–304.

[4] Winiwarter S, Ahlberg E, Watson E, Oprisiu I, Mogemark M, Noeske T, Greene N. In silico ADME in drug design–enhancing the impact. ADMET and DMPK, 2018, 6(1), 15–33.

[5] M Honorio K, L Moda T, D Andricopulo A. Pharmacokinetic properties and in silico ADME modeling in drug discovery. Medicinal Chemistry, 2013, 9(2), 163–76.

[6] Paul Gleeson M, Hersey A, Hannongbua S. In-silico ADME models: A general assessment of their utility in drug discovery applications. Current Topics in Medicinal Chemistry, 2011, 11(4), 358–81.

[7] Alqahtani S. In silico ADME-Tox modeling: Progress and prospects. Expert Opinion on Drug Metabolism & Toxicology, 2017, 13(11), 1147–58.

[8] Effinger A, O'Driscoll CM, McAllister M, Fotaki N. In vitro and in silico ADME prediction. ADME Processes in Pharmaceutical Sciences: Dosage, Design, and Pharmacotherapy Success, 2018, 301–30.

[9] Butina D, Segall MD, Frankcombe K. Predicting ADME properties in silico: Methods and models. Drug Discovery Today, 2002, 7(11), S83–8.

[10] Avdeef A. Absorption and Drug Development: Solubility, Permeability, and Charge State. John Wiley & Sons, 2012.

[11] Wagner JG. Pharmacokinetics for the Pharmaceutical Scientist. CRC Press, 2018.

[12] Benet LZ, Kroetz D, Sheiner L, Hardman J, Limbird L. Pharmacokinetics: The dynamics of drug absorption, distribution, metabolism, and elimination. Goodman and Gilman's the Pharmacological Basis of Therapeutics, 1996, 3, e27.
[13] Chillistone S, Hardman JG. Factors affecting drug absorption and distribution. Anaesthesia & Intensive Care Medicine, 2017, 18(7), 335–9.
[14] Schanker LS. Mechanisms of drug absorption and distribution. Annual Review of Pharmacology, 1961, 1(1), 29–45.
[15] Bharate S, Kumar V, Vishwakarma R. Determining partition coefficient (Log P), distribution coefficient (Log D) and ionization constant (pKa) in early drug discovery. Combinatorial Chemistry & High Throughput Screening, 2016, 19(6), 461–9.
[16] Csizmadia F, Tsantili-Kakoulidou A, Panderi I, Darvas F. Prediction of distribution coefficient from structure. 1. Estimation method. Journal of Pharmaceutical Sciences, 1997, 86(7), 865–71.
[17] Pathare BE, Tambe VR, Patil V. A review on various analytical methods used in determination of dissociation constant. International Journal of Pharmacy and Pharmaceutical Sciences, 2014, 6(8), 26–34.
[18] Babić S, Horvat AJ, Pavlović DM, Kaštelan-Macan M. Determination of pKa values of active pharmaceutical ingredients. TrAC Trends in Analytical Chemistry, 2007, 26(11), 1043–61.
[19] https://chemaxon.com/marvin
[20] Sim DS. Drug absorption and bioavailability. Pharmacological Basis of Acute Care, 2015, 17–26.
[21] Paul A. Drug absorption and bioavailability. Introduction to Basics of Pharmacology and Toxicology: Volume 1: General and Molecular Pharmacology: Principles of Drug Action, 2019, 81–8.
[22] Atkinson Jr AJ. Drug absorption and bioavailability. In: Atkinson's Principles of Clinical Pharmacology. Academic Press, 2022, 43–59.
[23] Mannhold R, Kubinyi H, Folkers G. Drug Bioavailability: Estimation of Solubility, Permeability, Absorption and Bioavailability. John Wiley & Sons, 2009.
[24] http://www.swissadme.ch/
[25] Øie S. Drug distribution and binding. The Journal of Clinical Pharmacology, 1986, 26(8), 583–6.
[26] Paixão P, Gouveia LF, Morais JA. Prediction of drug distribution within blood. European Journal of Pharmaceutical Sciences, 2009, 36(4–5), 544–54.
[27] Vay K, Scheler S, Frieß W. Application of Hansen solubility parameters for understanding and prediction of drug distribution in microspheres. International Journal of Pharmaceutics, 2011, 416(1), 202–9.
[28] Vendel E, Rottschäfer V, de Lange EC. The need for mathematical modelling of spatial drug distribution within the brain. Fluids and Barriers of the CNS, 2019, 16, 1–33.
[29] Gibaldi M, Koup JR. Pharmacokinetic concepts – drug binding, apparent volume of distribution and clearance. European Journal of Clinical Pharmacology, 1981, 20, 299–305.
[30] Gibaldi M, McNamara PJ. Apparent volumes of distribution and drug binding to plasma proteins and tissues. European Journal of Clinical Pharmacology, 1978, 13(5), 373–8.
[31] Savva M. On the origin of the apparent volume of distribution and its significance in pharmacokinetics. Journal of Biosciences and Medicines, 2022, 10(1), 78–98.
[32] Ghafourian T, Barzegar-Jalali M, Hakimiha N, Ghafourian T, Cronin MT. Quantitative structure-pharmacokinetic relationship modelling: Apparent volume of distribution. Journal of Pharmacy and Pharmacology, 2004, 56(3), 339–50.
[33] Bohnert T, Gan LS. Plasma protein binding: From discovery to development. Journal of Pharmaceutical Sciences, 2013, 102(9), 2953–94.
[34] Trainor GL. The importance of plasma protein binding in drug discovery. Expert Opinion on Drug Discovery, 2007, 2(1), 51–64.
[35] Olson RE, Christ DD. Plasma protein binding of drugs. In: Annual Reports in Medicinal Chemistry. Academic Press, 1996, 31, 327–336.

[36] Kratochwil NA, Huber W, Müller F, Kansy M, Gerber PR. Predicting plasma protein binding of drugs: A new approach. Biochemical Pharmacology, 2002, 64(9), 1355–74.
[37] Xiong G, Wu Z, Yi J, et al. ADMETlab 2.0: An integrated online platform for accurate and comprehensive predictions of ADMET properties. Nucleic Acids Research, 2021, 49(W1), W5–14.
[38] Wager TT, Hou X, Verhoest PR, Villalobos A. Moving beyond rules: The development of a central nervous system multiparameter optimization (CNS MPO) approach to enable alignment of drug-like properties. ACS Chemical Neuroscience, 2010, 1(6), 435–49.
[39] Wager TT, Hou X, Verhoest PR, Villalobos A. Central nervous system multiparameter optimization desirability: Application in drug discovery. ACS Chemical Neuroscience, 2016, 7(6), 767–75.
[40] Gunaydin H. Probabilistic approach to generating MPOs and its application as a scoring function for CNS drugs. ACS Medicinal Chemistry Letters, 2016, 7(1), 89–93.
[41] Gupta M, Lee HJ, Barden CJ, Weaver DF. The blood–brain barrier (BBB) score. Journal of Medicinal Chemistry, 2019, 62(21), 9824–36.
[42] Kirchmair J, Göller AH, Lang D, et al. Predicting drug metabolism: Experiment and/or computation? Nature Reviews Drug Discovery, 2015, 14(6), 387–404.
[43] Testa B, Krämer SD. The biochemistry of drug metabolism–an introduction: Part 1. principles and overview. Chemistry & Biodiversity, 2006, 3(10), 1053–101.
[44] Hutzler JM, Ring BJ, Anderson SR. Low-turnover drug molecules: A current challenge for drug metabolism scientists. Drug Metabolism and Disposition, 2015, 43(12), 1917–28.
[45] Smith DA. Physicochemical properties in drug metabolism and pharmacokinetics. Computer-Assisted Lead Finding and Optimization: Current Tools for Medicinal Chemistry, 1997, 265–76.
[46] Beaumont C, Young GC, Cavalier T, Young MA. Human absorption, distribution, metabolism and excretion properties of drug molecules: A plethora of approaches. British Journal of Clinical Pharmacology, 2014, 78(6), 1185–200.
[47] Kirchmair J, Howlett A, Peironcely JE, et al. How do metabolites differ from their parent molecules and how are they excreted? Journal of Chemical Information and Modeling, 2013, 53(2), 354–67.
[48] Talevi A, Bellera CL. Drug excretion. ADME Processes in Pharmaceutical Sciences: Dosage, Design, and Pharmacotherapy Success, 2018, 81–96.
[49] Smith DA, Beaumont K, Maurer TS, Di L. Clearance in drug design: Miniperspective. Journal of Medicinal Chemistry, 2018, 62(5), 2245–55.
[50] McLeay SC, Morrish GA, Kirkpatrick CM, Green B. The relationship between drug clearance and body size: Systematic review and meta-analysis of the literature published from 2000 to 2007. Clinical Pharmacokinetics, 2012, 51, 319–30.
[51] Sobol E, Bialer M. The relationships between half-life (t1/2) and mean residence time (MRT) in the two-compartment open body model. Biopharmaceutics & Drug Disposition, 2004, 25(4), 157–62.
[52] Sharma A, Slugg PH, Hammett JL, Jusko WJ. Estimation of oral bioavailability of a long half-life drug in healthy subjects. Pharmaceutical Research, 1998, 15, 1782–6.
[53] Guengerich FP. Mechanisms of drug toxicity and relevance to pharmaceutical development. Drug Metabolism and Pharmacokinetics, 2011, 26(1), 3–14.
[54] Park BK, Kitteringham NR, Powell H, Pirmohamed M. Advances in molecular toxicology–towards understanding idiosyncratic drug toxicity. Toxicology, 2000, 153(1–3), 39–60.
[55] Toropov AA, Toropova AP, Raska Jr I, Leszczynska D, Leszczynski J. Comprehension of drug toxicity: Software and databases. Computers in Biology and Medicine, 2014, 45, 20–5.
[56] Baillie TA. Metabolism and toxicity of drugs. Two decades of progress in industrial drug metabolism. Chemical Research in Toxicology, 2008, 21(1), 129–37.

[57] Lipinski CA, Lombardo F, Dominy BW, Feeney PJ. Experimental and computational approaches to estimate solubility and permeability in drug discovery and development settings. Advanced Drug Delivery Reviews, 2012, 64, 4–17.
[58] Muegge I, Heald SL, Brittelli D. Simple selection criteria for drug-like chemical matter. Journal of Medicinal Chemistry, 2001, 44(12), 1841–6.
[59] Muegge I. Selection criteria for drug-like compounds. Medicinal Research Reviews, 2003, 23(3), 302–21.

Chapter 11
De novo drug designing approaches

> De novo drug design in CADD relies on algorithms and computational models to predict the interactions between small molecules and biological targets, paving the way for the discovery of novel therapeutics with tailored properties. – Dr. Jane Smith

11.1 Introduction

De novo drug design represents a transformative approach in the development of novel pharmaceutical compounds [1]. Unlike traditional drug discovery methods, which rely on identifying existing molecules with therapeutic potential, de novo drug design starts from scratch, constructing entirely new molecules tailored to target specific diseases [2, 3]. This innovative approach leverages computational techniques and molecular modeling to design molecules with desired biological activities. De novo drug design is characterized by its computational prowess, drawing from the fields of bioinformatics, molecular modeling, and artificial intelligence to expedite the identification of promising drug candidates [4–6].

Central to de novo drug design is the concept of rational drug design, which relies on a deep understanding of the target biomolecule's structure and function [7]. Scientists employ various computational tools and algorithms to predict the three-dimensional structure of the target protein, usually a key player in a disease pathway [8–10]. This predictive modeling allows researchers to identify potential binding sites and interaction points on the target, providing invaluable insights for drug design. Once the target's structure is elucidated, de novo drug design algorithms come into play [11]. These algorithms utilize vast databases of chemical compounds to generate new molecular structures with the potential to interact effectively with the target protein. Computational chemistry techniques, such as molecular docking and virtual screening, assess the binding affinity and stability of these generated molecules, narrowing down the pool of candidates for further investigation [12].

Machine learning and artificial intelligence play a pivotal role in de novo drug design, as they can analyze vast datasets of known drug–target interactions, enabling the development of predictive models that guide the creation of novel compounds [13]. By learning from historical successes and failures, these algorithms become increasingly proficient at proposing molecules with higher probabilities of therapeutic efficacy and lower toxicity. Furthermore, de novo drug design embraces the principles of medicinal chemistry to fine-tune the properties of candidate molecules. Researchers aim to optimize factors like drug-likeness, bioavailability, and pharmacokinetics to ensure that the resulting compounds possess the desired characteristics for clinical use. This iterative

https://doi.org/10.1515/9783111434858-011

process often involves the synthesis and experimental validation of selected molecules to validate their effectiveness [14–16].

One of the advantages of de novo drug design is its potential to address previously intractable targets and diseases. By custom-designing molecules tailored to a specific target, this approach expands the scope of drug discovery beyond the limitations of natural compounds and existing drugs. It also offers the opportunity to tackle emerging diseases for which no treatments currently exist, including those caused by newly discovered pathogens. Despite its tremendous potential, de novo drug design faces several challenges. The computational power required for simulating complex molecular interactions and generating vast chemical libraries can be resource-intensive. Moreover, predicting a molecule's behavior in the intricate biological milieu of the human body remains a formidable task. Balancing the need for specificity and selectivity while minimizing off-target effects and toxicity poses a constant challenge in the development of any drug molecule [17–19].

Therefore, de novo drug design represents a paradigm shift in drug discovery, leveraging computational approaches, structural biology, and artificial intelligence to create novel therapeutic compounds. This innovative strategy holds great promise for addressing unmet medical needs, accelerating drug development timelines, and reducing the reliance on serendipity in the search for new treatments. As technology continues to advance, de novo drug design is poised to play an increasingly significant role in shaping the future of pharmaceutical research and healthcare [20].

In the previous chapters, we have studied different computer-aided approaches in which the probable potency of the known ligands is assessed either through structure-based molecular docking simulations or ligand ligand-based QSAR approach. Contrary to the above two approaches, the de novo (meaning "from the beginning") drug designing approach is the process of finding a totally novel drug molecule from scratch. Just like CADD, if a novel drug molecule is designed from the structure of the receptor, then it is categorized as structure-based de novo drug design, and similarly, if the reference ligand/ pharmacophore is considered for developing the novel drug molecule, then this approach is known as ligand-based de novo drug designing. The concept of the de novo drug design can be best understood as the inverse of the molecular docking and QSAR process. In both of these processes, the potency of the known ligands is estimated either by calculating their binding with the target receptor protein (in molecular docking) or by estimating their biological activity from a mathematical relationship, derived from a series of known ligands (in the QSAR study). However, in the case of de novo drug designing, instead of the known ligands, a series of totally novel drugs are designed either from the structure of the target receptor protein or the active known ligands against a particular disease (Fig. 11.1).

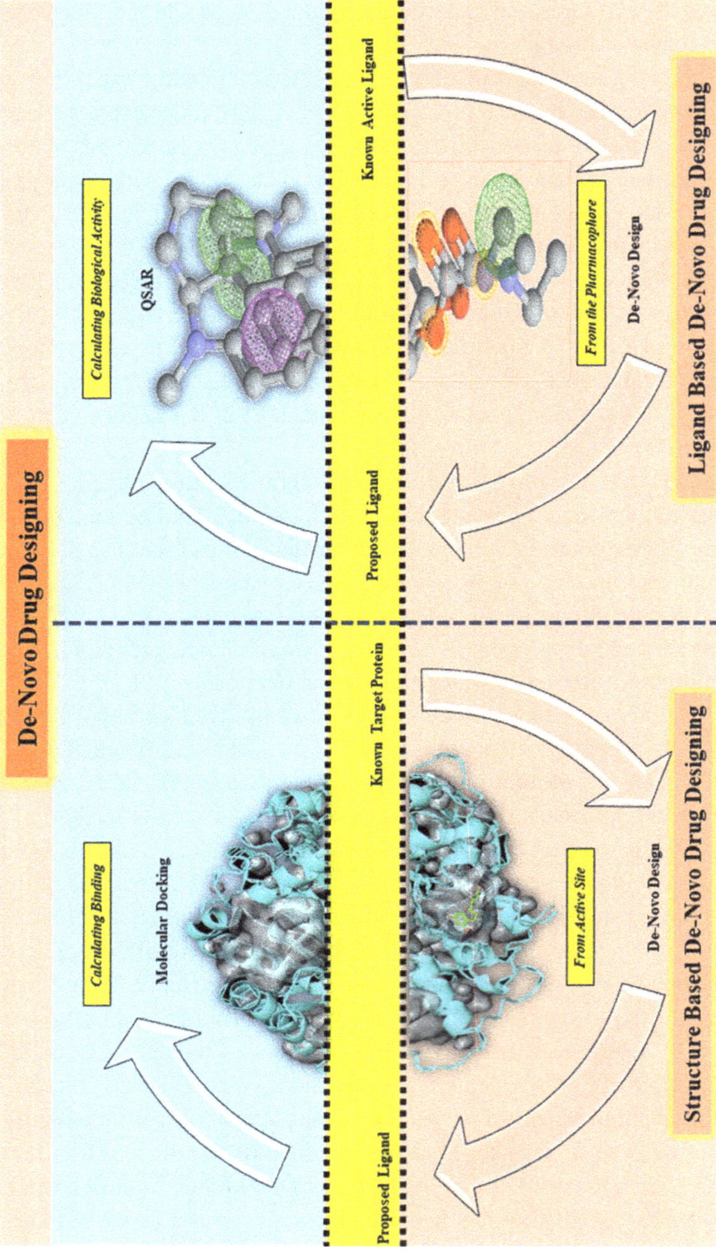

Fig. 11.1: Classification of the de novo drug design into structural-based and ligand-based de novo drug designing.

11.2 Classification of de novo drug designing

The de novo drug designing process can be classified based on the template required for the generation of the new drug molecules from scratch. Any de novo drug designing process can be classified as:
- Structure-based de novo drug designing
- Ligand-based de novo drug designing

11.2.1 Structure-based de novo drug designing

Structure-based de novo drug designing is a powerful strategy in the field of pharmaceutical research, offering a targeted and efficient approach for the creation of novel therapeutic compounds. At its core, this methodology relies on a deep understanding of the three-dimensional structure of a target biomolecule, often a protein, associated with a specific disease. With advances in structural biology techniques like X-ray crystallography and cryoelectron microscopy, scientists can elucidate the precise atomic-level details of these target proteins.

The key to structure-based de novo drug design lies in the rational design of molecules that can interact with the target protein in a highly specific manner. Computational tools and algorithms, such as molecular docking and virtual screening, are employed to predict how potential drug candidates will bind to the target. These simulations assess the binding affinity, geometry, and stability of the interactions, helping researchers identify promising molecules for further development. Machine learning and artificial intelligence also play a significant role, as they can analyze large datasets of known protein–ligand interactions to guide the creation of new molecules. By leveraging these technologies, researchers can generate novel compounds tailored to fit the target's active site and potentially disrupt its function, offering a precise and customizable approach to drug discovery. Ultimately, structure-based de novo drug design holds immense promise in expediting the development of innovative therapeutics, as it enables scientists to create compounds with a higher likelihood of efficacy and lower toxicity. This approach contributes to the ongoing pursuit of more effective and targeted treatments for a wide range of diseases.

Structure-based de novo drug designing process is the approach in which the 3D structure of the compounds is taken into account for the generation of novel drug molecules. The main objective of this process is to design novel drug molecules that interact with the target receptor in the desired and effective manner [21–23]. A complete flow diagram of the structure-based de-novo drug designing process is given in Fig. 11.2. The complete methodology of the structure-based de novo drug designing process is discussed below in detail:

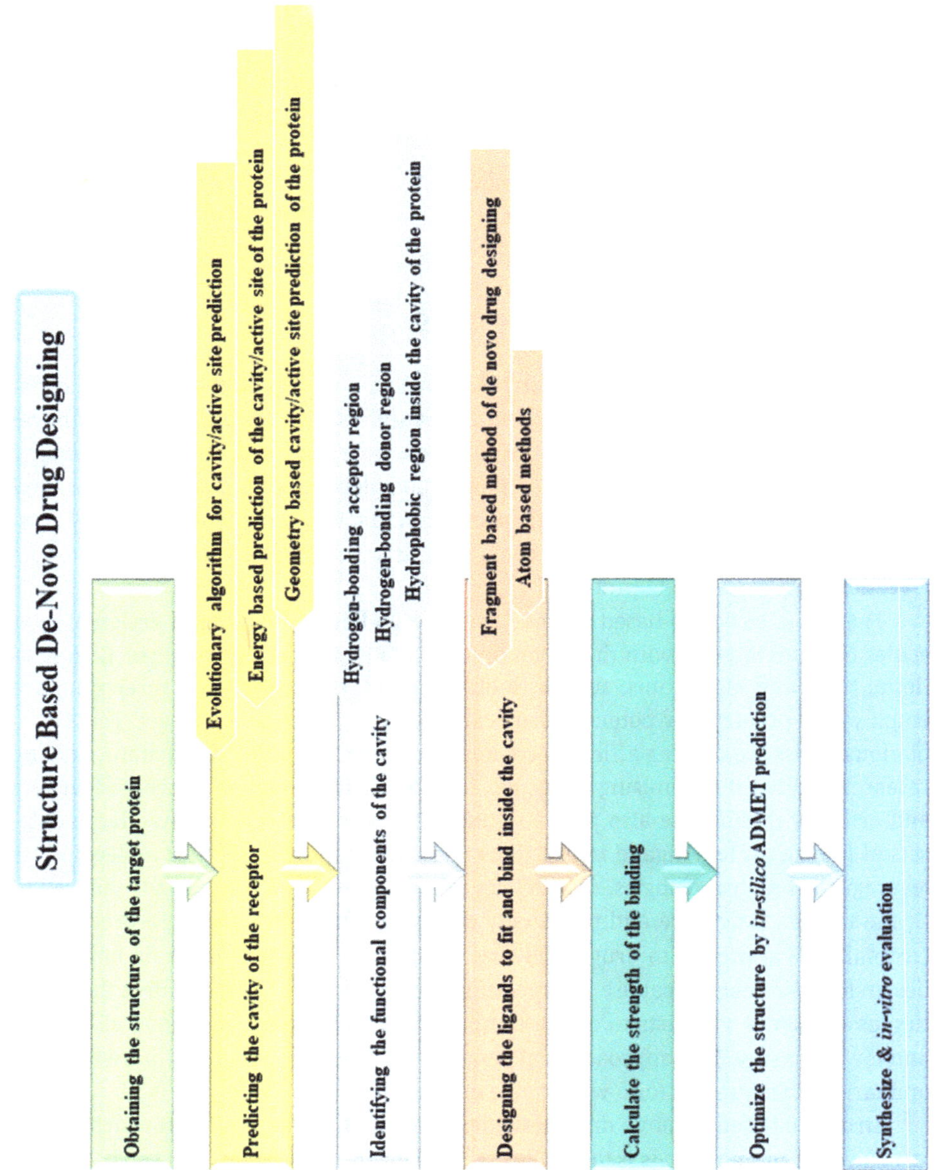

Fig. 11.2: Overview of the complete workflow of the steps involved in the structure-based de novo drug designing approach.

11.2.1.1 Obtaining the structure of the target protein

This is the very first step of any structure-based de novo drug designing approach. In this method, the 3D structures of the target protein against which new drugs are to be designed are obtained from the protein data bank. These structures present in the protein data bank are derived experimentally either through X-ray crystallography, NMR spectroscopy, or electron microscopy. The accuracy and quality of the 3D structure highly affect the accuracy of the structure-based de novo drug designing approach. Figure 11.3 depicts the structure of the human acetylcholinesterase enzyme (PDB ID: 4EY7) obtained from the protein data bank (www.rcsb.org). This structure is derived using the X-ray diffraction technique with a resolution of 2.35 Å.

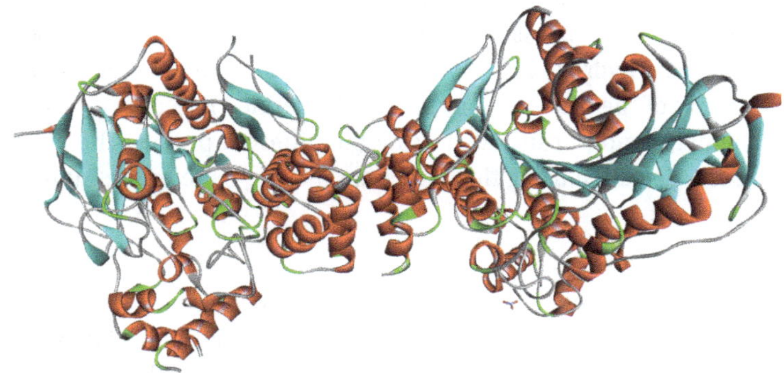

Fig. 11.3: The crystal structure of enzyme acetylcholinesterase (AChE) enzyme obtained from the X-ray diffraction technique obtained from the RCSB protein data bank (PDB ID: 4EY7).

11.2.1.2 Predicting the cavity/active site of the target protein

The cavity or the active site of any target receptor protein is an area of interest for any researcher as it is the region where potent drug molecules bind in a specific manner. Defining or predicting the cavity inside the target protein is the next step in the structure-based de novo drug-designing process [24]. There are various approaches through which the active site or cavity of the target protein can be calculated.

11.2.1.2.1 Evolutionary algorithm for cavity/active site prediction

This method is based on the principle that proteins having similar sequences of amino acids will fold similarly, and hence, will have a similar 3D structure. Likewise, the proteins that have similar structures will have similar active site pockets and therefore, if the active site of a protein is already known, then the cavity of all the proteins that have similar amino acid sequences can be determined using this information.

The similarity between the sequences of the amino acids among the proteins can be accessed using many tools. One of the most commonly employed tools for this purpose is BLAST, which is available open source. The most common in silico programs available for the detection of the cavity of the proteins based on evolutionary algorithms include ConSurf, Rate4site, and Garlig. The quality of this type of cavity detection relies highly on the quality of alignment of the amino acid sequence of the unknown protein and on the sequence of the protein of the known structure.

11.2.1.2.2 Energy-based prediction of the cavity/active site of the protein

This is another approach for determining the active site of the target protein. In this method, a probe (a simple ligand) is placed inside the different regions of the protein and the interaction energy of the amino acids of the protein with the probe is calculated. The regions that show minimum energy of the amino acid-probe interactions are considered favorable regions. Different probes are utilized for measuring different types of interactions, such as for assessing hydrophobic regions one type of probes will be used and for evaluating hydrogen bond acceptor area, different type of probes will be used. The final cavity of the protein molecule will be defined based on energetically favored and forbidden regions, calculated using the different probes. The most popular software for cavity detection using this algorithm include Qsitefinder, AutoLigand, and Grid.

11.2.1.2.3 Geometry-based cavity/active site prediction of the protein

This is the most common approach in predicting the cavity of the protein. Whenever a protein is crystallized to obtain its 3D structure, it is mostly crystallized with its internal ligand, which is the standard potent molecule against that receptor. Finding out the region where this internal ligand is bound inside the protein is the region of interest for any researcher and is defined as the active site or cavity of that protein. This is the most accurate method for calculating the active site of the target protein. In this approach, a cutoff value is put for the distance of the amino acids from the internal ligand, and the region that falls within the cutoff value is defined as the cavity or the active site of the protein. Table 11.1 depicts different geometry-based programs for the detection of active sites/cavities within the target protein:

The ultimate goal of any of the abovementioned algorithms for cavity site detection is to accurately find the region inside the protein where the ligand can bind to show the desirable pharmacological effect. The accuracy of this step is of utmost importance for successful structure-based de novo drug designing. Figure 11.4 depicts the active site of the human acetylcholinesterase enzyme predicted by the software, Discovery studio.

11.2.1.3 Identifying the functional component of the binding site of the protein

For designing any novel drug molecule, it is indispensable to have complete knowledge of the binding/active site of the target protein. For efficiently designing the novel drug molecules, the active site of the target protein is studied for mainly three regions that

Tab. 11.1: Commonly used geometry-based programs for the detection of active site/cavity within the target protein.

S. no.	Principle of the geometric algorithm program	Examples of the software
1	Sphere-based algorithm	– HOLE – SURFNET – PASS
2	GRID-based algorithm	– POCKET – LIGSITE – DoGSite
3	Surface-based methods	– NSA – SCREEN – MSPocket
4	Tessellation-based algorithms	– CAST – MOLE – APROPOS

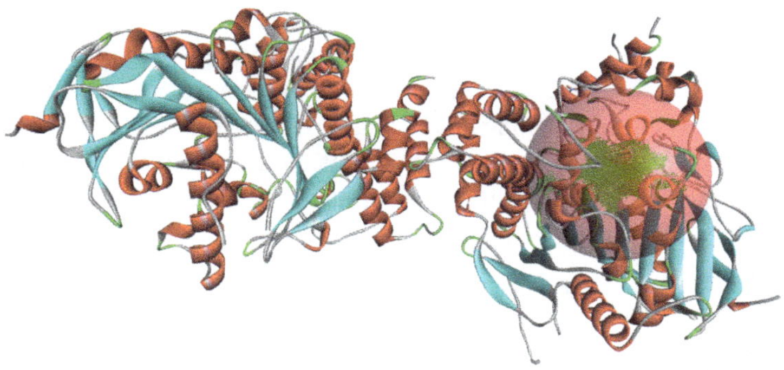

Fig. 11.4: The active site of the enzyme acetylcholinesterase (AChE) where the proposed ligand should bind effectively.

define the interaction of the ligand in most cases. These three regions include the hydrogen bond donor region, hydrogen bond acceptor region, and the hydrophobic area.

11.2.1.3.1 Hydrogen-bonding acceptor region

Hydrogen bonding is considered to be of the utmost importance whenever we define a protein–ligand interaction. Hydrogen bonding is an electrostatic force of attraction that comes into play when a hydrogen atom bound covalently to an electronegative atom is placed in the vicinity of another atom that has a lone pair of electrons (Fig. 11.5). The atom having a lone pair of electrons forms the dipole–dipole bond with the proton, known as the hydrogen-bond acceptor. In the protein molecules, amino

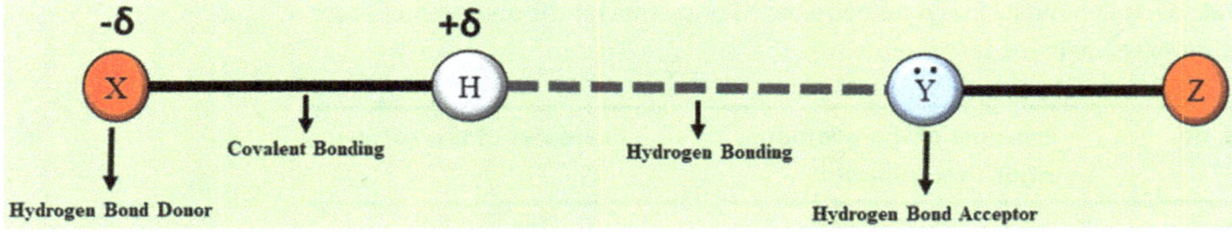

Fig. 11.5: The concept of hydrogen bonding inside the cavity of the receptor with the ligand.

acids aspartic acid (ASP) and glutamic acid (GLU) possess only the hydrogen-bond acceptor groups within them.

11.2.1.3.2 Hydrogen-bonding donor region

As discussed above, the hydrogen bond donor group is the electronegative atom that is covalently bound to the hydrogen atom. It is the electronegativity of the covalently bound atom that pulls the electron toward itself in the covalent bond and imparts a partial positive charge to the proton atom. The amino acids that only have the hydrogen bond donor groups inside the target protein are arginine (ARG), lysine (LYS), and tryptophan (TRP).

There are other amino acids inside the protein molecule that contain both hydrogen-bond acceptor and donor groups. These amino acids include asparagine (ASN), glutamine (GLN), histidine (HIS), serine (SER), threonine (THR), and tyrosine (TYR). Figure 11.6 depicts the hydrogen bonding acceptor and donor region inside the active cavity of the acetylcholinesterase enzyme. The green region on the surface of the cavity of the enzyme represents the hydrogen bond acceptor area whereas the purple color represents the hydrogen bond donor regions.

11.2.1.3.3 The hydrophobic region inside the cavity of the protein

Hydrophobic regions are also of utmost importance whenever we define or study the interaction of protein–ligand inside the cavity of the target protein. The hydrogen-bond regions are the polar regions within the protein and are considered the hydrophilic regions. Likewise, some regions contain nonpolar groups inside the cavity of the protein and are considered hydrophobic regions. Hydrophobic groups are mainly responsible for the Vander-Waals type of interaction between the ligand and the target protein. The amino acids that form the hydrophobic regions of the protein binding sites are phenylalanine (PHE), glycine (GLY), leucine (LEU), isoleucine (ILE), proline (PRO), methionine (MET), tryptophan (TRP), alanine (ALA), and valine (VAL). The very first method for defining the hydrophobic region inside the binding site of the target protein is the molecular lipophilic potential (MLP). Some of the most popular software for the determination of the hydrophobic regions of the binding site of the target protein are QGrid, PocketFinder, and PLIP.

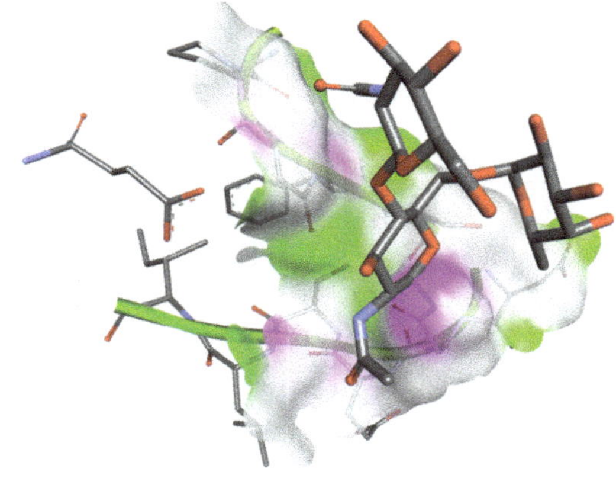

Fig. 11.6: The region of the active site of the enzyme acetylcholinesterase (AChE) where the green region on the surface represents the hydrogen bond acceptor area, whereas the purple color represents the hydrogen bond donor regions.

11.2.1.4 Designing of the molecule based on binding site analysis

De novo molecule designing based on binding site analysis is a specialized approach in drug discovery that focuses on creating entirely new molecular structures tailored to interact with specific target binding sites, typically on proteins involved in diseases. This method begins with a deep understanding of the 3D structure of the target protein and its active binding site, often determined through techniques like X-ray crystallography or cryoelectron microscopy. Computational tools and algorithms are then employed to design molecules that fit snugly into the binding site, forming strong and selective interactions. This process involves generating and evaluating various chemical structures to optimize binding affinity and other pharmacological properties. Machine learning and artificial intelligence techniques may assist in predicting and refining the designs.

Designing the novel molecule is the core of any de novo drug designing approach and is the next step of the structure-based de novo drug designing approach [25, 26]. Based on the analysis of the active site of the target protein, de novo drug designing is done. The novel molecule can be designed by using any one of the following approaches:

a. **Atom-based methods:** In this method, the starting point for the generation of the novel drug molecule is an atom. An atom is placed inside the cavity of the protein at a suitable point where it interacts favorably with the atoms of the amino acids of the receptor. This is a sequential approach in which each atom is added sequentially to the previous atom and then all the parameters such as optimization of the geometry, calculation of binding affinity, etc. are calculated at each step. A deep knowledge of chemistry is the prerequisite for any atom-based de novo designing of the drug mole-

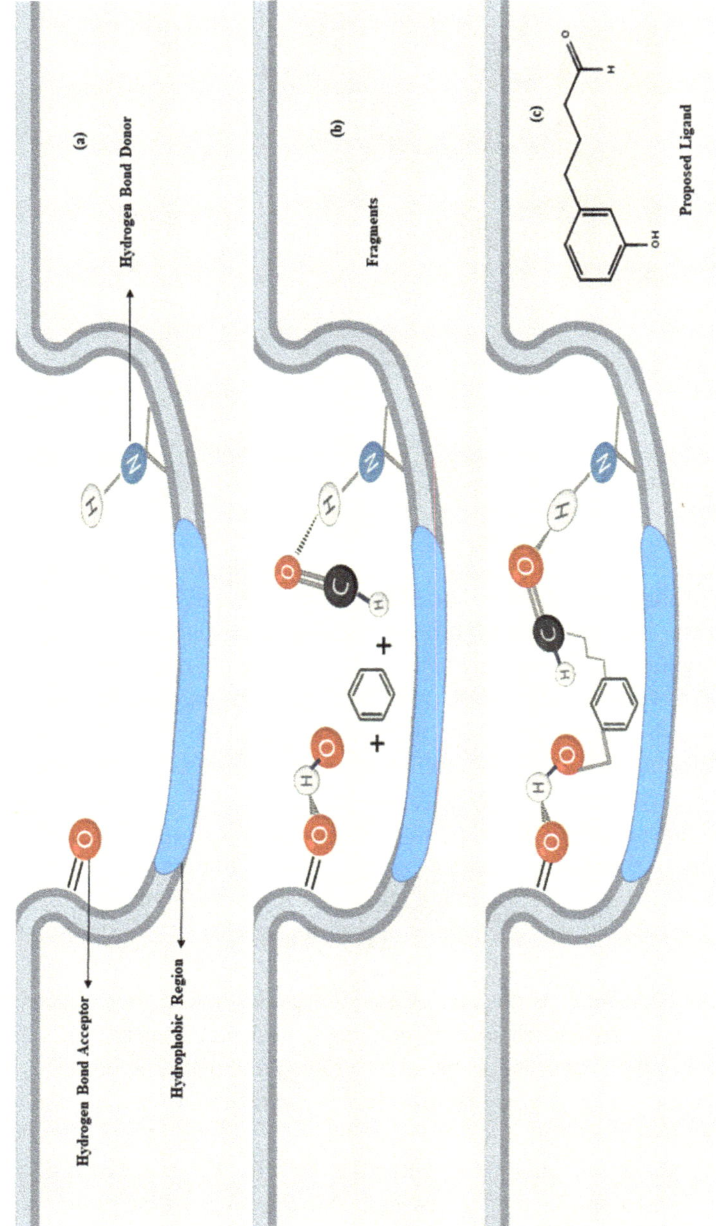

Fig. 11.7: Fragment-based approach for de novo designing of the ligand where different fragments are joined together as per the requirement of the active site of the receptor.

cule. The atom-based method can produce a large number of novel molecules that can act as potential drug molecules against the drug target. The only disadvantages of the method are its complexity, time, and cost due to which these methods are seldom used nowadays and are replaced by the fragment-based approach of de novo drug designing. Earlier, programs such as LEGEND, GENSTAR, and GROWMOL were employed for the atom-based de novo drug designing.

b. **Fragment-based method of de novo drug designing:** This is the most common and widely accepted method for the de novo designing of the drug molecules. In this approach, small fragments are placed inside the interaction sites of the cavity and then a novel molecule is designed based on these interaction points (Fig. 11.7).

Fragments are defined as small organic molecules with lower molecular weight, high solubility, and high affinity toward the interaction sites of the cavity of the target proteins. The fragments used in the molecule designing can either be in cyclic or ring form, depending upon the requirement of the binding site. As Lipinski's rule of five was used for the drug-likeness of the larger drug molecules, Congreve et al. in 2003 proposed the rule of three for defining the properties of the small fragments. As per the Congreve rule of three, smaller fragments should have the following properties:

- The molecular weight of the fragment should be equal to or less than 300 Da.
- The calculated $c \log P$ of the fragment should be equal to or less than 3.
- The number of hydrogen bond acceptors/donors should equal to or less than 3.

Along with these three rules, additional filters can also be applied to obtain more drug-like fragments which include:

- The number of rotatable bonds in the fragment should be equal to or less than 3.
- The polar surface area of the fragment should be equal to or less than 60 Å square.

For constructing the fragments along the binding site of the target protein to obtain potent novel drug molecules, different strategies are employed (Fig. 11.8). The main approaches employed for this purpose are:

- **Linking of the fragment**: In this approach, fragments are first placed inside the cavity as per their affinity with the interaction sites. They are then linked together with small linkers. Linkers generally consist of nonpolar organic alkyl groups such as methyl, propyl, and isopropyl groups.
- **Merging of the fragments**: This is another approach for creating the novel drug molecule from the small fragments. Contrary to the linking, in this approach, merging is performed, rather than linking, for creating the novel drug molecule. This can be best understood by considering one fragment having a carboxylic acid (COOH) group and another amino (NH_2) group is attached (Fig. 11.9). Instead of using the linker, merging will be done by the removal of a water molecule. The resultant molecule will have an amide group (CONH).

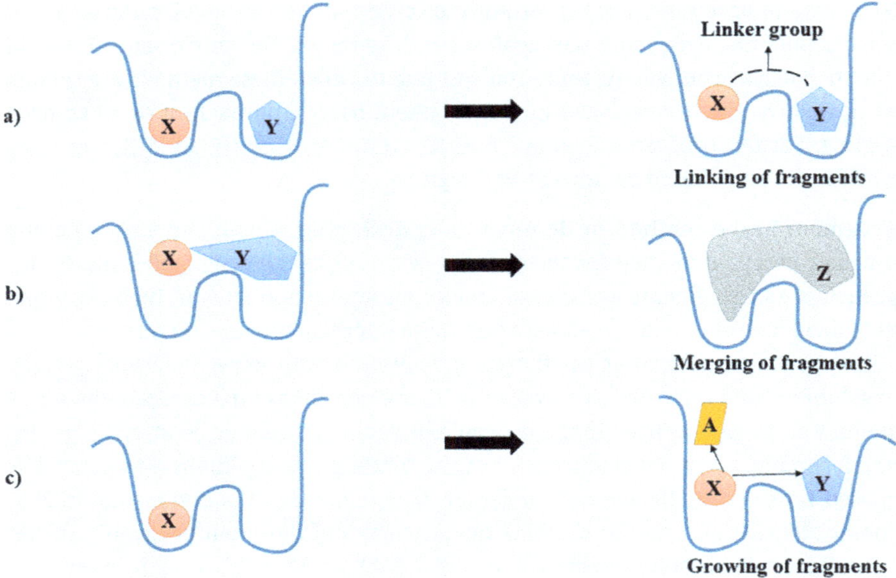

Fig. 11.8: Use of different approaches such as linking, merging, and growing of fragments to obtain a novel active drug against a particular drug based on the active cavity of the receptor.

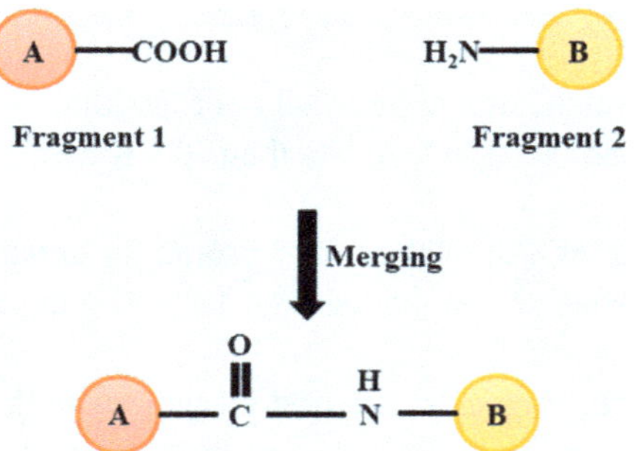

Fig. 11.9: The merging approach of the fragment-based method of the de novo drug designing.

- **Growing of the fragments**: In this approach, initially, a single fragment is placed at one interaction site of the cavity and then other fragments are grown along the binding site.

c. **Different strategies for finding the best fragments:** Screening of the fragments for the construction of a potent novel drug molecule is of utmost importance to get an efficient drug candidate molecule. This selection of the best fragments for the particular target protein is done from a pool or database of the fragments. There are several methods through which these fragment molecules are screened for their binding to the interaction sites of the cavity.

- **Performing direct binding assays of the fragments with the target protein**
  This is the straightforward pathway for assessing the binding of the fragments within the cavity of the target protein. The major limitation of this methodology is that these fragments have low affinity to the target protein when assessed individually. To overcome this issue, the binding of these fragments is evaluated at a very high concentration. Fragments that show the best binding at the interaction site of the cavity are selected for the further generation of the novel drug molecule.
- **Nuclear magnetic resonance (NMR)-based screening**
  In this approach ^{15}N-labeled protein molecule is used for screening purposes. A group of fragments is allowed to bind with this ^{15}N-labeled target protein and their position of binding is assessed through NMR spectroscopy. Fragments that bind inside the cavity of the target protein and are adjacent to each other are selected for further novel molecule generation.
- **Mass spectrometry-based screening**
  This is another approach for the selection of the best fragments for de novo designing of the molecules. Electrospray ionization mass spectrometry (ESIMS) is the choice of method for this type of screening of the fragments. With the ESIMS approach, fragments that bind covalently and noncovalently are distinguished from the pool of the fragments. The fragments showing the best binding in a noncovalent manner are selected further for the designing of the potent novel drug molecule.
- **X-ray crystallography-based screening**
  With the advancement in the technology of X-ray crystallography, it has become highly capable of assessing the binding of the small fragments inside the cavity of the target protein. These X-ray crystallographic methods have become so powerful that despite being small affinity of the fragments as compared to the lead molecules, they are still capable of finding the best bound fragments.

Fragment-based de novo drug designing is the most widely used method for novel potent drug molecule generation. The most common software used for performing this task is LUDI, which was first introduced in the year 1992.

11.2.1.5 Assessing the binding strength of the designed molecules

In this step, the molecules that are obtained from the previous step of designing are further evaluated for their binding in the active site of the target protein. The evaluation of the binding of these newly designed molecules is done with the help of various computational approaches such as molecular docking or molecular simulations. Out of these two methods, molecular docking finds a special place due to its high popularity and ease of performance along with being less time-consuming. The ranking of the

designed molecules is performed through the outcome of molecular docking results, based on any of these scoring algorithms:
- Physics-based scoring
- Empirical-based scoring
- Knowledge-based scoring
- Machine learning-based scoring

The choice of the type of scoring function is totally dependent on the choice of the individual researchers. Details of these scoring methodologies are already discussed in Chapter 6 of this book.

11.2.1.6 Optimizing the probable hits for the absorption, distribution, metabolism, excretion, and toxicology (ADMET) profile

Having high potency is not the sole requirement for any probable drug candidate. It is the pharmacokinetic profile of the molecule that plays an equally important role in a successful drug candidate. It is a fact that it is the concentration of the drug at the site of action that decides its pharmacological action; hence, having an optimum absorption, distribution, metabolism, excretion, and toxicology (ADMET) profile is of utmost importance for any potent drug candidate. Therefore in the next step, the ADMET profile of the novel molecules that have high affinity to the target receptor is assessed. Based on the ADMET profile, these molecules are further optimized or discarded from further study. The two most popular web servers for assessing the ADMET profile of any molecule are SwissADME and ADMETlab 2.0.

11.2.1.7 Synthesis and in vitro analysis of the most promising candidates

This is the final step of the de novo drug designing process in which the first synthetic route for the molecules that shows high affinity and good ADMET profile is investigated. Once the optimum synthetic route is decided, the synthesis of these molecules is performed. The synthesized molecules are then put under in vitro evaluation for the validation of the in silico results. The novel candidates that show good activity in the in vitro testing can be further evaluated for in vivo testing and clinical trials.

11.2.2 Ligand-based de novo drug designing approach

Ligand-based de novo drug designing is a strategic approach in drug discovery that focuses on developing new therapeutic compounds based on the characteristics and properties of known ligands or molecules that interact with a specific target of interest, typically a protein associated with a disease. Unlike structure-based methods that rely on the knowledge of a target's three-dimensional structure, ligand-based de novo drug design starts with the identification of a reference ligand or set of ligands known to bind to the target. The central concept behind this approach is to extract essential structural and chemical features from the reference ligands and use them as a template for generating entirely new molecules with the potential to interact effectively with the target. Computational tools, such as molecular modeling and chemoinformatics, play a pivotal role in this process, allowing researchers to analyze and manipulate molecular structures to design ligands that mimic the key interactions observed in the reference compounds.

Ligand-based de novo drug design is particularly valuable when the 3-D structure of the target protein is unknown or difficult to obtain. It leverages the wealth of existing ligand–target interaction data and can be a versatile strategy for optimizing drug candidates to improve binding affinity, selectivity, and pharmacokinetic properties. This approach significantly accelerates the drug discovery process by starting with validated ligands and tailoring new molecules based on their proven biological activity [27–30].

The goal of this approach is also the same as that of the structure-based approach, which is to develop a novel drug molecule. The only difference is that it is based on designing the novel drug molecule from the information gathered from the already active ligands against the particular drug target. The basic flow diagram of the ligand-based de novo drug designing approach is given in Fig. 11.10. All the steps in the ligand-based de novo drug design are similar to the structure-based de novo design except for the first two steps.

11.2.2.1 Searching for the active ligands

This is the very first step in the ligand-based de novo designing approach. The active ligands against the target disease are obtained from the literature. The active known compounds are selected based on their biological activity value that is already reported.

11.2.2.2 Obtaining the pharmacophore from the active compounds

This is the most crucial step of the ligand-based de novo drug designing approach. A pharmacophore is defined as an "ensemble of the essential features of the molecules in 3-D to produce a desired pharmacological response." There are several computational tools available for generating the pharmacophore from the active ligands. The most popular software and web servers for obtaining the pharmacophore from the ligands are PHASE, MOE, CATALYST, PHARMAGIST, PHARMER, and LIGANDSCOUT.

Figure 11.11 depicts a pharmacophore developed from the three known compounds (physostigmine, neostigmine, and rivastigmine) against acetylcholinesterase enzyme. The pharmacophore is developed from the online web server of Pharmagist and the visualization is done from the ZINCpharmer web server. The green ring in the pharmacophore represents the hydrophobic regions of the pharmacophore, the orange ring represents the hydrogen-bond acceptor area, and the purple ring represents the aromatic ring, essential for acting on the acetylcholinesterase enzyme.

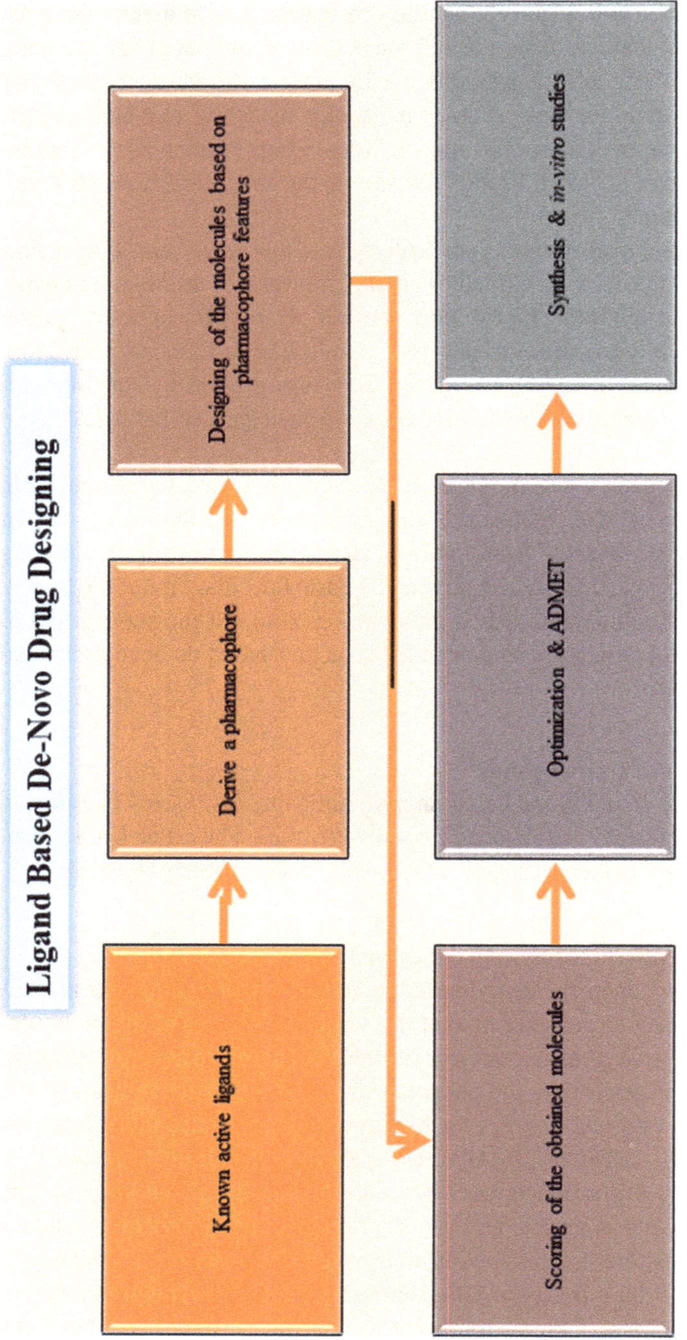

Fig. 11.10: The overview of the basic steps involved in the process of the ligand-based de novo drug designing approach.

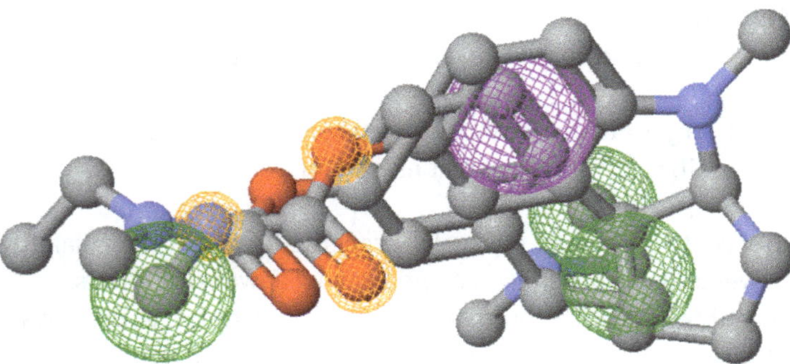

Fig. 11.11: An illustration of the pharmacophore developed through the ligand-based de novo drug designing approach.

The remaining steps in the ligand-based de novo drug designing from the designing of the novel molecules to the synthesis of these molecules are similar to that of the structure-based de novo drug designing already discussed above. The only difference in this approach is that instead of interaction sites of the cavity of the target protein, pharmacophore constrain are employed for the designing of the novel molecules.

11.3 Advantages of de novo drug designing approaches

De novo drug design approaches in CADD offer numerous advantages, contributing to the efficiency and effectiveness of the drug discovery process. Here are 10 major advantages of de novo drug design in CADD:

- **Novelty:** De novo design allows for the creation of entirely new drug candidates, avoiding reliance on existing compounds, which can be particularly beneficial when targeting novel or challenging diseases.
- **Target flexibility:** It can be applied even when the three-dimensional structure of the target biomolecule is unknown or challenging to obtain, as it relies on ligand-based or fragment-based approaches.
- **Cost-efficiency:** By reducing the need for high-throughput screening of large compound libraries, de novo design can save significant time and resources in the drug discovery process.
- **Customization:** Researchers can tailor drug candidates to specific targets, optimizing binding affinity, selectivity, and other pharmacological properties.
- **Rapid iteration:** De novo design allows for quick iterations and modifications of drug candidates in response to experimental results or changing project requirements.

- **Exploration of chemical space:** It enables the exploration of a broader range of chemical space, leading to the discovery of diverse and innovative drug candidates.
- **Minimization of off-target effects:** De novo design can focus on designing molecules that interact specifically with the target of interest, reducing the risk of off-target effects.
- **Reduced side effects:** By optimizing molecular properties, such as bioavailability and toxicity, de novo-designed drugs can potentially have fewer side effects.
- **Access to undruggable targets:** It opens up opportunities to target previously considered "undruggable" proteins or biomolecules by custom-designing ligands with the desired properties.
- **Intellectual property:** De novo-designed compounds are often entirely novel, which can lead to stronger intellectual property positions for pharmaceutical companies.

These advantages make de novo drug design approaches a valuable tool in the arsenal of CADD techniques, accelerating drug discovery, and enabling the development of innovative and effective pharmaceuticals.

11.4 Limitations of de novo drug designing approaches

De novo drug design approaches in CADD have made significant strides but still face several limitations:

- **Limited understanding of complex biology:** De novo methods heavily depend on our knowledge of target proteins and their interactions, which may be incomplete or inaccurate, especially for novel or poorly characterized targets.
- **Computational resources:** Performing extensive molecular simulations, virtual screening, and designing large chemical libraries can be computationally intensive, requiring high-performance computing infrastructure.
- **Accuracy of predictive models:** The accuracy of computational models used in de novo design is contingent on the quality and quantity of available data, leading to potential inaccuracies in predictions.
- **Chemical space exploration:** The vastness of chemical space makes it impossible to exhaustively search for all potential drug candidates, potentially missing promising molecules.
- **Protein flexibility:** Protein structures can undergo conformational changes, making it challenging to predict how they will interact with designed ligands under various conditions.
- **Lack of experimental validation:** Predicted drug candidates from de novo design must be synthesized and experimentally validated, which can be time-consuming and costly.

- **Toxicity and side effects:** De novo-designed compounds may have unforeseen toxicities or side effects due to challenges in predicting their interactions with off-target proteins.
- **Bioavailability and pharmacokinetics:** Predicting a molecule's behavior within a biological system, including absorption, distribution, metabolism, and excretion, remains a complex task.
- **Chemical synthesis challenges:** The synthesis of novel compounds designed through de novo methods can be intricate and may involve the development of new chemical routes, adding complexity and cost.
- **Regulatory hurdles:** Meeting regulatory requirements for drug approval can be challenging for de novo-designed drugs, as they often lack a historical safety and efficacy profile.

Despite these limitations, de novo drug design remains a promising approach in CADD, with ongoing advancements in computational techniques and an increasing understanding of molecular biology. Researchers continue to address these challenges to improve the efficiency and effectiveness of de novo drug design strategies.

References

[1] Mouchlis VD, Afantitis A, Serra A, et al. Advances in de novo drug design: From conventional to machine learning methods. International Journal of Molecular Sciences, 2021, 22(4), 1676.

[2] Tong X, Liu X, Tan X, et al. Generative models for de novo drug design. Journal of Medicinal Chemistry, 2021, 64(19), 14011–27.

[3] Hartenfeller M, Schneider G. De novo drug design. Chemoinformatics and Computational Chemical Biology, 2011, 299–323.

[4] Popova M, Isayev O, Tropsha A. Deep reinforcement learning for de novo drug design. Science Advances, 2018, 4(7), eaap7885.

[5] Devi RV, Sathya SS, Coumar MS. Evolutionary algorithms for de novo drug design–A survey. Applied Soft Computing, 2015, 27, 543–52.

[6] Gupta A, Müller AT, Huisman BJ, Fuchs JA, Schneider P, Schneider G. Generative recurrent networks for de novo drug design. Molecular Informatics, 2018, 37(1–2), 1700111.

[7] Schneider G. Future de novo drug design. Molecular Informatics, 2014, 33(6–7), 397–402.

[8] Skalic M, Jiménez J, Sabbadin D, De Fabritiis G. Shape-based generative modeling for de novo drug design. Journal of Chemical Information and Modeling, 2019, 59(3), 1205–14.

[9] Fischer T, Gazzola S, Riedl R. Approaching target selectivity by de novo drug design. Expert Opinion on Drug Discovery, 2019, 14(8), 791–803.

[10] Nicolaou CA, Apostolakis J, Pattichis CS. De novo drug design using multiobjective evolutionary graphs. Journal of Chemical Information and Modeling, 2009, 49(2), 295–307.

[11] Krishnan SR, Bung N, Bulusu G, Roy A. Accelerating de novo drug design against novel proteins using deep learning. Journal of Chemical Information and Modeling, 2021, 61(2), 621–30.

[12] Liu X, IJzerman AP, van Westen GJ. Computational approaches for de novo drug design: Past, present, and future. Artificial Neural Networks, 2020, 139–65.

[13] Ståhl N, Falkman G, Karlsson A, Mathiason G, Bostrom J. Deep reinforcement learning for multiparameter optimization in de novo drug design. Journal of Chemical Information and Modeling, 2019, 59(7), 3166–76.
[14] Xie W, Wang F, Li Y, Lai L, Pei J. Advances and challenges in de novo drug design using three-dimensional deep generative models. Journal of Chemical Information and Modeling, 2022, 62(10), 2269–79.
[15] Hartenfeller M, Schneider G. Enabling future drug discovery by de novo design. Wiley Interdisciplinary Reviews: Computational Molecular Science, 2011, 1(5), 742–59.
[16] Palazzesi F, Pozzan A. Deep learning applied to ligand-based de novo drug design. Artificial Intelligence in Drug Design, 2022, 273–99.
[17] Pellegrini E, Field MJ. Development and testing of a de novo drug-design algorithm. Journal of Computer-Aided Molecular Design, 2003, 17, 621–41.
[18] Rehman AU, Lu S, Khan AA, Khurshid B, Rasheed S, Wadood A, Zhang J. Hidden allosteric sites and De-Novo drug design. Expert Opinion on Drug Discovery, 2022, 17(3), 283–95.
[19] Schneider G, Baringhaus KH. De novo design: From models to molecules. De novo Molecular Design, 2013, 1–55.
[20] Jain SK, Agrawal A. De novo drug design: An overview. Indian Journal of Pharmaceutical Sciences, 2004, 66(6), 721–8.
[21] Li Y, Pei J, Lai L. Structure-based de novo drug design using 3D deep generative models. Chemical Science, 2021, 12(41), 13664–75.
[22] Krishnan SR, Bung N, Vangala SR, Srinivasan R, Bulusu G, Roy A. De novo structure-based drug design using deep learning. Journal of Chemical Information and Modeling, 2021, 62(21), 5100–9.
[23] Wang M, Hsieh CY, Wang J, et al. Relation: A deep generative model for structure-based de novo drug design. Journal of Medicinal Chemistry, 2022, 65(13), 9478–92.
[24] Yuan Y, Pei J, Lai L. Binding site detection and druggability prediction of protein targets for structure-based drug design. Current Pharmaceutical Design, 2013, 19(12), 2326–33.
[25] Seeliger D, de Groot BL. Ligand docking and binding site analysis with PyMOL and Autodock/Vina. Journal of Computer-Aided Molecular Design, 2010, 24(5), 417–22.
[26] Halgren TA. Identifying and characterizing binding sites and assessing druggability. Journal of Chemical Information and Modeling, 2009, 49(2), 377–89.
[27] Palazzesi F, Pozzan A. Deep learning applied to ligand-based de novo drug design. Artificial Intelligence in Drug Design, 2022, 273–99.
[28] Ajjarapu SM, Tiwari A, Ramteke PW, Singh DB, Kumar S. Ligand-based drug designing. In: Bioinformatics. Academic Press, 2022, 233–252.
[29] Huang HJ, Lee KJ, Yu HW, et al. Structure-based and ligand-based drug design for HER 2 receptor. Journal of Biomolecular Structure and Dynamics, 2010, 28(1), 23–37.
[30] Schneider G, Schneider P. Coping with complexity in ligand-based de novo design. In: Frontiers in Molecular Design and Chemical Information Science-Herman Skolnik Award Symposium 2015: Jürgen Bajorath. American Chemical Society, 2016, 143–158.

Chapter 12
Homology modeling

> Homology modeling is like molecular 3D printing, allowing us to create structural models of proteins using the blueprints of evolution. – Iddo Friedberg

12.1 Introduction

Homology modeling, also known as comparative modeling, is a computational technique used in structural biology to predict the 3D structure of a protein based on the known structure of a homologous protein [1]. Homology modeling plays a pivotal role in computer-aided drug designing (CADD) [2]. This technique is fundamental in the rational design of pharmaceutical compounds and the exploration of drug–protein interactions [3, 4]. At its core, homology modeling is a method used to predict the 3D structure of a protein based on its sequence similarity to a known protein with a resolved structure. In the context of CADD, this means that homology modeling enables researchers to create 3D models of target proteins, such as receptors or enzymes, when experimental structural data is unavailable [5]. These models serve as a foundation for virtual screening, de novo drug designing, ligand docking, and molecular dynamics simulations [6].

The process typically involves multiple steps, starting with the identification of a suitable template protein with a known structure. Sequences are aligned, and the target protein's structure is built based on the template [7]. Energy minimization and refinement follow to optimize the model's geometry. Once a reliable model is obtained, it can be employed to study how potential drug compounds interact with the target, aiding in the selection of promising drug candidates [8]. Homology modeling within CADD significantly accelerates the drug discovery process by reducing the reliance on costly and time-consuming experimental structure determination. It enhances our ability to understand the molecular basis of diseases and design drugs with high specificity and affinity, thereby contributing to the development of novel therapeutics [9, 10].

The 3D structure of a receptor protein is of indispensable requirement for adopting any structural-based drug-designing approach. This 3D structure of the target protein gives us an idea about the binding regions of that protein which is quite helpful in CADD approaches, whether molecular docking, molecular dynamics, or de novo drug designing. The 3D structures of proteins are available at the protein data bank (www.rcsb.org) of the Research Collaboratory for Structural Bioinformatics (RCSB). As per the data of RCSB, there are around 1.8 lakh entries of the proteins in this databank as of 2023 [11]. It is estimated that every year around 8,000 to 10,000 new entries of the protein structures are added to the RCSB protein data bank. Despite the availability of these huge numbers of proteins, there is still a large pool of protein molecules whose 3D structure is not known [12]. If a researcher is interested in performing a

https://doi.org/10.1515/9783111434858-012

structure-based drug-designing approach against any of the target receptors whose structure is not known or not available at the protein databank, then he will have no choice other than to cease his idea of research in the budding stage only. To rescue researchers from such a situation, a new technique, known as homology modeling or comparative modeling, has emerged. Homology modeling is based on the principle that similar proteins fold similarly [13]. This principle of homology modeling also finds a crucial application in dealing with the genetic mutations in the receptor protein, the structure of which is different from the natural protein.

12.2 Structure of a protein

To understand homology modeling, one should have a basic knowledge of the structure of the proteins so that each step of the homology modeling can be understood easily. Amino acids are considered as the building blocks of the protein. Amino acids are small organic molecules in which a carbon atom is present in the center. The valency of the carbon is 4, which enables it to form four covalent bonds with other atoms. In amino acids, the carbon atom forms one covalent bond each with a hydrogen atom, amino group (NH_2), carboxylic acid group (COOH), and one hetero group that is specific to the amino acid [14–17]. Each amino acid binds with another amino acid through a peptide bond (–CONH–) (Fig. 12.1).

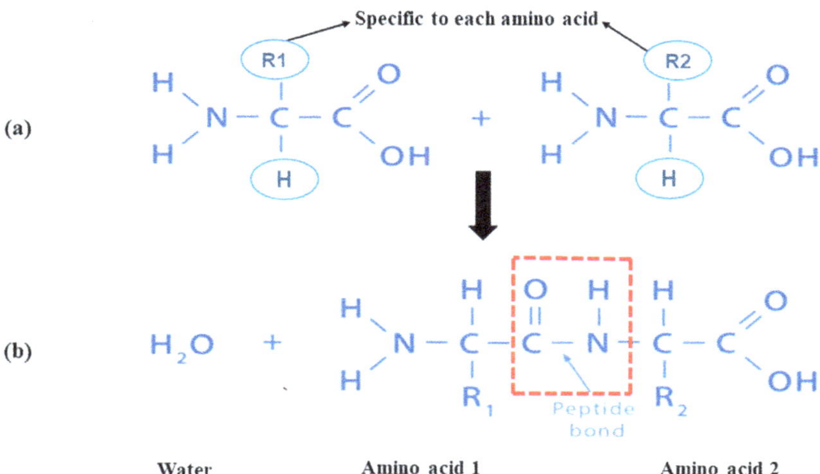

Fig. 12.1: Simple peptide bond between two amino acids.

12.2.1 Primary structure of the protein

The primary structure of a protein is its most fundamental level of structural organization. It refers to the linear sequence of amino acids that make up the protein [14]. These amino acids are linked together by peptide bonds, forming a chain-like structure. The primary structure is critical because it dictates all higher levels of the protein structure and ultimately determines the protein's function. The sequence of amino acids encodes the genetic information necessary for the protein's synthesis. Even a minor change, such as a single amino acid substitution, can have a profound impact on a protein's function and can be responsible for genetic disorders or diseases. In essence, the primary structure provides the blueprint for the protein's folding into its secondary, tertiary, and quaternary structures, which, in turn, determine the protein's unique shape and functionality [18]. Understanding the primary structure is a foundational aspect of biochemistry and molecular biology, as it holds the key to deciphering the complex functions and interactions of proteins in living organisms (Fig. 12.2).

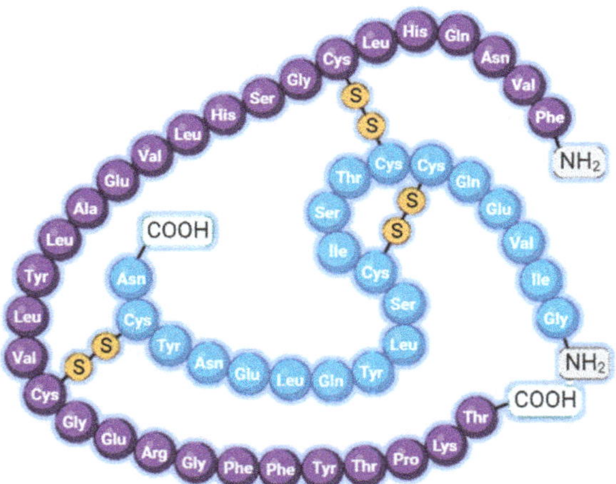

Fig. 12.2: Primary structure of the protein, represented by the linear sequence of the amino acids inside the protein.

12.2.2 Secondary structure of the protein

The secondary structure of a protein is a fundamental aspect of its 3D conformation and is primarily characterized by the local folding patterns within the polypeptide chain. The sequence of the amino acids does not exist as a straight chain inside the protein, instead they arrange themselves in a specific manner in the space. This arrangement of the chain of amino acids is referred to as the secondary structure of the protein. Secondary structures are essential in protein folding, as they determine the overall protein shape and its functional properties. Understanding the secondary structure of a protein is critical in predicting its biological activity, interactions with other molecules, and its role in various cellular processes. The two most common types of secondary structure elements in proteins are α-helices and β-strands [17] (Fig. 12.3).

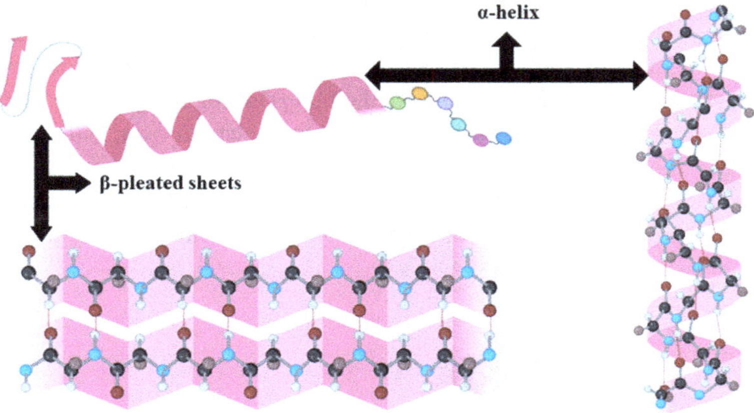

Fig. 12.3: Secondary structure of the protein, represented by the α-helixes and β-strands, formed from the linear sequences of the amino acids.

- **α-helixes**: This is the most common type of the secondary structure of most of the proteins. In this type of arrangement, the polypeptide chain of the amino acids exists in the form of a right-hand-sided coil. In α-helices, the peptide backbone forms a spiral, stabilized by hydrogen bonds between the carbonyl oxygen of one amino acid and the amide hydrogen of an amino acid further along the chain. These helices are stable and present in many proteins, providing structural integrity.
- **β-strands**: In this type of arrangement, two or more chains of the amino acid sequences lie parallel to each other in the form of a sheet. β-strands are extended segments where the peptide backbone forms a zigzag pattern. These strands can align and interact through hydrogen bonds to create β-sheets, contributing to the protein's stability and forming important structural motifs.

12.2.3 Tertiary structure of the protein

The tertiary structure of the protein arises from the further folding of the secondary structure of the protein (Fig. 12.4).

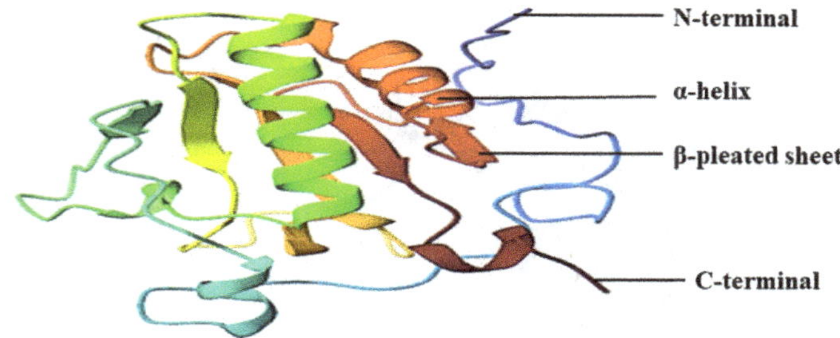

Fig. 12.4: Tertiary structure of the protein that is formed by the arrangement of different secondary structures of the protein.

The tertiary structure of a protein is the third and most complex level of protein structure, following the primary and secondary structures. It defines the precise 3D arrangement of a protein's atoms, including the spatial orientation of its secondary structural elements (α-helices and β-sheets). This intricate 3D conformation is crucial for a protein's function and biological activity. The tertiary structure is primarily stabilized by various noncovalent interactions, such as hydrogen bonds, van der Waals forces, electrostatic attractions, and hydrophobic interactions. These forces collectively contribute to the protein's stability and its ability to fold into a specific, functional shape. The unique tertiary structure of a protein determines its active sites, binding pockets, and overall architecture, allowing it to carry out its biological functions, whether as an enzyme catalyzing the chemical reactions, a receptor binding to specific molecules, or a structural component of cells and tissues. Alterations in the tertiary structure can lead to a loss of function and can be associated with diseases, emphasizing the critical role of protein folding in biology [15].

After the folding of the secondary structure of the protein, the tertiary protein either folds fibrous form or a globular form. Fibrous protein, such as keratin, is the tertiary structure in which polypeptide chains of the secondary run parallel to each other. Globular proteins are those in which polypeptide chains are folded in a coil-like structure. Human insulin is one such example of a globular protein.

12.2.4 Quaternary structure of the protein

The quaternary structure of a protein refers to its arrangement and interactions with other protein subunits, forming a complex, multi-subunit protein assembly (Fig. 12.5).

This level of structural organization is crucial in understanding a protein's overall function, as many biologically significant proteins exist as complexes rather than as solitary molecules. The quaternary structure is commonly observed in proteins with multiple polypeptide chains or subunits. These subunits may be identical (homomers) or different (heteromers) and are held together by various types of interactions, such

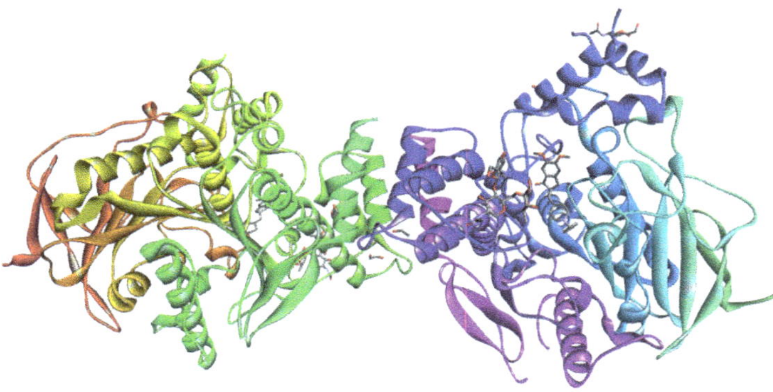

Fig. 12.5: Quaternary structure of protein formed by the assembly of two or more quaternary structures of the proteins.

as hydrogen bonds, ionic bonds, disulfide bridges, and hydrophobic interactions. The quaternary structure is responsible for stabilizing the protein complex and can have a profound impact on the protein's function. For example, hemoglobin, an essential oxygen-carrying protein in blood, exhibits a quaternary structure composed of four globin subunits. Each subunit binds to oxygen, and their arrangement allows efficient oxygen transport. Understanding the quaternary structure is vital in comprehending the cooperative behavior of such proteins and provides insights into their roles in biological processes.

One must understand that the overall structure of the protein is always decided by its sequence of amino acids, that is, from its primary structure [18].

12.3 Steps involved in the process of homology modeling

As already discussed, homology or comparative modeling is the process of modeling the 3D structure of a protein with the help of another protein whose 3D structure is known [19]. The basic flow diagram of the whole methodology of the homology modeling is given in Fig. 12.6.

12.3.1 Identification of the sequence of the amino acids of the query structure protein

Query protein is the protein whose 3D structure is not known and is to be modeled. To build or model the structure of the query protein, the very first step is to identify the sequence of the amino acids present in the protein. The structure of a protein is dependent on the amino acid sequence or the primary structure of the protein. There-

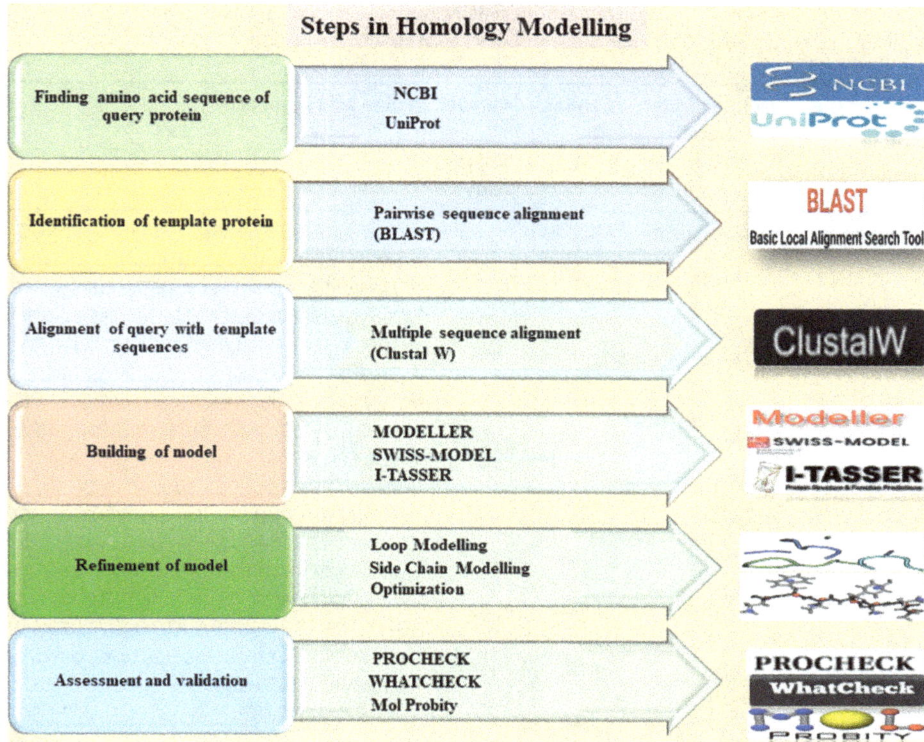

Fig. 12.6: Workflow of the steps involved in the process of homology modeling.

fore, this step is of utmost importance as a wrong sequence can lead to a false result in the structure generation of the protein.

There are several databases available for finding the amino acid sequence of the proteins and the most popular and commonly employed databases are Uniprot (www.uniprot.org) and National Center for Biotechnology Information (NCBI) protein databases [20, 21]. Let us understand this with an example of dehydrogenases. Dehydrogenases are enzymes employed for the removal of hydrogen in different biochemical reactions of the body. Table 12.1 depicts the amino acid sequence of one such dehydrogenase enzyme isolated from the species *Mucilaginibacter terrenus*.

This amino acid sequence is obtained from the protein databank of the NCBI. Here, every single letter represents an amino acid in the sequence, as each amino acid can be represented by its unique single-letter code.

Tab. 12.1: The amino acid sequence of the protein dehydrogenase enzyme isolated from the species Mucilaginibacter terrenus. One letter code represents each specific amino acid.

Name of the protein	Isolated from the species	GenBank ID	Number of amino acids	Amino acid sequence
Dehydrogenase	*Mucilaginibacter terrenus*	RFZ81413	658	MIFDRKNIDNDGLITFYKKLLLPRLVEEKMLILLRQG RISKWFSGIGQEAIAVGSTLAMQPDEYILPMHRNLG VFTARNIPISRLMAQWQGKMSGFTRGRDRSFHFGT QEYKIVGMISHLGPQLALADGIALADKLSGQQKVTL VYTGEGATSEGDFHEALNIAAVWDLPVIFLVENNGY ALSTPTNEQYRCEALADRAIGYGIEGRRINGNNILEV YDTVNELAQDMRRKPRPALLECMTFRMRGHEEASG TKYVPQHLFEEWRQHDPLDNFEQYLLDEQVLRKEW VPYLREEFTKLIDIEIEKVFNEPDIVPHAQTELNSVYRT YHKPEPKPYEVLSGKRYIDAISDGLRQSMRKYSNLVI MGQDIAEYGGAFKITQGFVEEFGKERVRNTPICESGI VGAAMGLALNGYKAVVEMQFADFVSSGFNQVVNN LAKTYYRWGQHVDVVIRMPTGAGTGAGPFHSQSNE AWFTKTPGLKVVYPAFPDDAKGLLMSAIEDPNPVMY FEHKYLYRSTTGNVPDDDHYVEIGKANIVRKGTKASI ITYGLGVHWAMEYADNHPEQSIEILDLRSLQPWDKE AVEKTVMKTSKVLILHEDTLTSGFGGELAAYIGEHLF KYLDAPVMRCGSLDTPIPMNKALEDQFLAKARLEEF MARLLSF

12.3.2 Identification of the template protein

Template protein is the protein whose 3D structure is known and has sequence similarity with the query protein. The next step in homology modeling is to find the template protein whose 3D structure is known and has the highest similarity with the query protein in terms of the amino acids. It is an unsaid acceptable criterion that there should be a minimum 45% similarity match of the template protein sequence with the query protein to obtain a 3D structure that can be further employed for structure-based drug designing. Here, one should also understand that the higher the similarity of the sequence, higher will be the accuracy in the determination of the 3D structure of the query protein. To identify the template protein, a process known as pairwise sequence alignment (PSA) is performed. When alignment is done between the two pairs of proteins, it is termed as PSA. To identify the template proteins that have similarities with the query protein, every protein sequence with known structures is put under PSA with the query protein. The template structures that show high similarity with the query protein are then selected for further preparation of the 3D structure of the query protein.

The template structures are identified from the Basic Local Alignment Tool (BLAST) server of the NCBI. Figure 12.7 depicts the top six results obtained for the template structure obtained from the BLAST server for the dehydrogenase enzyme, isolated from the species *Mucilaginibacter terrenus.*

It is evident from Fig. 12.7 that the lowest identity with the query is 39.68%, which covers 46% of the protein and 244 similar amino acids in similarity; whereas the highest identity of the query is 40% which also covers 46% of the protein and 248 of total amino acids. The graphical representation of these top six identical proteins is given in Fig. 12.8, which depicts the regions where the sequence of these template proteins matches with the query protein [22].

12.3.3 Alignment of the query protein with template protein structures

The next step in the homology modeling is performing the alignment of the selected template proteins with the query protein. When alignment is performed between more than two proteins, it is referred to as the multiple sequence alignment (MSA). MSA is performed to identify the regions of similarity between the query protein and the different template structures. In the MSA, regions of different template proteins that show similarity with the query protein are aligned with the amino acid sequence of the query protein. MUSCLES, ClustalW, and so on are the popular software or web servers for performing the MSA. The main objective of the MSA is to identify the regions of similarity between the query and template proteins, based on which the 3D structure of the query protein will be predicted [23].

Figure 12.8 already depicts the top six protein templates that show similarity with the query protein. By closely observing these top six results, we notice that all have demonstrated similarity in the same regions of the query protein. Figure 12.9 is the graphical representation of all the aligned target proteins with the query protein along with the regions of similarity. In Fig. 12.9, it is visible that the PDB ID 2J9F is showing high similarity in the starting region of the query protein and the PDB ID 1DTW is showing high similarity with the later regions of the query protein. Figure 12.10 depicts the MSA of the query protein with the two selected template proteins. The dots in the template protein sequence show the regions of no similarity with the query protein.

12.3.4 Building the model of the query protein

Once the alignment of the template proteins is done with the query protein, the next and the most important step is to create the alignment-based 3D model of the query protein. There are different algorithms available for creating the 3D model of the query protein.

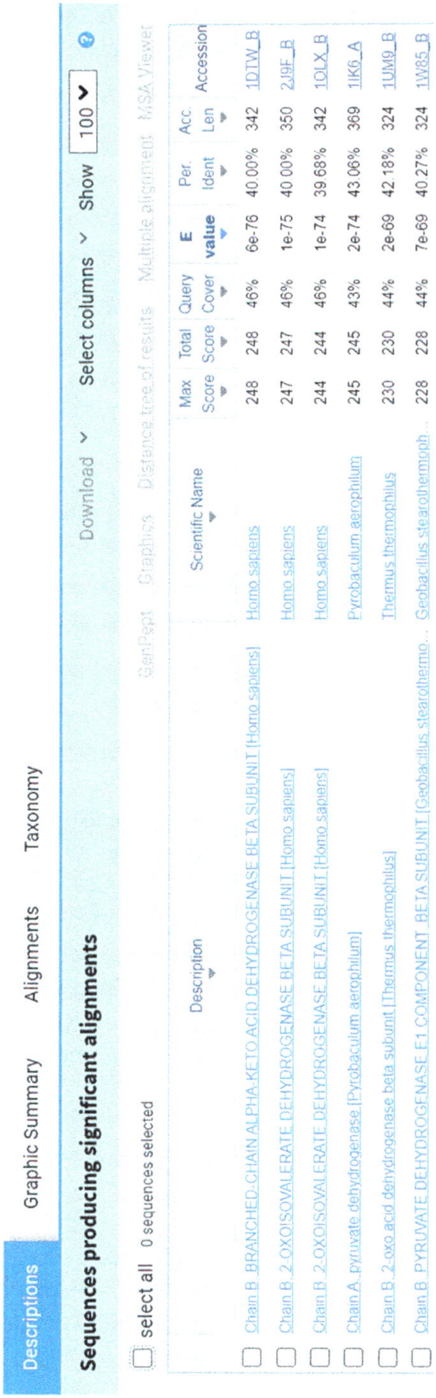

Descriptions | Graphic Summary | Alignments | Taxonomy

Sequences producing significant alignments

Download | Select columns | Show 100

select all 0 sequences selected

GenPept Graphics Distance tree of results Multiple alignment MSA Viewer

Description	Scientific Name	Max Score	Total Score	Query Cover	E value	Per. Ident	Acc. Len	Accession
Chain B, BRANCHED-CHAIN ALPHA-KETO ACID DEHYDROGENASE BETA SUBUNIT [Homo sapiens]	Homo sapiens	248	248	46%	6e-76	40.00%	342	1DTW_B
Chain B, 2-OXOISOVALERATE DEHYDROGENASE BETA SUBUNIT [Homo sapiens]	Homo sapiens	247	247	46%	1e-75	40.00%	350	2J9F_B
Chain B, 2-OXOISOVALERATE DEHYDROGENASE BETA SUBUNIT [Homo sapiens]	Homo sapiens	244	244	46%	1e-74	39.68%	342	1OLX_B
Chain A, pyruvate dehydrogenase [Pyrobaculum aerophilum]	Pyrobaculum aerophilum	245	245	43%	2e-74	43.06%	369	1IK6_A
Chain B, 2-oxo acid dehydrogenase beta subunit [Thermus thermophilus]	Thermus thermophilus	230	230	44%	2e-69	42.18%	324	1UM9_B
Chain B, PYRUVATE DEHYDROGENASE E1 COMPONENT, BETA SUBUNIT [Geobacillus stearothermo...	Geobacillus stearothermoph...	228	228	44%	7e-69	40.27%	324	1W85_B

Fig. 12.7: Top six proteins showing similarity with the amino acid sequence of our query protein.

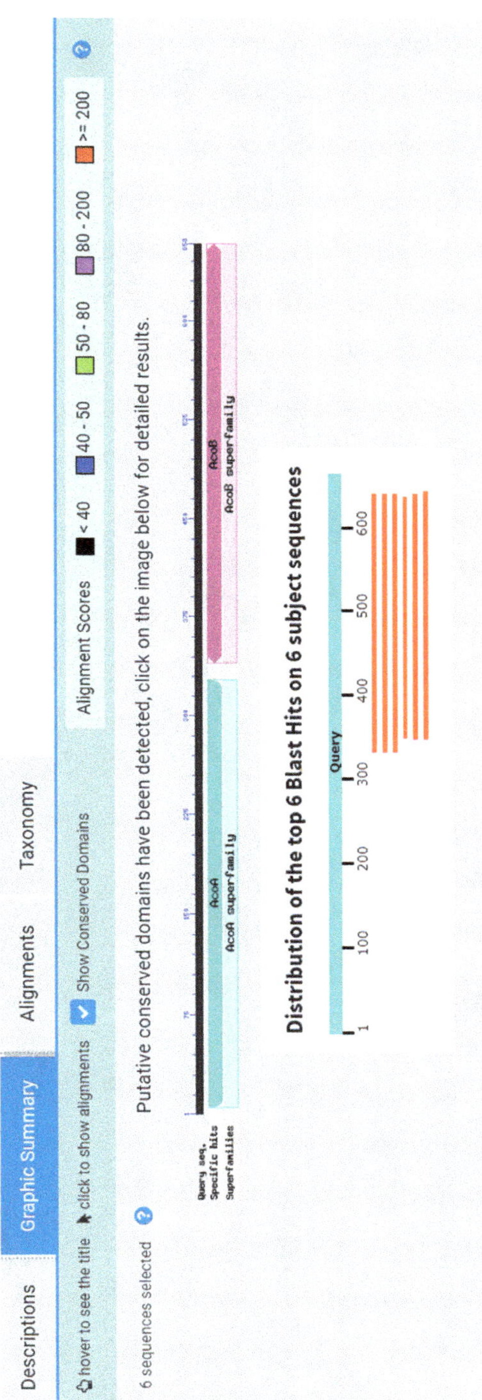

Fig. 12.8: The regions of the top six similar proteins where they show similarity with the query protein. The red regions of these proteins represent the regions of high similarity with our query protein.

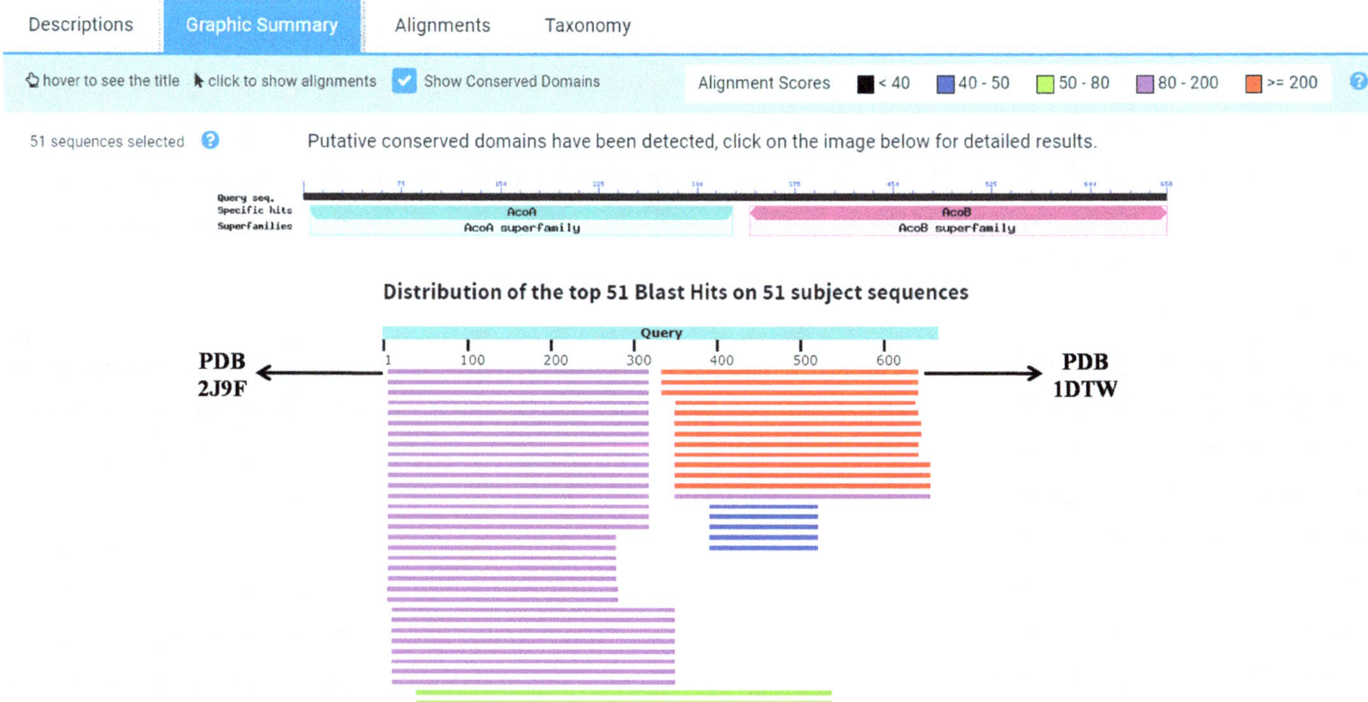

Fig. 12.9: Sequencing result of all the proteins similar to the query protein. The protein with PDB ID 2J9F shows similarity in the starting region and the protein with PDB ID 1DTW shows similarity in the later region of the query protein.

```
>P1;QUERY

MIFDRKNIDNDGLITFYKKLLLPRLVEEKMLILLRQGRISKWFSGIGQEAIAVGSTLAMQ
PDEYILPMHRNLGVFTARNIPISRLMAQWQGKMSGFTRGRDRSFHFGTQEYKIVGMISHL
GPQLALADGIALADKLSGQQKVTLVYTGEGATSEGDFHEALNIAAVWDLPVIFLVENNGY
ALSTPTNEQYRCEALADRAIGYGIEGRRINGNNILEVYDTVNELAQDMRRKPRPALLECM
TFRMRGHEEASGTKYVPQHLFEEWRQHDPLDNFEQYLLDEQVLRKEWVPYLREEFTKLID
IEIEKVFNEPDIVPHAQTELNSVYRTYHKPEPKPYEVLSGKRYIDAISDGLRQSMRKYSN
LVIMGQDIAEYGGAFKITQGFVEEFGKERVRNTPICESGIVGAAMGLALNGYKAVVEMQF
ADFVSSGFNQVVNNLAKTYYRWG---QHVDVVIRMPTGAGTGAGPFHSQSNEAWFTKTPG
LKVVYPAFPDDAKGLLMSAIEDPNPVMYFEHKYLYRSTTGNVPDDDHYVEIGKANIVRKG
TKASIITYGLGVHWAMEYADNHPEQS---IEILDLRSLQPWDKEAVEKTVMKTSKVLILH
EDTLTSGFGGELAAYIGEHLFKYLDAPVMRCGSLDTPIPMNKALEDQFLAKARLEEFMAR
LLSF
*
>P1;2J9F

------HLPKEKVLKLYKSMTLLNTMDRILYESQRQGRISFYMTNYGEEGTHVGSAAALD
NTDLVFGQYREAGVLMYRDYPLELFMAQCYGNISDLGKGRQMPVHYGCKERHFVTISSPL
ATQIPQAVGAAYAAKRANANRVVICYFGEGAASEGDAHAGFNFAATLECPIIFFCRNNGY
AISTPTSEQYRGDGIAARGPGYGIMSIRVDGNDVFAVYNATKEARRRAVAENQPFLIEAM
TYRIGHHSTSDDSSAYRPVDEVNYWDKQDHPISRLRHYLLSQGWWDEEQEKAWRKQSRRK
VMEAFEQAERKPKPNPN-------------------------------------------
------------------------------------------------------------
------------------------------------------------------------
------------------------------------------------------------
------------------------------------------------------------
------------------------------------------------------------
*
>P1;1DTW

------------------------------------------------------------
------------------------------------------------------------
------------------------------------------------------------
------------------------------------------------------------
------------------------------------------------------------
-----------------------------PEPREYGQTQKMNLFQSVTSALDNSLAKDPT
AVIFGEDVAFGG-VFRCTVGLRDKYGKDRVFNTPLCEQGIVGFGIGIAVTGATAIAEIQF
ADYIFPAFDQIVNEAAKYRYRSGDLFNCGSLTIRSPWGCVGHGALYHSQSPEAFFAHCPG
IKVVIPRSPFQAKGLLLSCIEDKNPCIFFEPKILYRAAAEEVPIEPYNIPLSQAEVIQEG
SDVTLVAWGTQVHVIREVASMAKEKLGVSCEVIDLRTIIPWDVDTICKSVIKTGRLLISH
EAPLTGGFASEISSTVQEECFLNLEAPISRVCGYDTPFP---------------------
----
*
```

Fig. 12.10: Multiple sequence alignment results of query protein with the top two similar target proteins (with the proteins having PDB ID 2J9F and 1DTW).

12.3.4.1 Rigid body assembly method

In this approach, rigid regions are first identified inside the aligned template structures. These rigid regions of the template structures that show similarity with the query protein are then used for building a backbone of the query structure. Once the backbone is created, other nonconserved regions, which are loop and side chains, are assembled on this backbone iteratively to obtain the final 3D structure of the query protein. This algorithm is used by programs that were developed initially for homology modeling. 3D-JIGSAW, SWISS-MODEL, and BUILDER are the most widely used and popular programs for model building based on this algorithm [24].

12.3.4.2 Segment matching approach

This method is slightly different as compared to the conventional methods of model building. In this approach, first, the query protein sequence is divided into small segments. The structures of these small segments are then searched in the databases of known structures, based on the energy function, geometry constraints, or a combination of both [25]. Once the structure of the small segments is obtained from the database, then these segments are combined to form the backbone of the query protein, followed by developing loop and side chains on the backbone. SegMod/ENCAD is one such program used for the segment-matching-based modeling of the proteins.

12.3.4.3 Satisfaction of spatial restraint method

Based on the structure of the template protein, certain spatial restraints are applied to the query protein for building its 3D structures. The restrains that can be applied in this type of modeling include:

- **Angle restraints**: This is the allowed angles between the atoms of a molecule. If the template structure follows certain fixed bond angles, then this restrain is applied to the query protein model building.
- **Distance restraints**: This restrain is applied when a desired bond length is required between the atoms of query protein based on the pattern observed from the template protein.
- **Dihedral angle restrains**: The dihedral angle is another important criterion that is taken into consideration while developing the query protein model. A certain value of the torsion angle is set that must be satisfied by the query protein while building its 3D structure.

Once the model is created based on the spatial requirement elucidated from the template structure, further energy minimization of the 3D structure of the query protein is performed, based on molecular mechanics. MODELLER is the most common and very popular software for the homology modeling works using this algorithm [26].

12.3.4.4 Artificial evolution approach

Artificial evolution approach is a hybrid approach to homology modeling in which both the rigid body assembly method and stepwise template evolutionary mutation are employed simultaneously. The mutation in the template is done until it matches with the sequence of the query protein. NEST is the program that is based on this algorithm for generating the structure of the unknown proteins [27].

Figure 12.11 depicts the 3D model developed of the earlier employed query protein, dehydrogenase enzyme isolated from the species *Mucilaginibacter terrenus*. This model is developed using the SWISS-MODEL program as explained above.

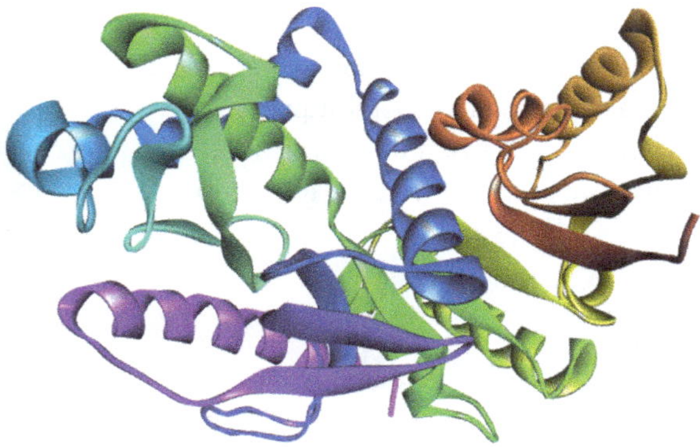

Fig. 12.11: The protein tertiary structure derived from homology modeling of our query protein dehydrogenase enzyme isolated from the species *Mucilaginibacter terrenus*.

12.3.5 Refinement of the model developed

This is a very important and crucial step that should be given proper attention for the development of a high-quality model of the query protein. The main objective of this step is to provide further refinement to the generated structure through different approaches. Basically, three steps are performed during the refinement of the generated model, including modeling of the loop, side chain modeling, and tuning of the final structure of the query protein [28–30].

12.3.5.1 Modeling of the loops

Loops are the regions of the query protein that are not conserved. The sequence of amino acids in these loops has no evolutionary similarity with any template and hence their alignment cannot be performed. Loops are the regions of the structure that combine two secondary structural parts of the protein (α-helix and β-strand). On average, about one-third region of the total structure of the protein corresponds to the region of the loop. Modeling of the loops is also of utmost importance as sometimes these regions encode for a special feature of the query protein. In drug discovery, loops play an important role when the loop region coincides with the active binding site of the query protein. For the prediction of the loops, the following two approaches are employed.

12.3.5.1.1 Database search approach

This is one of the most common methods employed for loop modeling and is also known by the name knowledge-based, template-based, or homology-based approach. In this method, the sequence of amino acids of the query protein is compared with the structures of loops having similar sequences available in the database. This method is suitable for modeling small-length loops. The most widely acceptable loop modeling servers, based on this approach, are GalaxyLoop (https://galaxy.seoklab.org/cgi-bin/submit.cgi?type=LOOP), DaReUS Loop (https://bioserv.rpbs.univ-paris-diderot.fr/services/DaReUS-Loop/), and Fread (https://opig.stats.ox.ac.uk/webapps/sabdab-sabpred/sabpred/more/#FREAD).

12.3.5.1.2 Ab initio method of loop modeling

Ab initio method of loop modeling is an exhaustive method for modeling the loop region of the query proteins. In this approach, a conformational search of the loops is performed based on the force field-based scoring. The conformation that shows the least energy is considered the most acceptable loop. RCD+ (https://rcd.iqfr.csic.es/), LOOPY (https://doi.org/10.1073/pnas.102179699), MODELLER, and LoopBuilder are programs and web servers that employ the ab initio method of loop modeling.

12.3.5.1.3 Hybrid methods of loop modeling

In the hybrid methods of loop modeling, a hybrid algorithm is employed for generating the loops of the query protein. This type of program is developed on both the database and the ab initio algorithm for performing successful modeling of the loops. These types of programs offer high speed and can also be performed on a large number of residues containing loops. Sphinx (https://opig.stats.ox.ac.uk/webapps/sabdab-sabpred/sabpred/more/#Sphinx) is one such popular web server that is developed on the hybrid-based approach.

12.3.5.2 The modeling side chain of the model

Once the loops are constructed on the backbone of the structure of the query protein, the next computational challenge is to derive the side chains of amino acids on this backbone. Side-chain amino acids can exist in a limited number of possible conformations. Therefore, instead of considering all the possible conformational space, only possible limited conformations of the amino acids are evaluated. These conformers of the amino acid side chains are termed rotamers. Side chain predictability of the side chain of the backbone of the model developed is based on two components:

12.3.5.2.1 Rotamer library

With an increasing number of entries in the database of the protein data bank (PDB), it has made it possible for researchers to create separate libraries of the rotamers from this database. Two types of rotamers libraries are developed by different researchers, the backbone-dependent and the backbone-independent libraries. Out of these two, backbone-dependent library developed by Shapovalov and Dunbrack is the most reliable and widely accepted library.

12.3.5.2.2 Search methods and scoring

A search method is employed for searching the optimal rotamers of the amino acid chains. Accuracy and speed are two criteria based on which search methods are selected for finding out the best rotamers. The most popular methods employed for this purpose include simulated annealing-based search, dead-end elimination (DEE), and fast sparse phase retrieval (FASP).

Energy-based scoring of the rotamers is then performed to select the energetically most favorable rotamer. For this purpose, force field-based scoring is the most widely accepted method for scoring the rotamers. Programs such as RAMP (http://www.ram.org/computing/ramp), CIS-RR (http://jianglab.ibp.ac.cn/lims/cisrr/cisrr.html), and SCWRL (http://dunbrack.fccc.edu/Software.php) are used widely for the amino acid side chain modeling of the developed model.

12.3.5.3 Optimization of the developed model

The final developed model, after the insertion of loops and side chains into the backbone of the initial model, required further optimization to obtain the most energetically acceptable model. The initial structure of the query protein obtained through comparative modeling contains many errors in terms of the spatial arrangement of the atoms, bond angles, bond lengths, and torsion angles. This correction is obtained through various approaches to obtain a refined 3D model of the query protein. The main objective of this step is to obtain a model close to nearly experimental values in terms of spatial arrangement. The various approaches that are used for model optimization include:

12.3.5.3.1 Force field-based energy minimization

Force field-based energy minimization is the most popular approach for optimization of the structure of the query protein. The energy of the 3D structure of the query protein is minimized further with the help of physics-based/knowledge-based force fields to obtain an energetically favored structure. 3Drefine, i3Drefine, and Seok servers are the most popular tools or web servers employed for energy minimization of the initial model generated by the query protein.

12.3.5.3.2 Molecular dynamics (MD)-based optimization

Molecular dynamics-based optimization of the 3D structure of the query protein has gained a lot of popularity in the last decade. In this approach, Newton's law of motion is employed for refining the structure further using atom-based force fields. The first refinement of the approximate 3D structure of the query protein was reported in 2008 by Jiang Zhu. Replica-exchange molecular dynamics (REMD) technique is the most commonly employed method for the MD-based optimization of the raw structure of the query protein. Galaxyrefine (https://galaxy.seoklab.org/cgi-bin/submit.cgi?type=REFINE) is one such web server that does this MD-based refinement with good accuracy. Figure 12.12 depicts the 3D refined structure of the generated model of the dehydrogenase enzyme isolated from the species *Mucilaginibacter terrenus* using the Galaxyrefine server. It is evident by comparing Figs. 12.11 and 12.12 that there is a significant difference in the folding of the earlier and refined structures of the same query protein.

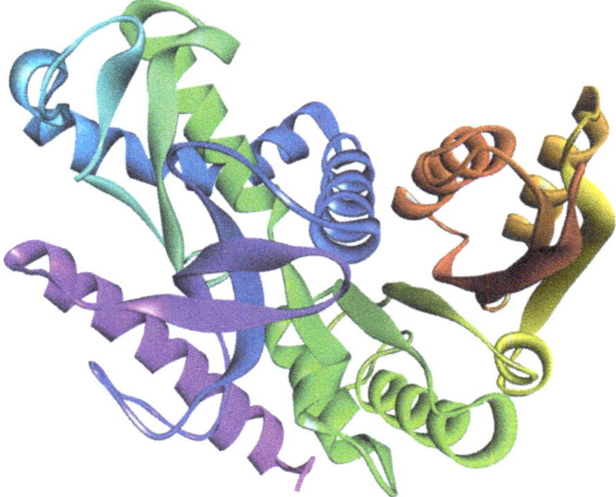

Fig. 12.12: Secondary structure of query protein after performing the refinement steps to the previously derived model.

12.3.5.3.3 Monte Carlo-based optimization of the structure

The Monte Carlo-based optimization approach is used as a supplement for further refinement of the model that is already optimized through any of the above two methods. Monte Carlo is a probability-based simple approach in which a new model is proposed from the initial model and the new model is accepted or rejected based on certain criteria. This whole process is repeated until the most acceptable model of the query protein is obtained.

12.3.6 Assessment and validation of the developed model

The whole process of homology modeling is based on combining many small processes to obtain a 3D structure of the protein. Inaccuracy in any of the processes or steps can lead to errors in the finally generated model of the query protein. Hence, assessment and validation of the final model of the query protein is of immense necessity. The developed model of the query protein is assessed and validated for its spatial aspects, which are done by analyzing the bond lengths, rotational angles, and torsion angles of the developed model. WhatCheck9 (https://swift.cmbi.umcn.nl/gv/whatcheck/), PROCHECK (https://www.ebi.ac.uk/thornton-srv/software/PROCHECK), and Molprobity (http://molprobity.biochem.duke.edu/) are three most important web servers for performing the assessment and validation of the final model. Table 12.2 depicts some of the important validation parameters of the generated 3D model calculated using the Molprobity server.

Tab. 12.2: Validation parameters of the generated 3D model calculated using the Molprobity server.

S. no.	Parameter	Value of our model	Acceptance criteria
1	Molprobity score	1.68 (90th percentile)	The structure in the 0th percentile is considered the worst predicted and the one in the 100th percentile is considered the best among the structures with similar resolution
2	Favored rotamers	275 (98.57%)	Should be greater than 98%
3	Amino acids in Ramachandran's most favored region	325 (94.50%)	Should be greater than 90%
4	Rama distribution *Z*-score	0.29 ± 0.42	Should be less than 02

All these parameters are crucial in assessing the quality of the generated model. Out of these, Ramachandran plot is of utmost importance showing us the number of amino acids falling in the energetically favorable and forbidden regions. Figure 12.13 depicts the Ramachandran plot of the generated 3D model of the dehydrogenase enzyme isolated from the species *Mucilaginibacter terrenus*. In this plot, we can see four regions, red-colored (most favored region), yellow-colored (additional favored region), light yellow-colored (generously allowed region), and white-colored (forbidden or disallowed region) [31].

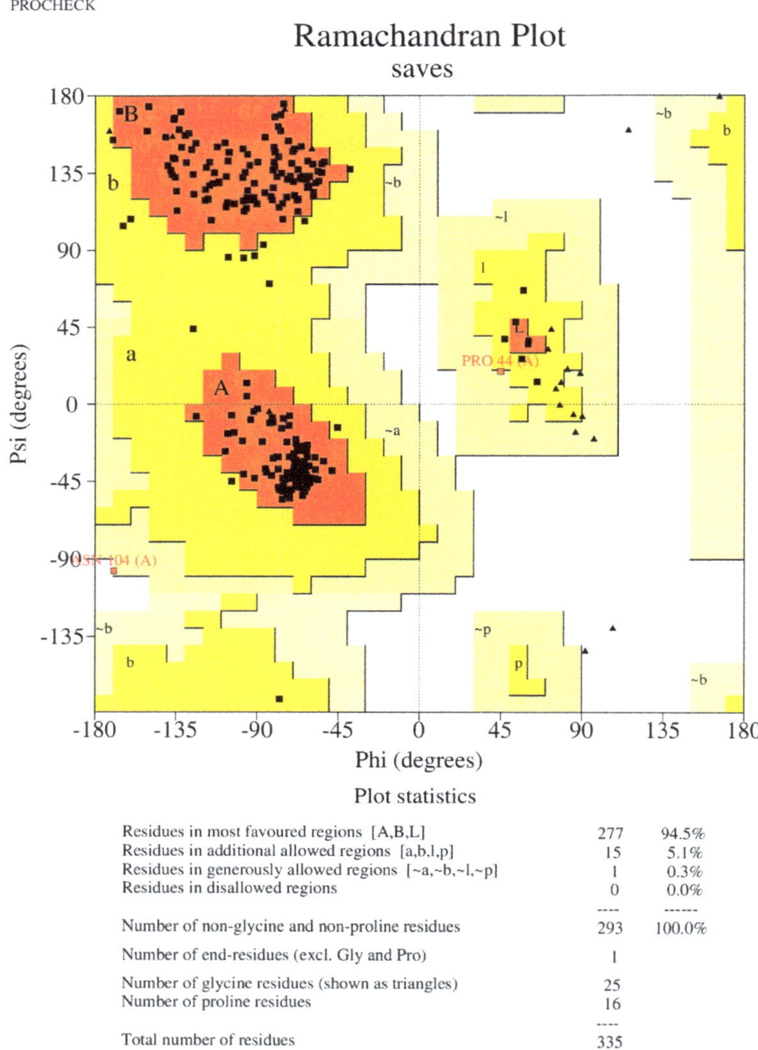

Based on an analysis of 118 structures of resolution of at least 2.0 Angstroms and R-factor no greater than 20%, a good quality model would be expected to have over 90% in the most favoured regions.

Fig. 12.13: Ramachandran plot of the query protein with 94.5% amino acids in the most favorable region.

12.4 Applications of homology modeling in CADD

The primary application of homology modeling lies in the prediction of 3D protein structures based on the knowledge of the structures of homologous proteins. Homology modeling is a versatile tool in CADD that aids in the rational design of drugs, understanding molecular interactions, and optimizing the drug development processes. It is particularly valuable when experimental protein structures are scarce or difficult to obtain. The following are some major applications of homology modeling in CADD.

- **Structure prediction:** Homology modeling is used to predict the 3D structures of target proteins, especially when experimental structures are unavailable. These predicted structures serve as a basis for understanding protein–ligand interactions.
- **Rational drug design:** Once the 3D structure of the target protein is predicted, it can be used to design small molecules (ligands) that interact with the protein's active site. This is crucial for the rational design of drugs and drug candidates.
- **Virtual screening:** Predicted protein structures can be employed in virtual screening experiments to identify potential drug candidates from large compound libraries. This helps in narrowing down the search for lead compounds.
- **Binding mode analysis:** Homology models enable the visualization and analysis of the binding modes between ligands and target proteins. This information is vital for optimizing the binding affinity and specificity of drug candidates.
- **Mutagenesis studies:** Homology modeling can be used to predict the effects of mutations in the target protein. This is valuable for understanding how genetic variations may impact drug responses and personalized medicine.
- **Protein–protein interaction studies:** Homology modeling can also be applied to predict the structures of protein complexes and protein–protein interactions. This is important in understanding various signaling pathways and designing drugs that target protein–protein interactions.
- **Drug resistance mechanisms:** Homology modeling helps in studying drug resistance mechanisms. By modeling the mutated target protein structures, researchers can design new drugs that can overcome resistance.
- **Enzyme mechanism elucidation:** It is used to elucidate the catalytic mechanisms of enzymes, which is important for designing enzyme inhibitors and studying metabolic pathways.
- **Biosimilars development:** In the development of biosimilars, homology modeling is used to ensure that the 3D structure of the biosimilar closely matches that of the original biological drug, ensuring similar efficacy and safety profiles.
- **Protein function prediction:** Homology modeling can provide insights into the function of uncharacterized proteins by comparing them to homologous proteins with known functions.

12.5 Advantages of homology modeling

Homology modeling is a versatile and valuable tool in computer-aided drug design, allowing researchers to make informed decisions and streamlining the drug development process. Homology modeling offers some key advantages over the conventional drug designing process, which are depicted below.

- **Rapid structure prediction:** Homology modeling allows for the quick prediction of protein structures, saving time in drug development.
- **Cost-effective:** It is a cost-effective alternative to experimental X-ray crystallography or NMR spectroscopy for structure determination.
- **Facilitates rational drug design:** Knowing the 3D structure of a target protein enables rational drug design, optimizing the development process.
- **Structural insights:** It provides insight into the active site and binding pocket of a target protein, which is crucial for designing effective drugs.
- **Polypharmacology:** It aids in the identification of potential off-target effects, helping to design safer drugs with fewer side effects.
- **Virtual screening:** Homology models can be used in virtual screening to identify potential drug candidates for a specific target.
- **Prediction of protein mutations:** It can predict the effects of mutations in a protein, aiding in the understanding of drug resistance and personalized medicine.
- **Advantageous for membrane proteins:** It is particularly useful for predicting the structures of membrane proteins, which are challenging to study experimentally.
- **High accuracy:** With high sequence similarity between the target and template, homology models can be remarkably accurate in predicting the overall structure.
- **Enhanced drug optimization:** It can guide structure-based drug optimization by identifying potential binding sites and interactions for lead compounds.

12.6 Limitations of homology modeling

Homology modeling, while a valuable tool in computer-aided drug design, has several limitations that researchers should be aware of. It is important to consider these limitations when using homology modeling in computer-aided drug design and validate the resulting models through experimental data and other computational methods, when possible. The following are some of the major limitations of homology modeling in CADD:

- **Sequence similarity requirement:** Homology modeling relies on the availability of a sufficiently similar template protein structure. If a suitable template is not available, modeling becomes challenging or unreliable.
- **Inaccurate modeling for highly divergent sequences:** Homology modeling is less effective when modeling proteins with low sequence similarity to known structures, leading to potentially inaccurate models.

- **Incorrect loop modeling:** Loops, especially long and flexible ones, can be challenging to model accurately, often resulting in inaccuracies in the final structure.
- **Domain and multidomain proteins:** Homology modeling struggles with proteins that consist of multiple domains, as the arrangement and orientation of these domains may not be predicted accurately.
- **Posttranslational modifications and nonstandard residues:** Homology modeling does not account for posttranslational modifications or nonstandard amino acids, limiting its applicability in certain cases.
- **Incomplete or incorrect template structures:** If the selected template structure contains errors, gaps, or missing regions, these inaccuracies can be propagated to the homology model.
- **Limited applicability for membrane proteins:** Membrane proteins have distinct structural features, such as hydrophobic transmembrane domains, which are challenging to model accurately using standard homology modeling techniques.
- **Dynamic or disordered regions:** Proteins with highly dynamic or disordered regions may not be accurately represented in the homology model, leading to potential inaccuracies.
- **Ligand interactions:** Homology modeling does not predict ligand interactions or binding sites accurately, making it less suitable for drug design without additional information.
- **Challenging cases:** There are some proteins for which homology modeling may not be suitable due to unique structural features, and alternative modeling techniques or experimental methods are required.

References

[1] Summers NL, Karplus M. Construction of side-chains in homology modelling: Application to the C-terminal lobe of rhizopuspepsin. Journal of Molecular Biology, 1989, 210(4), 785–811.

[2] Hillisch A, Pineda LF, Hilgenfeld R. Utility of homology models in the drug discovery process. Drug Discovery Today, 2004, 9(15), 659–69.

[3] E Lohning A, M Levonis S, Williams-Noonan B, S Schweiker S. A practical guide to molecular docking and homology modelling for medicinal chemists. Current Topics in Medicinal Chemistry, 2017, 17(18), 2023–40.

[4] Launay G, Simonson T. Homology modelling of protein-protein complexes: A simple method and its possibilities and limitations. BMC Bioinformatics, 2008, 9(1), 1–6.

[5] Aszódi A, Taylor WR. Homology modelling by distance geometry. Folding and Design, 1996, 1(5), 325–34.

[6] Vyas VK, Ukawala RD, Ghate M, Chintha C. Homology modeling a fast tool for drug discovery: Current perspectives. Indian Journal of Pharmaceutical Sciences, 2012, 74(1), 1.

[7] Wiltgen M. Algorithms for Structure Comparison and Analysis: Homology Modelling of Proteins. Elsevier, Cambridge, 2018.

[8] Valanciute A, Nygaard L, Zschach H, Jepsen MM, Lindorff-Larsen K, Stein A. Accurate protein stability predictions from homology models. Computational and Structural Biotechnology Journal, 2023, 21, 66–73.
[9] Cavasotto CN, Phatak SS. Homology modeling in drug discovery: Current trends and applications. Drug Discovery Today, 2009, 14(13–14), 676–83.
[10] França TC. Homology modeling: An important tool for the drug discovery. Journal of Biomolecular Structure and Dynamics, 2015, 33(8), 1780–93.
[11] https://www.rcsb.org/
[12] Hillisch A, Pineda LF, Hilgenfeld R. Utility of homology models in the drug discovery process. Drug Discovery Today, 2004, 9(15), 659–69.
[13] E Lohning A, M Levonis S, Williams-Noonan B, S Schweiker S. A practical guide to molecular docking and homology modelling for medicinal chemists. Current Topics in Medicinal Chemistry, 2017, 17(18), 2023–40.
[14] Kyte J. Structure in Protein Chemistry. Garland Science, 2006.
[15] Chothia C. Principles that determine the structure of proteins. Annual Review of Biochemistry, 1984, 53(1), 537–72.
[16] Robertson AD, Murphy KP. Protein structure and the energetics of protein stability. Chemical Reviews, 1997, 97(5), 1251–68.
[17] Schulz GE, Schirmer RH. Principles of Protein Structure. Springer Science & Business Media, 2013.
[18] Branden CI, Tooze J. Introduction to Protein Structure. Garland Science, 2012.
[19] Pitman MR, Menz RI. Methods for protein homology modelling. In: Applied Mycology and Biotechnology. Elsevier, 2006, 6, 37–59.
[20] UniProt Consortium. UniProt: A hub for protein information. Nucleic Acids Research, 2015, 43(D1), D204–12.
[21] https://www.ncbi.nlm.nih.gov/
[22] Ye J, McGinnis S, Madden TL. BLAST: Improvements for better sequence analysis. Nucleic Acids Research, 2006, 34(suppl_2), W6–9.
[23] Thompson JD, Gibson TJ, Higgins DG. Multiple sequence alignment using ClustalW and ClustalX. Current Protocols in Bioinformatics, 2003, 1, 2–3.
[24] Schwede T, Kopp J, Guex N, Peitsch MC. SWISS-MODEL: An automated protein homology-modeling server. Nucleic Acids Research, 2003, 31(13), 3381–5.
[25] Pitman MR, Menz RI. Methods for protein homology modelling. In: Applied Mycology and Biotechnology. Elsevier, 2006, 6, 37–59.
[26] Fiser A, Šali A. Modeller: Generation and refinement of homology-based protein structure models. In: Methods in Enzymology. Academic Press, 2003, 374, 461–91.
[27] Xiang Z. Advances in homology protein structure modeling. Current Protein and Peptide Science, 2006, 7(3), 217–27.
[28] Claude JB, Suhre K, Notredame C, Claverie JM, Abergel C. CaspR: A web server for automated molecular replacement using homology modelling. Nucleic Acids Research, 2004, 32(suppl_2), W606–9.
[29] Raval A, Piana S, Eastwood MP, Dror RO, Shaw DE. Refinement of protein structure homology models via long, all-atom molecular dynamics simulations. Proteins: Structure, Function, and Bioinformatics, 2012, 80(8), 2071–9.
[30] Kairys V, Gilson MK, Fernandes MX. Using protein homology models for structure-based studies: Approaches to model refinement. The Scientific World Journal, 2006, 6, 1542–54.
[31] Laskowski RA, Furnham N, Thornton JM. The Ramachandran plot and protein structure validation. InBiomolecular forms and functions: A celebration of 50 years of the Ramachandran map, 2013, 62–75.

Chapter 13 Pharmacophore mapping in computer-aided drug designing (CADD)

In the realm of drug design, pharmacophore mapping is the blueprint of hope for a healthier future.
– Dr. Jane Pharmaceutica

13.1 Introduction

Pharmacophore mapping is a pivotal technique within the realm of CADD. It serves as a virtual compass guiding researchers through the complex landscape of molecular interactions, with the ultimate goal of discovering novel therapeutic agents [1]. At its core, pharmacophore mapping seeks to identify and define the essential structural and chemical features that a molecule must possess to interact with a specific biological target, such as a receptor or enzyme. The process begins by analyzing the 3D structures of known ligands that bind to the target of interest. These ligands' common features, including key functional groups and spatial arrangements, are extracted and abstracted into a pharmacophore model. This model represents the essential "pharmacophoric" elements required for effective binding [2, 3].

Pharmacophore mapping is invaluable for various aspects of drug discovery. It aids in virtual screening, allowing researchers to sift through vast chemical libraries to identify compounds that fit the pharmacophore, thereby expediting the lead discovery process [4]. Furthermore, it is a critical tool for optimizing drug candidates by guiding structural modifications to enhance their affinity and specificity for the target [5]. Pharmacophore mapping is a cornerstone of CADD, enabling the rational design of drugs with increased precision and efficiency. It accelerates the identification of promising leads and aids in the development of innovative therapies, contributing significantly to the field of pharmaceutical research and in the quest for improved healthcare solutions [6, 7].

The concept of pharmacophore emerged in the last decade of the nineteenth century. It took almost one century for it to emerge as a computational tool in the CADD, when in the early 1990s, many commercial programs were introduced for the development of 3D pharmacophore models [8, 9].

To gain insight and to develop a complete knowledge of the pharmacophore concept, it is of utmost importance to dive into the historical perspective of this concept.

https://doi.org/10.1515/9783111434858-013

13.2 Historical perspective of the pharmacophore concept

The credit for coining the term "pharmacophore" is given to Paul Ehlrich (father of chemotherapy) [10]. Ironically, he never used the term "pharmacophore" in any of his publications regarding his research related to this concept. It was Marshall who used the term pharmacophore [11] in his publication *On the physiological action of the alkaloids of jaborandi leaves*, based on Paul Ehrlich's side-chain theory published in 1897 [10].

The role of Paul Ehlrich in the development of the pharmacophore concept is very interesting and his attempts laid the foundation of modern day concepts of pharmacophore modeling in CADD. In his research, he introduced the terms "haptophore" and "toxophore." As per the interpretation from Ehlrich's work, he defined haptophores as the chemical groups that are responsible for binding the drug to the side chains (in the present scenario receptors). As he was working on the development of the chemotherapeutic agents, he defined toxophores as the chemical groups that show their toxic effect for alleviating the growth of microorganisms.

Some of the researchers argue that he did not use the term "pharmacophore," so he could not be given the credit for coining the term. Though he did not use this term in his publication, he was well aware of the same, as officially he used this term "pharmacorum" that later became "pharmacophore" in his Nobel Prize acceptance speech of 1908. This can be related well to the "particle nature of the wave" that was defined by Albert Einstein in 1906. He termed the particles of light as "quanta" in 1906 and got the Nobel Prize for the same in 1921 but a more acceptable term for the same is "photon," which was coined by Gilbert Lewin in 1926. This is an unarguable fact that credit for the concept of photon is attributed to Albert Einstein, although he never used this term. On the same analogy, credit for the concept of the term "pharmacophore" is attributed to Paul Ehlrich.

At the time of Paul Ehlrich, weak chemical interactions such as hydrogen bonding and Van der Waals interactions were not discovered, and hence the concept of haptophore and toxophore was limited to defining the role of the strong chemical interactions, such as covalent bonds, on the biological activity. With the passage of time and with more relevant research in the field of chemistry, the concept of pharmacophore gained popularity. It has been envisaged that for a molecule, it is not mandatory to have separate haptophore and toxophore; instead, a single group can perform both tasks, leading to the common term "pharmacophore." Thus, the modern definition of pharmacophore that came into the picture was given by Shueler in 1960 as the pattern of forces in the space that are required for the desired pharmacological response. As per Schueler, pharmacophore was defined as "a molecular framework that carries (phoros) the essential features responsible for a drug's (pharmacon) biological activity."

Beckett and his coworkers proposed the pharmacophore models for the muscarinic agents for the first time [12]. They gave the desired approximate distances between the different features that are essential for a molecule to act as an analog of muscarine (Fig. 13.1). Their work was based on approximations. The first commuted pharmacophore model was developed by Kier and his coworkers in 1967 (Fig. 13.2) for the muscarinic agents [13]. Ironically, both of the above models were developed for the same receptors and were similar in nature but were developed in different labs.

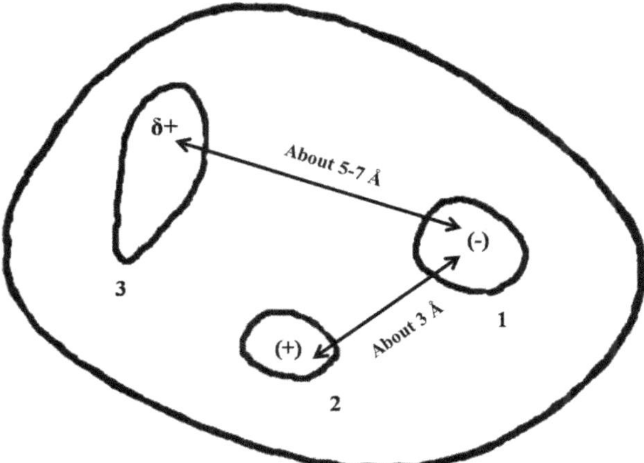

Fig. 13.1: The first pharmacophore developed by Becket and coworkers for a muscarinic receptor agonist. They give the desired approximate distance between the required substituent for a molecule to act as a potent muscarinic receptor agonist.

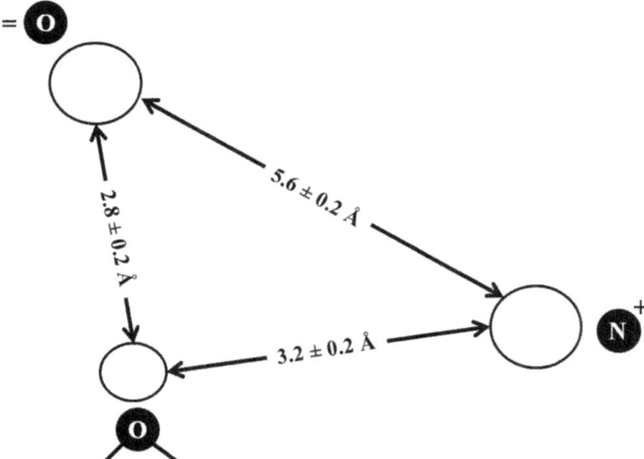

Fig. 13.2: Pharmacophore, developed by Keir and his coworkers, for the muscarinic agonist.

With the emergence of highly sophisticated computational tools with the passage of time, it was from the early 1990s that 3D pharmacophore modeling became a possibility that laid down the foundation of highly helpful pharmacophore-based drug designing approaches such as virtual screening, de novo drug designing, etc. Considering the importance of pharmacophore, the IUPAC defined this concept in the year 1998 [14] as "A pharmacophore is the ensemble of steric and electronic features that is necessary to ensure the optimal supramolecular interactions with a specific biological target structure and to trigger (or to block) its biological response."

13.3 Understanding the concept: pharmacophore

Pharmacophore is not just limited to identifying the chemical or functional groups essential for biological activity, the notion which evolved from the lab of Paul Ehlrich. Instead, it is the much broader term that includes every feature that is essential for imparting pharmacological response to receptors. Considering the above idea, the most complete and simplest definition for the pharmacophore can be given as "A pharmacophore is the pattern of features of a molecule that is responsible for a biological effect."

Here, this pattern of features includes everything, from the earlier concept of the "arrangement of chemical groups" to the modern perspective of hydrogen bonding, hydrophobicity, etc. The following are the most important features that are taken into consideration while defining a pharmacophore for any particular target (Fig. 13.3):

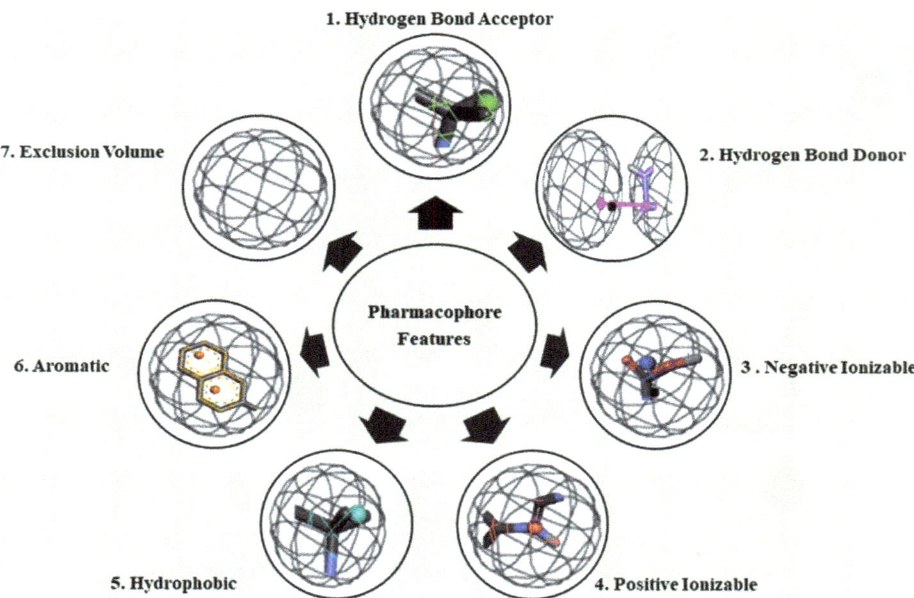

Fig. 13.3: The most common attributes of a pharmacophore to be developed.

- Hydrogen bond acceptors (HBAs)
- Hydrogen bond donors (HBDs)

- Positive ionizable groups
- Negative ionizable groups
- Hydrophobic regions
- Metal binding regions
- Aromatic groups

These pharmacophoric features have to be arranged in a 3D space to obtain a clear idea of how a new molecule should be designed, based on the pharmacophore. This is termed "pharmacophore modeling or mapping."

13.4 Types of pharmacophore modeling (mapping)

Pharmacophore modeling, similar to the analogy to CADD, is classified as ligand-based and structure-based pharmacophore modeling (Fig. 13.4). Both of these approaches differ from each other in generating the model, however, the application of both remains the same [2].

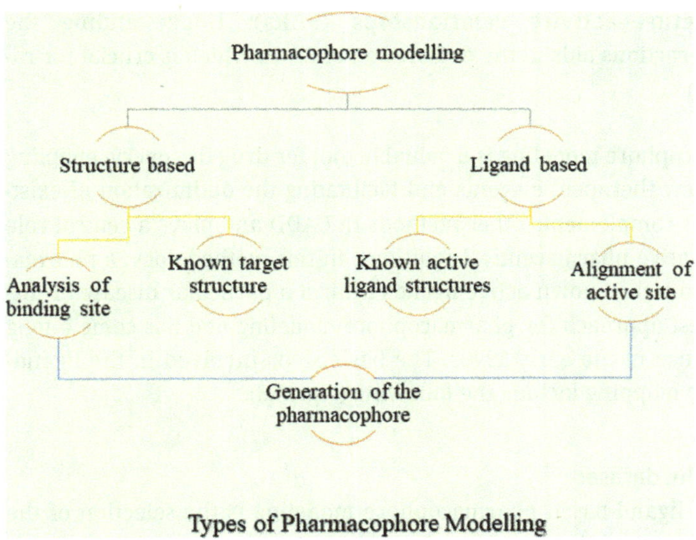

Fig. 13.4: Classification of the pharmacophore modeling, based on the approach used for the development of the pharmacophore.

13.4.1 Ligand-based pharmacophore modeling

Ligand-based pharmacophore modeling is a powerful method in drug discovery and CADD that focuses on characterizing the common structural and chemical features of active ligands or molecules known to interact with a specific biological target, such as a receptor or enzyme [15]. This approach is particularly valuable when detailed structural information about the target is limited or unavailable. The process begins by compiling a set of structurally diverse ligands that are known to bind to the target of interest. These ligands are then aligned based on their common features, and a pharmacophore model is constructed [16–18]. The pharmacophore model encapsulates critical spatial arrangements and functional groups essential for binding and biological activity, essentially representing the molecular "key" to the target's "lock."

Ligand-based pharmacophore modeling is a versatile tool in drug discovery for:

- **Lead identification:** It helps identify potential lead compounds by screening large chemical databases to find molecules that match the pharmacophore model.
- **Lead optimization:** Pharmacophore models guide medicinal chemists in modifying lead compounds to enhance their affinity, selectivity, and other desirable properties.
- **Predicting biological activity:** These models can provide insights into the pharmacological activity of new compounds or compounds with features similar to known actives.
- **Exploring structure–activity relationships (SARs):** Understanding the ligand–target interactions aids in the exploration of SARs, which is crucial for rational drug design.

Ligand-based pharmacophore modeling is a valuable tool for drug discovery, enabling the development of new therapeutic agents and facilitating the optimization of existing drug candidates. It complements other methods in CADD and plays a central role in the quest for innovative pharmaceutical solutions. In this methodology, a pharmacophore is deduced from the known active ligands against a particular disease or disorder. This is the oldest approach for pharmacophore modeling and has come a long way since its emergence in the early 1990s. The basic steps involved in the ligand-based pharmacophore mapping include the following (Fig. 13.5).

13.4.1.1 Curation of the dataset

The very first step of ligand-based pharmacophore modeling is the selection of the dataset for the generation of the pharmacophore. In this step, known active compounds that are diverse chemically but share a common target biological activity are selected. The dataset is then further divided into the training and test datasets. The training dataset is used for the generation of the pharmacophore model and the test dataset is used for the validation of the generated pharmacophore model. The dataset can be divided manually, randomly, or through machine learning approaches such as Kennard stone or Euclidean-based algorithms.

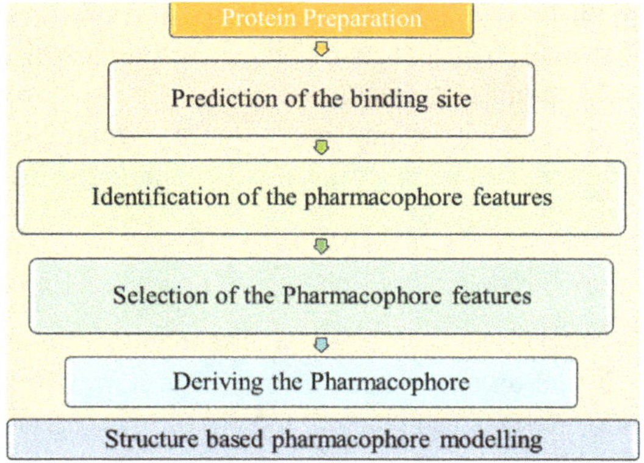

Fig. 13.5: Basic steps involved in the development of the ligand-based pharmacophore model.

13.4.1.2 Conformational search in ligand-based pharmacophore modelling

The quality of any ligand-based developed pharmacophore totally depends on this step. Any drug molecule that interacts with the receptor is in its 3-D conformation, which is known as the bioactive conformation of that drug molecule. Hence, the conformation of the ligand plays a crucial role in the generation of any ligand-based pharmacophore model. The goal of any conformation search algorithm is to search for the different conformations of each molecule of the training dataset that are close to the bioactive conformations from all the possible conformations (total conformational space) [19].

The search for the different conformations can be done through any of the mentioned methods:

- Systematic search algorithms
- Stochastic or random search algorithms
- Simulation method-based search algorithms

Each of the algorithms is discussed in detail in Chapter 9 of the molecular docking under the heading "Conformational search." A generalized workflow of the conformational search in the pharmacophore modeling is as follows:

a) **Finding the number of rotatable bonds:** The number of rotatable bonds present in the molecule is of utmost importance as it decides the possible conformations of the molecule. All the single bonds present between two atoms, other than hydrogen inside the molecule, are considered rotatable, excluding the terminal bonds (groups such as methyl and amino attached to the molecule are not considered for the calculation of the rotatable bond).
b) **Generation of the possible conformations:** Depending upon the conformational search algorithm applied, all the possible conformations are generated for the molecule under study and are used for the generation of the pharmacophore

model. Using a systematic search algorithm can lead to a combinatorial blast, i.e., a very huge number of conformations can be generated, hence consuming more time. If a molecule has more rotatable bonds, an exhaustive search can be more challenging due to combinatorial blast, as with every increase in the rotatable bond, the number of conformations increases exponentially as per the following equation:

$$N = (360/k)^n \tag{13.1}$$

where N is the number of possible conformations, k is every incremental increase in the torsion angle from one conformation to the next, and n is the number of rotatable bonds. The generated conformations using any of the algorithms should also be geometrically and energetically minimized using any of the optimum force field applications.

c) **Removal of duplicate and similar conformations:** In this step, closely related conformations or duplicate conformations are removed. To identify such conformations, the root mean square deviation (RMSD) parameter is employed:

$$\text{RMSD} = \sqrt{\Sigma(\Delta d)^2/N} \tag{13.2}$$

where Δd is the difference in the distance between the two same atoms of different conformations and N represents the number of atoms, other than hydrogen, present in the molecule. The lower the value of the RMSD between the two conformations, it is more likely they are similar in the conformation.

d) **Selection of the conformations close to the bioactive conformation:** The main purpose of the conformational search in the pharmacophore modeling is to identify those conformations that are close to the ligand-bound conformation so that a robust pharmacophore model can be generated. Hence in this step, from all the conformations available from the previous step, only those that are close to the bioactive conformation are kept.

The popular programs available for the conformational search for pharmacophore modeling are given in Tab. 13.1.

13.4.1.3 Generation of the ligand-based pharmacophore

This is the next step of the pharmacophore modeling in which a desired ligand-based pharmacophore model is generated. The two most common algorithms for the generation of the pharmacophore models are HipHop and Hypogen of Catalyst, currently owned by Biovia of Discovery Studio [20, 21]. The generation of the pharmacophore model is performed in three steps:

Tab. 13.1: List of popular programs available for the conformational search for pharmacophore modeling.

S. no.	Name of program	Description
1	Best	This is the program of the software package catalyst. This is an extensive method of conformational search, based on the CHARMm force field.
2	Fast	This program also belongs to the Catalyst software package and is also based on the CHARMm force field. It is less extensive and more approximate, hence, performs conformational search faster.
3	Chem-X	This program works on the rule-based conformational analysis and employs a "bump check" algorithm to remove duplicate conformations.
4	Omega	This program works on the deterministic rule-based conformational analysis. It works on the Clear Force Field for all the energy-related calculations.
5	Macromodel	Conformational analysis is performed using the Monte Carlo approach of the search algorithm. The MMFF force field is used in this program for performing all the energy calculations.

a) **Constructive phase:** It is the first step of the pharmacophore model generation in which all possible features of the pharmacophore are included in the pharmacophore. In the HipHop methodology, the activity of the molecules is not taken into consideration; instead, construction of the pharmacophore is done based on common features among the drug like active ligands of the training dataset. On the contrary, the hypogen algorithm constructs the pharmacophore model based on the activity data and can be used for the further prediction of the biological activity of the unknown compounds. In both of the approaches, multiple pharmacophore models are generated in this step.
b) **Subtractive phase:** In this step, the pharmacophores that are not useful are removed. Here, pharmacophore models from the least active compounds of the training set are assessed and if they are found similar to that obtained from step a), then they are removed from the dataset of generated models of step a).
c) **Optimizing the final pharmacophore models:** Pharmacophore models that came after step *b)* are further optimized for obtaining better models, employing the simulated annealing algorithm. The simulated annealing process starts with the initial pharmacophore model and then small changes are made in the pharmacophore model. The new model is evaluated based on correlation coefficient and cost function values. If the new model is found better than the previous one, then it is accepted as the new model, and if not, the process is stopped and the previous model is selected as the final optimized pharmacophore model.

13.4.1.4 Validation of the pharmacophore model

This is a crucial step in the ligand-based pharmacophore modeling in which the robustness of the developed model is evaluated. Validating the ligand-based model ensures that the model is not generated by chance and it can be used for getting reproducible results. Various approaches employed for the validation of the ligand-based pharmacophore model include the following:

13.4.1.4.1 Cost analysis approach

In the cost analysis approach, three cost parameters are calculated and based on their values, the model is validated. These cost parameters are calculated from the values of the weight component (*W*), error component (*E*), and configuration component (*C*). The ideal value of the weight component is 2 and deviation from this value is evaluated in the pharmacophore models. The error component represents the difference between the observed and predicted values of the training dataset of the molecule [22]. The configuration component represents the entropy values of the pharmacophore model inside the conformational space.

a) **Fixed cost:** The fixed cost value of the pharmacophore model represents the lowest possible cost of the hypothetical model that can fit all the data perfectly. Fixed cost is calculated by adding the minimum possible error, weight, and configuration components in the following equation:

$$\text{Fixed cost} = eE(x = 0) + wW(x = 0) + cC \tag{13.3}$$

   where *e*, *w*, and *c* are the coefficients of the error, weight, and configuration components, respectively.

b) **Null cost:** The null cost value of the model represents the maximum cost of the pharmacophore model when no feature of the same is selected and the activity of the model is approximated to the average activity of the training set molecules.

c) **Total cost:** The total cost of the pharmacophore model represents the actual cost of each of the pharmacophore hypotheses calculated by the summation of the error, weight, and configuration component:

$$\text{Total cost} = eE + wW + cC \tag{13.4}$$

   where *e*, *w*, and *c* are the coefficients of the error, weight, and configuration components, respectively.

   For a robust pharmacophore model, the value of total cost should be close to the fixed cost and there should be a considerable difference between the total cost and null cost. When the difference between the total cost and null cost is greater than 60, the developed model is considered to be highly robust. If the difference is between 40 and 60, then it is estimated that the developed model has a prediction accuracy between 75–90%. If the difference between the total cost and null cost is below 40, the developed model is not considered reliable.

13.4.1.4.2 Fisher's randomization test

In Fisher's randomization test validation approach, the biological activity of all the training dataset molecules is scrambled and then the pharmacophore model is generated, keeping other parameters and pharmacophore features the same as that used for the development of the original pharmacophore model. Now, if the newly developed model is found to be better or similar to the original data, we consider that the original pharmacophore is developed only by chance and is not a robust pharmacophore model. A cross-validation approach, such as the leave-one-out method is used for calculating the correlation values of the developed models through scrambling of the activity [23].

13.4.1.4.3 Test set prediction method for the validation

In the initial step of the pharmacophore modeling, a dataset of the molecules is divided into training and test datasets. With the training dataset molecules, the pharmacophore model is developed and the test dataset molecules are utilized for the validation of the pharmacophore model. In this approach, the activity of the dataset molecules is calculated through the developed pharmacophore model and compared with the actual experimental activity value. Here also, cross-validation approach is used for correlating the results obtained of the predicted biological activity and the experimental biological activity of the test dataset molecules [24].

13.4.1.4.4 Goodness-of-Hit (GH) scoring approach for the pharmacophore model validation

In Goodness-of-hit (GH) scoring approach, a database of molecules that contain actives and decoys against the pharmacophore is used for the validation of the model [25, 26]. Let us consider an example to understand this methodology. Suppose we used a database of 500 molecules for the validation of the developed pharmacophore model. Out of these 500 molecules, 15 actives (A) are present, and the remaining are decoys. After performing the virtual screening of these 500 molecules on the pharmacophore model that is to be validated, let us suppose that the generated model has given 18 hit compounds (Ht), and out of these 13 (Ha) are from the known 15 active compounds. Then, for evaluating the robustness of the model, the GH score is considered, which is calculated as follows:

- Total compounds present in the database (D) = 500
- Total actives present in the database (A) = 15
- Total hits obtained (Ht) = 18
- Total active detected (Ha) = 13
- % yield of the actives = (Ha/Ht) × 100 = 72.2%
- % ratio of actives = (Ha/A) × 100 = 86.67%
- Enrichment factor (E) = (Ha × D)/(Ht × A) = 24.07
- False negative = A − Ha = 02

- False positives = Ht – Ha = 05
- GH score = (Ha(3A + Ht)/4HtA)(1 – (Ht – Ha)/ (D – A) = 0.75

The value of the GH score should be between 0.7 and 0.8 for a robust pharmacophore model.

13.4.1.5 Advantages of ligand-based pharmacophore modeling

Ligand-based pharmacophore modeling is a valuable approach in CADD that focuses on the characteristics and properties of known active ligands (molecules that bind to a specific target). It is a versatile and powerful technique in CADD, particularly when dealing with targets where structural information is limited or when optimizing leads for drug development. It offers a practical means of identifying potential drug candidates and understanding the pharmacological requirements for binding to a specific target. It offers several advantages, including:

- **Data-driven approach:** Ligand-based pharmacophore modeling is based on experimental data from known ligands, making it a data-driven method that leverages existing knowledge.
- **Identification of common features:** It identifies common structural and chemical features critical for binding to a specific target, providing insights into the key interactions required for biological activity.
- **Lead optimization:** It is useful for lead optimization, allowing researchers to modify existing ligands to improve binding affinity, selectivity, and other properties.
- **Virtual screening:** Ligand-based pharmacophore models can be used in virtual screening to filter large compound libraries quickly, identifying potential hits that fit the pharmacophore.
- **Scaffold hopping:** It enables researchers to explore diverse chemical scaffolds that share the same pharmacophoric features, leading to the discovery of structurally distinct but biologically active compounds.
- **No structural information required:** Unlike structure-based methods, ligand-based pharmacophore modeling does not require detailed structural information about the target protein, making it applicable in cases where protein structures are unavailable or challenging to obtain.
- **Fast and cost-effective:** It is relatively fast and cost-effective, making it a practical choice for early-stage drug discovery and lead identification.
- **Applicability to multiple targets:** Ligand-based pharmacophores can be applied to multiple targets within a target family, facilitating the discovery of multi-target drugs.
- **Dynamic ligand–target interactions:** It takes into account the dynamic nature of ligand–target interactions, allowing for the consideration of conformational flexibility and binding kinetics.

- **Validation and benchmarking:** The method is easily validated and benchmarked against known active and inactive compounds, enhancing its reliability.
- **Complementary to structure-based approaches:** It can be used in combination with structure-based methods to gain a more comprehensive understanding of the ligand–target interactions.

13.4.1.6 Limitations of ligand-based pharmacophore modeling

Ligand-based pharmacophore modeling is a valuable technique in CADD, but it has several limitations that should be considered:

- **Dependence on known ligands:** Ligand-based pharmacophore modeling relies on a set of known active ligands. If these ligands are not representative of the full chemical space of potential ligands for a target, the resulting pharmacophore model may not be comprehensive.
- **Lack of information on protein structure:** This approach does not take into account the 3D structure of the target protein. As a result, it may not provide insights into the binding site's structural features and interactions, which can be critical for designing drugs that specifically interact with the target.
- **Inability to predict novel ligands:** Ligand-based pharmacophore models are only as good as the ligands used to create them. They cannot predict the activity of entirely new or structurally dissimilar compounds that were not part of the training set.
- **Limited applicability to different targets:** A pharmacophore model developed for one target may not be applicable to other targets. It lacks the target specificity and transferability of structure-based approaches.
- **Overfitting:** There is a risk of overfitting, where the pharmacophore model captures noise or irrelevant features from the training ligands, leading to poor predictive performance for new ligands.
- **Variability in ligand conformation:** Ligands can adopt different conformations, and the model may not account for this variability. It may not accurately represent the dynamic nature of ligand binding.
- **Lack of information on mechanism of action:** Ligand-based pharmacophore models do not provide insights into the mechanism of action at the molecular level, limiting the ability to design mechanism-specific drugs.
- **Challenging for multitarget drug design:** Developing pharmacophore models for multitarget drug design is complex and may not always yield accurate results.
- **Limited insights into** absorption, distribution, metabolism, excretion, and toxicity **(ADMET) properties:** This method does not inherently provide information about a compound's ADMET properties, which are crucial in drug development.
- **Data quality and quantity:** The quality and quantity of available ligand data can significantly impact the reliability of the pharmacophore model. Incomplete or noisy data can lead to inaccurate models.

To mitigate these limitations, researchers often combine ligand-based pharmacophore modeling with other techniques, such as structure-based methods, molecular dynamics simulations, or machine learning approaches, to gain a more comprehensive understanding of drug-target interactions and improve the drug discovery process.

13.4.2 Structure-based pharmacophore modeling

Structure-based pharmacophore modeling is an advanced and powerful technique used in drug discovery and design [27, 28]. It combines structural biology data with pharmacophore modeling to help identify crucial interactions between ligands (small molecules or drug candidates) and their target proteins. This approach has proven to be instrumental in rational drug design, enabling the development of more effective and specific therapeutic agents [29].

The process begins by determining the 3D structure of the target protein using techniques like X-ray crystallography or NMR spectroscopy. Once the protein structure is available, specific binding sites or pockets within the protein are identified. These binding sites are crucial for the protein's function and are often associated with the active site where ligands bind. Structure-based pharmacophore modeling leverages this structural information to create a pharmacophore model that accurately represents the key interactions between the ligand and the protein's binding site. This model considers features like hydrogen bond donors and acceptors, aromatic regions, and hydrophobic pockets, which are essential for ligand–protein interactions. The resultant pharmacophore model serves as a valuable tool for virtual screening and lead optimization. Researchers can use it to rapidly filter through compound libraries and identify potential drug candidates that fit the binding site's specific requirements. Moreover, it facilitates the design of new ligands with improved affinity and selectivity for the target protein, which can lead to more successful drug development [30, 31].

Structure-based pharmacophore modeling harnesses the power of structural biology to create accurate representations of ligand–protein interactions, enhancing the efficiency and effectiveness of drug discovery and design processes. It plays a pivotal role in the development of targeted therapies and is a cornerstone of modern pharmaceutical research. This method is possible only when the 3D structure of the target protein is known. In some approaches, the pharmacophore model is also obtained from the structural features of the internal ligand bound to the target protein. The only limitation to this approach is that a known 3D structure of the target protein should be available and with the continuous increase in the protein structures in the protein data bank, this approach is gaining immense popularity [32, 33]. The basic steps involved in the structure-based pharmacophore modeling are discussed below (Fig. 13.6).

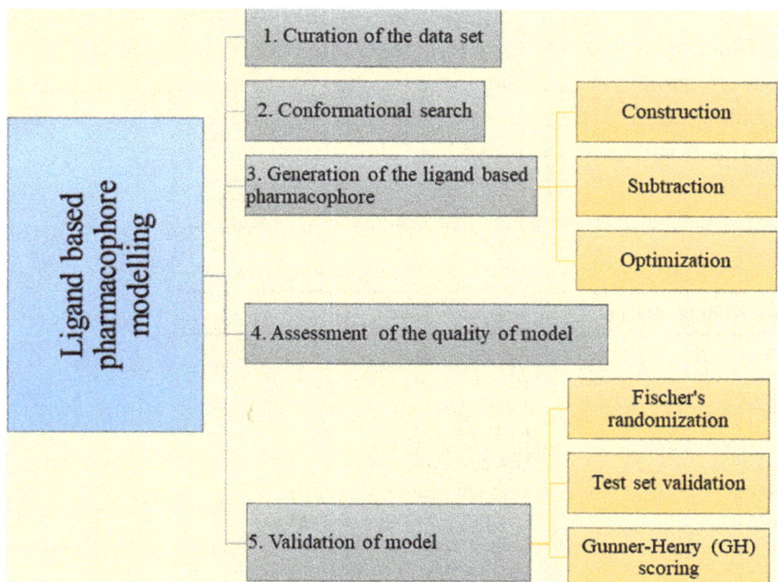

Fig. 13.6: Basic steps involved in the development of the structure-based pharmacophore model.

13.4.2.1 Preparation of the protein

Protein preparation is the initial step of the structure-based pharmacophore modeling. The 3D structure of the target protein that is obtained from the protein data bank is retrieved in its raw form and contains certain errors. To obtain an error-free and significant pharmacophore model from the structure of the target protein, it becomes necessary to prepare the target protein before performing any modeling step on it. The main considerations during the protein preparation steps include:

- The protein structure present in the protein data bank is deposited with the missing hydrogen, hence before pharmacophore modelling, all the missing hydrogen atoms should be added to the structure of the target protein.
- The target protein structure should also be checked for any missing amino acid residue and should be corrected.
- The protonation state of amino acid residues such as asparagine, glutamine, and histidine should also be solved as these residues play a crucial role in hydrogen bonding.

13.4.2.2 Prediction of the binding site of the protein

This is the second step of the structure-based pharmacophore modeling in which the active site of the target protein is detected. There are various algorithms for performing this step and are broadly classified as the energy-based and geometry-based detection of the active binding site of the target protein. It has been found that target proteins that contain bound internal targets depict the pharmacophore with approximately 95% accuracy, however, the target proteins that do not contain internal ligands are difficult to handle for deducing the active, site hence lowering the accuracy of the pharmacophore. Basic programs, along with their algorithm used for detecting the active site of the target protein, are given in Tab. 13.2.

Tab. 13.2: List of basic programs and their algorithm used for detecting the active site of the target protein.

S. no.	Name of the program	Algorithm used
1	PocketFinder	Energy-based
2	SiteD	Energy-based (solvation based)
3	SuperStar	Energy-based
4	MUSIC	Energy-based
5	SiteMap	Energy-based (solvation based)
6	Insight II	Geometry-based (grid-based)
7	GRID	Geometry-based (grid-based)
8	Snooker	Geometry based (Delaunay tessellation)
9	SiteFinder	Geometry-based (use of alpha complexes)
10	PASS	Geometry-based (use of alpha complexes)
11	SURFNET	Geometry-based (use of alpha complexes)
12	LIGSITE	Geometry-based (use of alpha complexes)
13	APROPOS	Geometry-based (use of alpha complexes)
14	CAST	Geometry-based (use of alpha complexes)

13.4.2.3 Identification of pharmacophore features

Identification of the pharmacophore feature is the next and crucial one in the modeling of the pharmacophore. There are two approaches employed for this purpose. In the first approach, all the pharmacophoric features are deduced from the internal ligand that is bound to the target protein's active cavity. The second approach is used if the internally bound ligand is not present in the active cavity of the target protein. The pharmacophoric feature in this second approach is identified based on amino acid residues present inside the cavity, which is based on their requirement for chemical interactions with the potent ligand.

Out of these two approaches, the elucidation of pharmacophore features, based on the internal ligand, is considered more accurate and fast. In this approach, features of pharmacophore are placed on all the functional groups present on the ligands. If more than one structure of the target protein with different bound internal ligands is available, then this step resembles the ligand-based pharmacophore modeling in which initially all the internally bound ligands are extracted and their alignment is performed. Based on the molecular alignment, all the pharmacophoric features are deduced. LigandScout and MOE are programs that identify the pharmacophore features using this approach.

The second approach is to deduce the pharmacophore features from the apo-structure of the target protein. It is a tedious task as compared to the previous one. In this approach, initially, the active cavity of the target protein is identified, followed by the clustering of the hotspots inside the binding cavity where interaction with the ligand can occur. These hotspots are then combined to identify all the pharmacophoric features required of a potent ligand. FLAP, GBPM, Pocket V2, Snooker, and Sybyl are popular programs that elucidate the pharmacophore features from the active site of the target protein.

13.4.2.4 Selection of the pharmacophoric features

Once all the possible pharmacophore features are identified, the next step is to select only those features that are important and necessary for the possibly potent ligand. Careful selection of only those features is required at this step that correlate well with the biological activity. The different approaches used for this purpose include the following.

a) **Energy-based selection:** Energy-based selection is one of the most popular methods for the selection of the desired pharmacophoric features in the modeling of the pharmacophore. In this approach, if the ligand-bound inside the active cavity of the protein is known, then probe docking or molecular dynamics-based scoring is performed for selecting the desired pharmacophore features. This energy-based scoring based on probe docking or dynamics simulations is performed at different sites of interaction of the bound ligand within the cavity of the protein. The features or points that show highly favorable interactions, based on energy calculations, are then selected for developing the pharmacophore.

   If the protein structure is devoid of internally bound ligands, and pharmacophore features are identified from the cavity of the apo-protein, then the hydration site analysis approach is applied. This approach is based on the principle that functional groups that can replace water from the active cavity of the protein structure contribute significantly to the generation of the potent ligand and thus become an indispensable member of the pharmacophore model.

b) **Selection if more than one internally bound ligand is available for the target protein:** This approach is employed if there is more than one internal ligand available for the same target protein. In this approach, the individual pharmacophore obtained from each of the internal ligands is superimposed and aligned with each other, and the common features are selected for obtaining the significant pharmacophore model. SQUID and FLAP are the two most popular algorithms that employ this approach for the selection of the pharmacophore features.

c) **Selection based on making changes in the amino acid residues inside the binding cavity of the protein:** In this approach, amino acids present inside the binding cavity of the protein are changed with other amino acids and its impact on the binding of the ligand is then assessed. Amino acids that contribute significantly to the ligand binding are then studied further and only those pharmacophoric features that satisfy the requirement of those amino acids are selected to obtain a prominent pharmacophore.

13.4.2.5 Refinement of the pharmacophore model

This is the last step of the structure-based pharmacophore modeling approach. Till now, we have studied the development of the pharmacophore model by selecting only those features that correlate well with the biological activity. In this approach, the final refinement of the developed model is performed to develop a robust and significant pharmacophore model. Approaches employed for the refinement of the pharmacophore include the following.

a) **Use of active ligands for the training of the structure-based pharmacophore model:** The active ligand approach is very useful for the further refinement of the pharmacophore model. This not only enhances the robustness of the pharmacophore model but also reduces the detection of false positives through virtual screening. In this methodology, screening of the known active ligands is performed on the developed pharmacophore and the ability of the pharmacophore is assessed to identify them as active. Any refinement, if required, is done on the pharmacophore model, based on the screening results of the active ligands. The Ligand profiler of the Acclerys program performs this task very efficiently.
b) **Applying shape constraints to the pharmacophore model:** Shape, volume, and geometry are crucial parameters in any of the CADD approaches. Based on the shape and geometry requirement of the binding site of the target protein, a shape restrain can be applied to the final pharmacophore model developed. The pharmacophore features that are in accordance with the shape restraints are included in the final refined pharmacophore and those that are outside the shape restrains are excluded.

13.4.2.6 Advantages of structure-based pharmacophore modeling

Structure-based pharmacophore modeling is a powerful approach in CADD, offering a structured and rational methodology for designing drugs with improved specificity, affinity, and overall efficacy. It has become an essential tool in the arsenal of drug discovery, contributing to the development of more effective and safer medications:

- **Informed drug design:** Structure-based pharmacophore modeling leverages the 3D structure of a target protein, enabling a more informed and precise drug design process. This approach can account for the specific structural features and binding sites of the target.
- **Understanding binding interactions:** It provides insights into the key interactions between ligands and target proteins. This understanding is crucial for designing molecules that can mimic or disrupt these interactions effectively.

- **Enhanced specificity:** By considering the target's structure, structure-based pharmacophore modeling can lead to the development of highly specific ligands, minimizing off-target effects, and improving the safety profile of drugs.
- **Rational ligand optimization:** This technique allows for the rational optimization of lead compounds. Researchers can modify existing ligands to enhance binding affinity and selectivity by matching them with the target's 3-D structure.
- **Virtual screening efficiency:** It significantly improves the virtual screening processes by eliminating compounds that do not fit the pharmacophore model, reducing the number of compounds that need to be experimentally tested.
- **Binding site prediction:** Structure-based pharmacophore modeling can help predict potential binding sites on the target protein, which is valuable for identifying allosteric binding sites or other druggable regions.
- **Ligand–receptor flexibility:** This approach accounts for the flexibility of both ligands and target proteins, which is critical in understanding how ligands can adapt to the binding pocket.
- **Acceleration of lead discovery:** It accelerates the lead discovery process by providing a structural basis for identifying potential drug candidates. This saves time and resources in the drug development pipeline.
- **Personalized medicine:** Structure-based pharmacophore modeling can be applied in the context of patient-specific target proteins, contributing to personalized medicine and tailored drug therapies.
- **Diverse applications:** Beyond drug discovery, this technique finds applications in understanding protein–protein interactions, enzyme–substrate interactions, and other molecular recognition events.

13.4.2.7 Limitations of structure-based pharmacophore modeling

Structure-based pharmacophore modeling, while a valuable technique in CADD, has its limitations that should be considered. Some of these limitations include:

- **Dependence on available structural data:** Structure-based pharmacophore modeling relies on the availability of experimentally determined protein–ligand complex structures. If such structures are lacking for the target of interest, the method cannot be applied effectively.
- **Static representation:** It typically provides a static representation of the ligand-binding site. It may not capture the conformational flexibility of the binding site or ligands, which can be crucial for accurately predicting ligand–receptor interactions.
- **Inaccurate complex structures:** Experimental structures can sometimes contain inaccuracies, especially when obtained through X-ray crystallography or NMR. Inaccurate structural data can lead to flawed pharmacophore models.
- **Limited coverage:** Structure-based pharmacophore modeling is limited to known ligands and target structures. It may not identify novel ligands or alternative binding modes that were not present in the training set.

- **Sensitivity to water molecules:** Water molecules in the binding site can influence ligand–receptor interactions. However, these are often neglected in pharmacophore modeling, potentially leading to inaccurate predictions.
- **Challenging for multitarget pharmacophores:** Creating pharmacophores for multiple targets (polypharmacophores) can be complex and may not capture all the nuances of different binding sites.
- **Computationally intensive:** Generating accurate structure-based pharmacophores can be computationally intensive, especially for large protein–ligand complexes. This can limit its practicality for high-throughput virtual screening.
- **Limited to protein targets:** Structure-based pharmacophore modeling is primarily applicable to protein targets, and it may not be suitable for other biomolecular targets, such as RNA or DNA.
- **Ignoring protein dynamics:** It does not account for dynamic protein–ligand interactions or induced-fit effects, which are critical for understanding binding affinity and specificity.
- **Overlooking solvent effects:** It often simplifies the binding site to a dry environment, neglecting the influence of solvent molecules and their role in ligand binding.

Despite these limitations, structure-based pharmacophore modeling remains a valuable tool in CADD. Researchers should be aware of its constraints and consider complementing it with other methods to enhance the accuracy and reliability of drug discovery efforts.

13.5 Applications of the pharmacophore modeling approaches

Pharmacophore modeling is a valuable tool in CADD, providing insights into molecular interactions and accelerating the drug discovery process. Its diverse applications make it a fundamental component of modern drug development efforts. Pharmacophore modeling is a versatile and powerful approach in CADD, offering various applications in the drug discovery process. Below are some key applications of pharmacophore modeling in CADD:

- **Virtual screening:** Pharmacophore models are used to screen large chemical databases, identifying compounds that match the pharmacophoric features. This accelerates the identification of potential drug candidates.
- **Lead compound identification:** Pharmacophore modeling helps in the discovery of lead compounds by specifying the key interactions required for a molecule to bind to a specific target.
- **Structure-based drug design:** Pharmacophore models can be integrated with 3D structures of target proteins to design new compounds that fit the pharmacophore and interact effectively with the target.

- **Hit optimization:** Pharmacophore models guide medicinal chemists in optimizing the properties of lead compounds, ensuring they maintain the essential pharmacophoric features while improving their drug-like properties.
- **Bioisosteric replacement:** Pharmacophore modeling assists in identifying suitable bioisosteric replacements for certain functional groups, allowing for the design of analogs with improved properties.
- **Polypharmacology studies:** Pharmacophore models can be employed to predict the interactions of compounds with multiple targets, which is essential for understanding potential off-target effects and designing multi-target drugs.
- **Toxicity prediction:** Pharmacophore modeling is used to predict the potential toxicity of compounds by identifying features associated with toxic effects.
- **Pharmacokinetics and ADMET prediction:** Pharmacophore models can help estimate a compound's absorption, distribution, metabolism, excretion, and toxicity (ADMET) properties.
- **Fragment-based drug design:** Pharmacophore models can guide the assembly of fragments into larger compounds with the desired pharmacophoric features.
- **Personalized medicine:** Personalized pharmacophore models can be developed, based on individual patient data, allowing for the design of patient-specific therapies.
- **Chemical probe design:** Pharmacophore modeling assists in the development of chemical probes used to investigate the biological function of a specific target.
- **Natural product-based drug discovery:** Pharmacophore models can be applied to identify natural product compounds with pharmacophoric features that make them suitable for drug development.
- **G Protein-coupled receptor (GPCR) ligand discovery:** Pharmacophore modeling is used in the discovery of ligands for GPCRs and other membrane proteins.

References

[1] Gao Q, Yang L, Zhu Y. Pharmacophore based drug design approach as a practical process in drug discovery. Current Computer-Aided Drug Design, 2010, 6(1), 37–49.

[2] Khedkar SA, Malde AK, Coutinho EC, Srivastava S. Pharmacophore modeling in drug discovery and development: an overview. Medicinal Chemistry, 2007, 3(2), 187–97.

[3] Pathak DV, Vyas A, Sagar SR, Bhatt HG, Patel PK. Pharmacophore Mapping: An Important Tool in Modern Drug Design and Discovery. Applied Computer-Aided Drug Design: Models and Methods, 2023, 57.

[4] Humblet C, Marshall GR. Pharmacophore identification and receptor mapping. In: Annual Reports in Medicinal Chemistry. Academic Press, 1980, 15, 267–276.

[5] Swaminathan P. Advances in pharmacophore modeling and its role in drug designing. Computer-Aided Drug Design, 2020, 223–43.

[6] Tyagi R, Singh A, Chaudhary KK, Yadav MK. Pharmacophore modeling and its applications. In: Bioinformatics. Academic Press, 2022, 269–289.

[7] Yang SY. Pharmacophore modeling and applications in drug discovery: challenges and recent advances. Drug Discovery Today, 2010, 15(11–12), 444–50.
[8] Shanmugarajan D, Akkiraju LJ, Panda S, Hazra S. Pharmacophore mapping and modeling approaches for drug development. In: Computational Approaches for Novel Therapeutic and Diagnostic Designing to Mitigate SARS-CoV2 Infection. Academic Press, 2022, 171–89.
[9] Qing X, Yin Lee X, De Raeymaeker J, et al. Pharmacophore modeling: advances, limitations, and current utility in drug discovery. Journal of Receptor, Ligand and Channel Research, 2014, 81–92.
[10] Ehrlich P. Die Wertbemessung des Diphtherieheilserums und deren theoretische Grundlagen. G. Fischer, 1897.
[11] Marshall CR. On the physiological action of the alkaloids of jaborandi leaves. The Journal of Physiology, 1904, 31(2), 120.
[12] Beckett AH, Harper NJ, Clitherow JW. The importance of stereoisomerism in muscarinic activity. Journal of Pharmacy and Pharmacology, 1963, 15(1), 362–71.
[13] KIER LB. Molecular orbital calculation of preferred conformations of acetylcholine, muscarine, and muscarone. Molecular Pharmacology, 1967, 3(5), 487–94.
[14] Wermuth CG, Ganellin CR, Lindberg P, Mitscher LA. Glossary of terms used in medicinal chemistry (IUPAC Recommendations 1998). Pure and Applied Chemistry, 1998, 70(5), 1129–43.
[15] Ajjarapu SM, Tiwari A, Ramteke PW, Singh DB, Kumar S. Ligand-based drug designing. In: Bioinformatics. Academic Press, 2022, 233–52.
[16] Engels M, BB S, Divakar S, Geetha G. Ligand based pharmacophore modeling, virtual screening and molecular docking studies to design novel pancreatic lipase inhibitors. International Journal of Pharmacy and Pharmaceutical Sciences, 2017, (4.33), 29.
[17] Kandakatla N, Ramakrishnan G. Ligand-based pharmacophore modeling and virtual screening studies to design novel HDAC2 inhibitors. Advances in Bioinformatics, 2014, 2014.
[18] Pal S, Kumar V, Kundu B, Bhattacharya D, Preethy N, Reddy MP, Talukdar A. Ligand-based pharmacophore modeling, virtual screening and molecular docking studies for discovery of potential topoisomerase I inhibitors. Computational and Structural Biotechnology Journal, 2019, 17, 291–310.
[19] Acharya C, Coop A, Polli J, MacKerell A. Recent advances in ligand-based drug design: relevance and utility of the conformationally sampled pharmacophore approach. Current Computer-Aided Drug Design, 2011, 7(1), 10–22.
[20] Clement OO, Mehl AT. HipHop: Pharmacophores. Pharmacophore Perception, Development, and Use in Drug Design, 2000, 2, 69.
[21] https://www.3ds.com/products/biovia/discovery-studio/ligand-and-pharmacophore-based-design
[22] Pal S, Kumar V, Kundu B, Bhattacharya D, Preethy N, Reddy MP, Talukdar A. Ligand-based pharmacophore modeling, virtual screening and molecular docking studies for discovery of potential topoisomerase I inhibitors. Computational and Structural Biotechnology Journal, 2019, 17, 291–310.
[23] John S, Thangapandian S, Arooj M, Hong JC, Kim KD, Lee KW. Development, evaluation and application of 3D QSAR Pharmacophore model in the discovery of potential human renin inhibitors. InBMC bioinformatics. BioMed Central, 2011, 12(14), 1–14.
[24] Schuurmann G, Ebert RU, Chen J, Wang B, Kuhne R. External validation and prediction employing the predictive squared correlation coefficient– Test set activity mean vs training set activity mean. Journal of Chemical Information and Modeling, 2008, 48(11), 2140–5.
[25] Gupta N, Sitwala N, Patel K. Pharmacophore modelling, validation, 3D virtual screening, docking, design and in silico ADMET simulation study of histone deacetylase class-1 inhibitors. Medicinal Chemistry Research, 2014, 23, 4853–64.

[26] Castleman P, Szwabowski G, Bowman D, Cole J, Parrill AL, Baker DL. Ligand-based G Protein Coupled Receptor pharmacophore modeling: Assessing the role of ligand function in model development. Journal of Molecular Graphics and Modelling, 2022, 111, 108107.

[27] Sanders MP, McGuire R, Roumen L, de Esch IJ, de Vlieg J, Klomp JP, de Graaf C. From the protein's perspective: the benefits and challenges of protein structure-based pharmacophore modeling. MedChemComm, 2012, 3(1), 28–38.

[28] Steindl TM, Schuster D, Wolber G, Laggner C, Langer T. High-throughput structure-based pharmacophore modelling as a basis for successful parallel virtual screening. Journal of Computer-Aided Molecular Design, 2006, 20, 703–15.

[29] Szwabowski GL, Daigle Jr BJ, Baker DL, Parrill AL. Structure-based pharmacophore modeling 2. Developing a novel framework for structure-based pharmacophore model generation and selection. Journal of Molecular Graphics and Modelling, 2023, 122, 108488.

[30] Pirhadi S, Shiri F, Ghasemi JB. Methods and applications of structure based pharmacophores in drug discovery. Current Topics in Medicinal Chemistry, 2013, 13(9), 1036–47.

[31] Szwabowski GL, Cole JA, Baker DL, Parrill AL. Structure-based pharmacophore modeling 1. Automated random pharmacophore model generation. Journal of Molecular Graphics and Modelling, 2023, 121, 108429.

[32] Gao Q, Yang L, Zhu Y. Pharmacophore based drug design approach as a practical process in drug discovery. Current Computer-Aided Drug Design, 2010, 6(1), 37–49.

[33] Shiri F, Pirhadi S, Ghasemi JB. Dynamic structure based pharmacophore modeling of the Acetylcholinesterase reveals several potential inhibitors. Journal of Biomolecular Structure and Dynamics, 2019, 37(7), 1800–12.

Chapter 14
Virtual screening approaches in the computer-aided drug designing

> Virtual screening is now an integral part of the drug discovery process, offering rapid and cost-effective method for identifying potential drug candidates. – Wolfgang Sippl

14.1 Introduction

The cornerstone of any drug design approach is the development of a potent chemical entity with the desired pharmacological profile. In the preceding chapters, various methodologies of computer-aided drug design (CADD) have been explored, including molecular docking, quantitative structure–activity relationship (QSAR, 2D and 3D), pharmacophore modeling, and de novo design. These methodologies are instrumental in achieving the aforementioned objective. Virtual screening can be understood as the application of these established CADD methodologies to identify potent drug molecules.

Most simply, virtual screening can be defined as the in silico process used in the early stages of drug discovery for finding out or identifying potential lead molecules or compounds from the pool of a chemical database by employing any knowledge-based screening approaches such as molecular docking, pharmacophore-based similarity, and QSAR-based similarity. The concept of virtual screening can be best understood by comparing it with the process of filtration using a funnel (Fig. 14.1). A large pool of molecules is allowed to pass through a funnel having certain filters and the molecules that do not hold good with the applied filters are excluded and those that match the filter requirement pass and emerge out as the possible hits [1–3]. The term "virtual screening" was first used by Hovarth D in the year 1997 in his publication [4]. Before diving deep into the concept and methodology of virtual screening, one must have clear knowledge of the different chemical databases available for performing the virtual screening process.

14.2 Chemical databases for the virtual screening in CADD

Chemical databases are the libraries of either small fragments, lead-like, nearly drug-like, or drug-like compounds that are either available commercially or in the open-source domain. Irrespective of the virtual screening technique, these chemical databases can be used for finding our leads for a particular drug target. The detail of the different types of chemical databases available is given in Tab. 14.1.

https://doi.org/10.1515/9783111434858-014

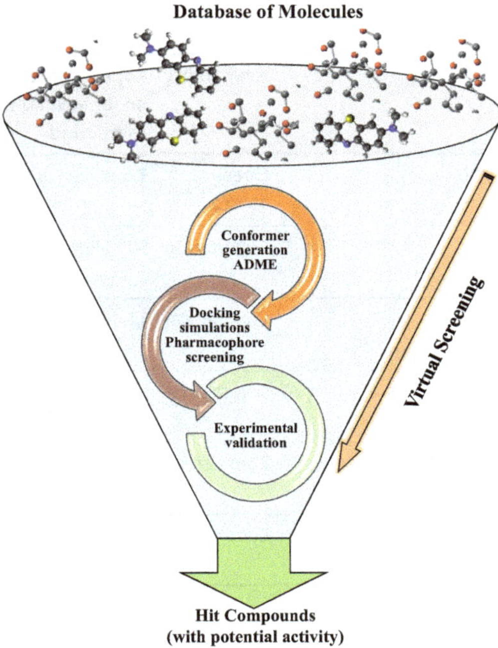

Fig. 14.1: Overview of the concept of virtual screening for finding the active hit compounds from a database.

Tab. 14.1: List of available databases.

S. no.	Name of the database	Link	Availability	Approximate no. of available compounds
1	Zinc database	https://zinc.docking.org/	Open-source	230 million [5]
2	ChemDB database	https://cdb.ics.uci.edu/	Open-source	5 million [6]
3	Pubchem database	https://pubchem.ncbi.nlm.nih.gov/	Open-source	115 million [7]
4	Chemspider database	https://www.chemspider.com/	Open-source	100 million [8]
5	Drugbank database	https://go.drugbank.com/	Open-source	0.5 million [9]
6	NCI database	https://cactus.nci.nih.gov/download/roadmap/	Open-source	0.46 million [10]

Tab. 14.1 (continued)

S. no.	Name of the database	Link	Availability	Approximate no. of available compounds
7	Chembl Database	https://www.ebi.ac.uk/chembl/	Open-source	2.3 million [11]
8	eMolecules database	https://www.emolecules.com/	Commercial	50 million [12]
9	Chemspace database	https://chem-space.com/search	Commercial	11.46 billion [13]
10	Maybridge database	https://www.thermofisher.in/chemicals/en/brands/maybridge.html	Commercial	52,160 [14]
11	ChemDiv database	https://www.chemdiv.com/drug-discovery-services/in-silico-cadd-drug-discovery-services/	Commercial	1.6 million [15]
12	Asinex database	https://www.asinex.com/screening-libraries-(all-libraries)	Commercial	5.7 million [16]

14.3 Classification of virtual screening

Virtual screening is also classified similar to the classification of the CADD process [17–19]. Broadly, virtual screening is classified into two types: structure-based virtual screening and ligand-based virtual screening (Fig. 14.2).

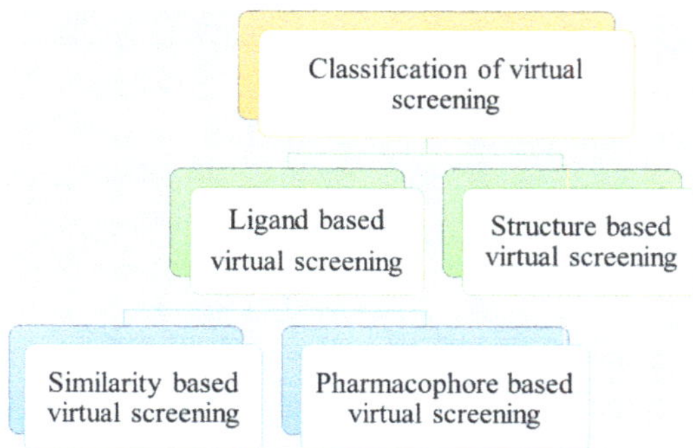

Fig. 14.2: Classification of the virtual screening technique into the structure- and ligand-based virtual screening methods.

14.3.1 Structure-based virtual screening

With the emergence of the high-quality 3D structure of proteins in the last few decades, the process of structure-based virtual screening has gained immense popularity. In this approach, the large database of the compounds is screened based on the active site properties of the target protein. Molecular docking is the most commonly employed methodology for performing the structure-based virtual screening of chemical databases. Only those docking algorithms that perform fast docking are employed for the virtual screening due the large number of compounds to be screened [20–22]. The workflow of the structure-based virtual screening algorithm is given in Fig. 14.3.

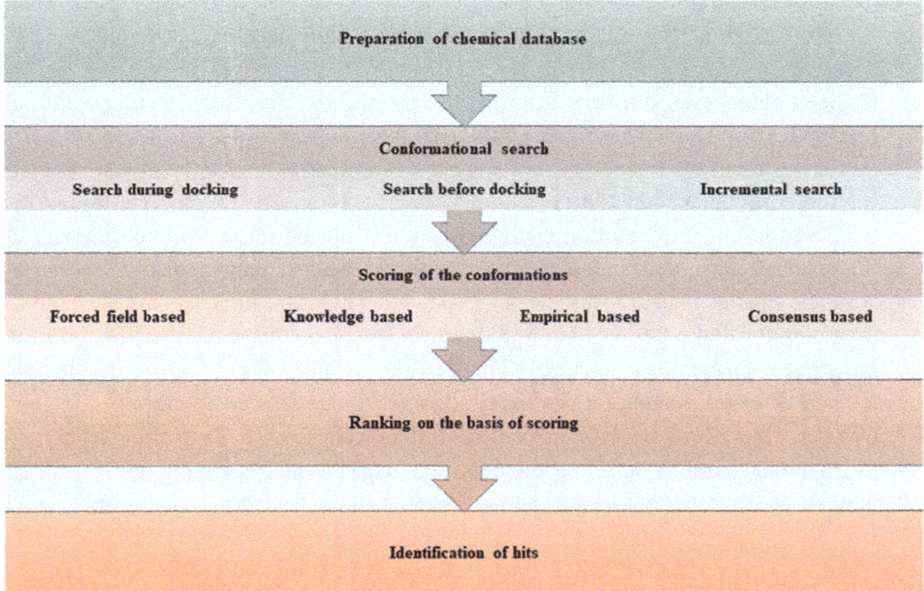

Fig. 14.3: Basic steps involved in the process of the structure-based virtual screening process of the compounds.

A detail of the steps performed in the structure-based virtual screening is given below:

14.3.1.1 Preparation of the chemical database

For performing any virtual screening process, the biggest prerequisite is the selection of the chemical database. Various databases that are available commercially or in the public domain are already given in Tab. 14.1. Depending upon the requirement of the virtual screening, one can choose from different databases such as chemical databases of probable cancer molecules and databases of herbal molecules.

14.3.1.2 Performing the conformational search

This is the next step of performing the structural-based virtual screening. Broadly, conformational search algorithms employed by the docking programs in structural-based virtual screening are of three types [23]:

a. **Conformational search performed during docking**: This is the most common type of search algorithm used for structural-based virtual screening. As in the virtual screening approach, it is required that process is performed rapidly, hence only stochastic conformational search algorithms are employed here. Monte Carlo- and genetic algorithm-based search algorithms are the two most common stochastic approaches used for this purpose.

- **Monte Carlo-based conformational search**: Monte Carlo-based conformational search is gaining popularity. Initially, this approach was used in molecular dynamics simulations for performing energy minimization, but later it was adopted for the conformational search in molecular docking also. In this methodology, first, a random conformation of the molecule is placed inside the binding site of the protein target, and then a random change in the conformation of the molecule is made. Energy-based parameters of both the conformations are calculated and it is then decided whether to accept or reject the new conformation, based on the Boltzmann probability function. This randomization in the conformation is performed when conformations of certain threshold values are obtained.
- **Genetic algorithms-based conformational search**: Genetic algorithm-based conformational search is based on the adaptive heuristic search technique focused on the process of natural selection and genetics. Initially, conformations of the ligand are generated based on certain parameters and these parameters are considered analogous to the genes. From the generated population of different conformations, a process known as "crossover" of the natural selection is performed to evolve out those conformations that have the lowest energy parameters.

b. **Conformation search performed before docking**: In the above approach, conformational search is performed by placing the ligand inside the binding cavity of the target protein. In this approach, all the energy calculations for obtaining the bioactive conformations are performed first and only then low-energy conformations in rigid form are placed inside the active site of the target protein for performing the scoring of those poses.

c. **Incremental conformational search**: Incremental conformational search is also a rapid conformational search approach employed in structural-based virtual screening. In this approach, initially, the molecule is cleaved into small fragments around the rotatable bonds present inside it. Then, the largest rigid fragment obtained is first placed inside the binding cavity of the protein. After that, each fragment is added, one by one, to the initial fragment in different orientations and scored based on a certain docking score. The fragment orientation that gives the best docking score is attached to the initial fragment and this whole process is repeated until the whole molecule is created.

Based on these above approaches, different molecular docking programs used in the virtual screening are given in Tab. 14.2.

Tab. 14.2: List of different molecular docking programs used in virtual screening.

S. no.	Conformational search approach	Molecular docking program
1.	Conformational search during docking (Monte Carlo approach)	Glide, MCDOCK, ICM, Dock Vision, and ProDock
	Conformational search during docking (genetic algorithm approach)	AutoDock, Gold, MolDock, FlopDock, and EADock
2.	Conformational search before docking	FRED and SLIDE
3.	Incremental search algorithm	FLEXx, DOCK, FLOG, and SurFlex

14.3.1.3 Scoring functions in the structure-based virtual screening

This is the next step of the structure-based virtual screening that is performed to assess the binding of the different conformations with the target protein [24]. Detailed descriptions of the different scoring functions employed for this purpose are already discussed in Chapter 9 (scoring of the ligand–receptor binding).

14.3.1.4 Ranking based on scoring

In the previous step, various conformations obtained from the conformational search were assessed for the binding affinity with the binding site of the target protein. As we already know, ligand–receptor binding is evaluated based on Gibb's free energy (ΔG) and the ligand that has a lower ΔG value for the binding is considered to have better binding. The molecules with lower ΔG values are ranked better compared to other molecules in the chemical database.

14.3.1.5 Identification of the hits

Hits are those molecules that are considered to possess good pharmacological activity against a particular target. This is the last step of the structural-based virtual screening in which first a threshold value of ΔG for the ligand–receptor binding is selected and then the molecules that possess a ΔG value better than the threshold value are considered as the hits against the desired target protein [25, 26]. These hit molecules are further selected for the further drug discovery process.

14.3.1.5.1 Advantages of structure-based virtual screening

- **Target specificity:** Structure-based virtual screening allows for the direct targeting of specific biomolecular targets, such as proteins or enzymes, involved in disease pathways, enhancing the likelihood of identifying compounds with high affinity and selectivity.
- **Rational design:** By utilizing the three-dimensional structure of the target protein or receptor, structure-based virtual screening enables rational drug design, wherein potential ligands can be designed or modified to fit into the binding site and interact optimally with the target, increasing the chances of success in drug discovery.
- **Speed and efficiency:** Virtual screening techniques can rapidly screen large compound libraries against target structures, enabling the evaluation of thousands to millions of compounds in silico, significantly reducing the time and resources, compared to traditional experimental screening methods.
- **Cost-effectiveness:** Virtual screening reduces the need for expensive and labor-intensive experimental assays by prioritizing compounds with favorable binding interactions for further experimental validation, thereby minimizing costs and increasing efficiency in the drug discovery process.
- **Hit expansion:** Structure-based virtual screening can identify structurally diverse hits with varying scaffolds that interact with different regions of the target protein, facilitating hit expansion and diversification of the chemical space for lead optimization and drug development.
- **Identification of novel binding sites:** Virtual screening can reveal potential binding sites on the target protein that may not have been previously characterized, leading to the discovery of allosteric modulators or alternative binding pockets for drug targeting, and expanding the scope of drug discovery efforts.
- **Integration with experimental data:** Virtual screening results can be integrated with experimental data, such as biochemical assays or structure–activity relationship (SAR) studies, to validate and prioritize hit compounds, guiding further optimization and lead identification in the drug design process.
- **Enhanced hit rates:** Structure-based virtual screening often yields higher hit rates compared to other screening methods, as it focuses on compounds with favorable binding interactions and structural complementarity with the target, increasing the likelihood of identifying potent and selective lead compounds.
- **Flexible workflow:** Virtual screening workflows are flexible and adaptable, allowing researchers to customize parameters, scoring functions, and screening protocols based on the specific requirements of the target and project goals, optimizing the screening process for maximum effectiveness.

- **Insight into SAR:** Virtual screening provides valuable insights into SAR by elucidating the molecular interactions between ligands and target proteins, guiding the rational design and optimization of lead compounds with the desired pharmacological properties.

14.3.1.5.2 Limitations of structure-based virtual screening

- **Dependence on protein structure availability:** Structure-based virtual screening relies heavily on the availability of high-quality protein structures. The limited availability of experimentally determined protein structures for the target of interest can constrain the application of structure-based virtual screening.
- **Inaccuracies in protein structure prediction:** Predicting protein structures computationally can introduce inaccuracies and uncertainties, affecting the reliability of structure-based virtual screening results. Inaccurate or incomplete protein structures may lead to false-positive or false-negative predictions.
- **Binding site flexibility:** Structure-based virtual screening often assumes a rigid protein–ligand binding site, neglecting the inherent flexibility and dynamics of protein structures. This oversimplification can result in overlooking potential ligand-binding modes or interactions, leading to suboptimal ligand selection.
- **Scoring function limitations:** The accuracy of structure-based virtual screening results heavily depends on the scoring functions used to evaluate ligand–protein interactions. Existing scoring functions may not fully capture the complexities of molecular recognition, leading to inaccuracies in ranking and prioritizing ligands.
- **Chemical diversity of compound libraries:** Structure-based virtual screening requires large compound libraries for screening, but the diversity of available chemical compounds may be limited. Inadequate chemical diversity in compound libraries can restrict the identification of novel lead compounds with diverse chemical scaffolds.
- **Computational resources and time constraints:** Performing structure-based virtual screening involves computationally intensive tasks, such as molecular docking and scoring, which require significant computational resources and time. Limited computational resources and time constraints may hinder the screening of large compound libraries or thorough exploration of the conformational space.
- **Overlooking water-mediated interactions:** Structure-based virtual screening often neglects water-mediated interactions within the protein–ligand binding site. Water molecules play a crucial role in mediating ligand binding and influencing binding affinity, but their consideration adds complexity to the docking process.
- **Ignoring protein flexibility during docking:** Structure-based virtual screening typically assumes a static protein conformation during docking, overlooking the conformational flexibility of protein structures. Ignoring protein flexibility may

lead to inaccurate predictions of ligand binding and affinity, particularly for flexible binding sites.

- **Validation and experimental confirmation:** Structure-based virtual screening predictions require experimental validation to confirm the binding affinity and biological activity of the identified lead compounds. Experimental validation is essential but time-consuming and costly, limiting the throughput and efficiency of structure-based virtual screening.
- **Addressing ligand-specific issues:** Certain ligand-specific issues, such as covalent binding, metal coordination, or allosteric modulation, may pose challenges for structure-based virtual screening. Specialized docking protocols or scoring functions may be required to address these ligand-specific interactions accurately.

14.3.2 Ligand-based virtual screening

When virtual screening is performed based on known active ligands present against the target disease, it is known as the ligand-based virtual screening approach. In this technique, a chemical database is searched for the potent molecules that are closely related to the active known ligands [27]. The screening for the search of potent molecules is done based on either the similarity with known ligands (2D and 3D) or in close relation with the pharmacophore developed from the active ligand. Ligand-based virtual screening works on the principle of similarity, i.e., compounds that possess similarity probably will have similar biological activity [28]. The complete workflow of the ligand-based virtual screening is given in Fig. 14.4.

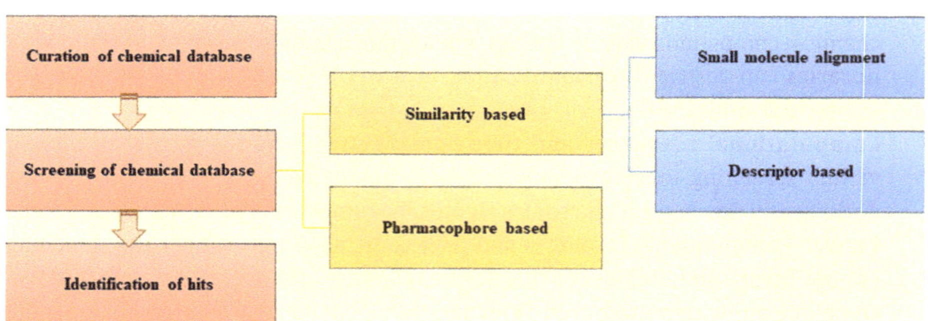

Fig. 14.4: Basic steps involved in the process of the ligand-based virtual screening process of the compounds.

14.3.2.1 Curation of the chemical dataset

This is the very first step of any virtual screening approach, whether it is structure-based or ligand-based. Details of the chemical database are already discussed in the earlier section of the chapter and various databases are given in Tab. 14.1.

14.3.2.2 Screening of the chemical database

This is the next step of the ligand-based virtual screening methodology. Based on the different screening methodologies, the ligand-based virtual screening is classified further. The most common screening algorithms employed in ligand-based virtual screening are similarity-based and pharmacophore-based screening.

a. **Similarity search-based screening**: Similarity search-based screening is the most common method employed for the ligand-based virtual screening technique. The chemical database in this approach is screened based on certain similarities. Similarity-based screening is done by one of the following methods.

i. **Small- molecule alignment-based similarity search**: In small-molecule alignment-based similarity search, each molecule of the chemical database is aligned over the reference active known ligand. There are several molecular alignment programs available that perform this task based on various algorithms. In this approach, the test molecule is always kept flexible whereas the reference active ligand can either be kept rigid or flexible during the alignment. The most popular programs available for performing the molecular-based similarity search are given in Tab. 14.3.

Tab. 14.3: List of popular programs available for performing the molecular-based similarity search.

S. no.	Name of the software/ program	Principle	Flexibility of the reference active ligand	Description
1.	FlexS	Molecular superposition	Rigid	– The test molecule is divided into small fragments. – Individual fragments are aligned on the reference ligand incrementally.
2.	GASP	Molecular superposition	Flexible	– This program is based on the Genetic algorithm. – Alignment is based on the mapping of the features (Hydrogen bond acceptor, ring center, etc.) between the reference and test molecule.

Tab. 14.3 (continued)

S. no.	Name of the software/ program	Principle	Flexibility of the reference active ligand	Description
3.	MEP	Alignment based on force fields	Flexible	– It is also based on the Genetic algorithm approach. – Alignment of the molecular fields of the test and reference molecule is performed to assess the similarity between them.
4.	MIMIC	Alignment based on force fields	Rigid	– This is more advanced than the MEP program. – Gaussian function is applied in the alignment, in addition to the MEP parameters in this program.
5.	fFLASH	3D fragment-based molecular superposition	Rigid	– This program is very fast for small- and medium-sized molecules. – The alignment is based on the graph-based clique detection technique.

ii. **Descriptor-based similarity search**: The one major drawback of the alignment-based similarity search is that each test molecule is to be aligned with the reference molecule, which is a very time-consuming affair. To overcome this, descriptor-based similarity search approach is employed. In this approach, the descriptor value of the reference molecule is compared with the values of each molecule of the chemical database to search for similarity. The descriptor can be defined as the representation of the properties of the molecule in a mathematical term, which is calculated by a certain algorithm. For performing the similarity-based search based on the descriptors, mostly following types of descriptors are employed:
   - **1D descriptors**: These are the descriptors that can be calculated easily from the structural properties of the molecule. The most common examples of the 1D descriptors are the number of hydrogen bond acceptors/donors, the number of rotatable bonds, log *P*, and so on. The role of the 1D descriptors in the virtual screening is mainly limited to the initial filtering of the chemical database (such as Lipsinki's Rule of Five).
   - **2D descriptors**: These are the descriptors that can be obtained from the graphical 2D representation of the molecule. These descriptors are most commonly employed for performing ligand-based virtual screening processes. Virtual screening, with the use of these descriptors, is based on either the classical linear 2D descriptors or tree-type 2D descriptors.

- **Classical linear 2D descriptors**: These are the classical 2D descriptors that are calculated using traditional algorithms such as topological or structural parameters. Examples of these types of descriptors include the number of aromatic rings available in the molecule, connectivity index and carbo index, eigenvalue-based descriptors, etc. The one major drawback with the use of linear 2D descriptors is that it can perform a similarity search between compounds that share common structural features. Molecules that have structural differences but have similar biological activity cannot be screened using the classical linear 2D descriptors.
- **Tree-type descriptors**: To overcome the drawback of the similarity search using linear 2D descriptors, tree-type 2D descriptors were introduced [29]. These descriptors were first introduced by Rarey and Dixon in their approach of "Free tree (Ftree)" descriptors (Fig. 14.5). These descriptors are represented by a tree-like structure in which the nodes of the tree represent the fragments or functional groups of the molecule and the graph (tree) represents the overall topology of the molecule. When searching for similarity between the two molecules, their Ftree descriptors are compared – the comparison between the nodes finds similarity between the functional groups, and the comparison between the overall trees finds similarity between the overall molecular structures of the two compounds.

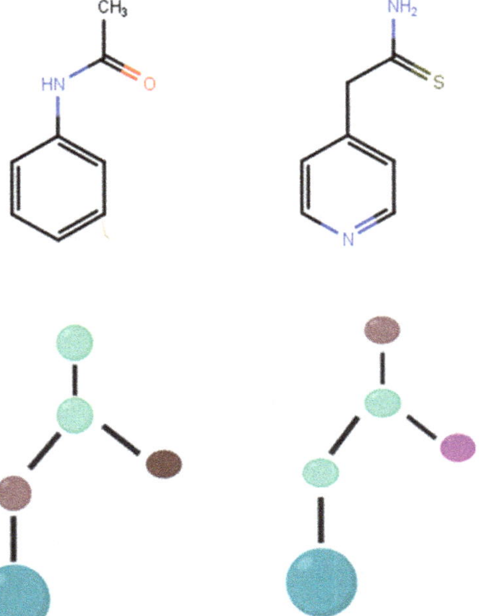

Fig. 14.5: Description of the free tree (Ftree) type of descriptors.

iii. **Similarity coefficients for finding the similarity between two molecules**: When 2D descriptors are employed for virtual screening, the similarity between the reference and test molecule is evaluated based on various similarity coefficients. The most commonly employed similarity coefficient for this purpose is the Tanimoto coefficient, given in the following equation. Tanimoto coefficient is defined as the ratio of the common features between the two molecules to the total number of features between those two molecules. Mathematically,

$$\text{Tanimoto coefficient}\,(T) = c/(a+b-c) \tag{14.1}$$

where c is the features set that is common in both molecules A and B, a is the feature set of molecule A, and b is the feature set of molecule B. The two molecules that show a Tanimoto coefficient value greater than 0.85 are considered similar and a value below this leads to the inference that the two molecules are not similar:

- **3D descriptors**: These descriptors are based on or calculated from the 3D spatial arrangement of the molecules. In the ligand-based virtual screening, linear 3D descriptors play a role, which includes distance between the functional groups in space and basic molecular shape properties. These 3D linear descriptors are evaluated for a fixed conformation but it is a well-established fact that a ligand interacts with the target protein in a flexible conformation and also the conformation that binds with the binding site of the target protein can have a totally different conformation. To overcome this, there is another approach of employing multiple conformations of a single molecule for generating the 3D linear descriptors but this leads to huge assumptions in the calculations. Due to all these reasons, ligand-based virtual screening, based on these 3D descriptors, is of limited use.

b. **Pharmacophore-based screening**: Apart from similarity-based screening, another important approach for screening the chemical database is pharmacophore-based screening. A complete workflow of a pharmacophore-based screening is given in Fig. 14.6.

i. **Generation of the pharmacophore**: This is the very first step of the pharmacophore-based approach in which a query pharmacophore is generated. This query pharmacophore can be generated either from the active ligands against a particular target or from the internal ligand present in the binding site of the target protein. The complete methodology of the generation of these pharmacophores is already discussed in Chapter 13 of this book.

ii. **Database searching**: Database searching is the second step of this process in which a curated chemical database is searched for probable actives. Database searching is a two-step process:

- **Pre-filtering of the chemical database**: This is the initial step of the database searching in which quick pre-filters are applied to the chemical database,

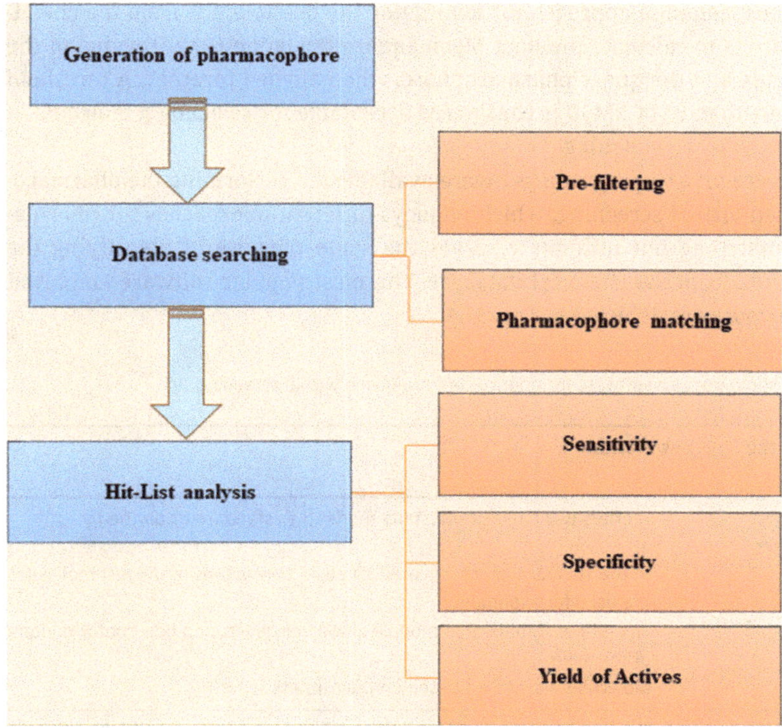

Fig. 14.6: Basic steps involved in the process of pharmacophore-based virtual screening.

which quickly excludes those compounds that are not considered for further searching. The most commonly employed pre-filtering approach is feature-based (0D descriptor-based) matching as the calculation of these types of descriptors is very fast and simple, yet very effective. It is performed by first calculating the features of the query pharmacophore and then calculating those features of the molecules of the chemical database. Molecules that contain features similar to the query pharmacophore are retained and the rest are excluded.

– **Pharmacophore matching**: Once the pre-filtering is applied to the chemical database, the next step of the database searching is to perform the 3D alignment of the molecules on the query pharmacophore. The molecules that align well with the query pharmacophore are considered hits and the rest are discarded. Pharmacophore matching is the most crucial step of any pharmacophore-based virtual screening as the accuracy of this step determines the quality of the overall screening process.

The most common approach for identifying the possible hits from the chemical database is to calculate the Root Mean Square Deviation (RMSD) value of the test molecule and the query pharmacophore, when aligned together. A threshold value of less than 01 of RMSD is considered acceptable for selecting a molecule in the pharmacophore matching.

There are several programs/software available for performing the pharmacophore-based virtual screening, which employs different approaches for pharmacophore matching but ultimately serves the same purpose of identifying the probable hits from the chemical database. The most popular software employed for this purpose are depicted in Tab. 14.4.

Tab. 14.4: List of some popular software for pharmacophore-based virtual screening.

S. no.	Name of the program	Description
1	Phase of Schrodinger	– It offers two types of searching modes, i.e., standard or on-the-fly approach. – In the standard approach, predefined conformations of the test molecules are used for alignment. – The on-the-fly approach module allows for generating new conformations of the molecules. – Molecules are ranked based on their fitness score.
2	Catalyst	– Similar to the phase program, it also offers standard and best flexible search mode for the generation of conformations. – The RSMD calculation is employed for the selection of hits from the chemical database, based on the alignment.
3	LigandScout	– All conformations that can match with the query pharmacophore are considered. – A pattern-matching approach is used for the molecular alignment process. – Molecules are ranked based on the RMSD score or molecular volume overlap score.
4	MOE	– Alignment is done with predefined conformations of the dataset as no new conformation can be generated. – Annotation points are created from the pharmacophore generated from the database molecules. – The ranking is done based on the matching of the database molecule's annotation points with that of the query pharmacophore.

c. **Hit-list analysis**: The overall quality of the pharmacophore-based screening is assessed through certain parameters, which include:

i. **Sensitivity**: The sensitivity of the screening process is defined as the ratio of the true positives to the total active compounds as per the following equation. Its value ranges from 0 to 1, where a value of 0 indicates that the screening process is not sensitive while a value of 1 indicates that the process of virtual screening is highly sensitive in retrieving the actives:

$$\text{Sensitivity} = \frac{\text{TP}}{\text{TP}} + \text{FN} \tag{14.2}$$

where TP represents the true positives detected by the software and FN is the false negatives detected by the screening algorithm.

ii. **Specificity**: It is another parameter for the analysis of the quality of the screening algorithm, which is defined as the ratio of the detected true negative to the sum of the detected true negative and false positives by the algorithm (equation (14.3)). Its value also ranges from 0 to 1, where 1 represents the highly specific algorithm:

$$\text{Sensitivity} = \frac{\text{TN}}{\text{TN}} + \text{FP} \tag{14.3}$$

where TN is the detected true negative and FP is the false positive detected by the algorithm.

iii. **Yield of actives**: Yield of actives is the ratio of the detected true positives detected by the program to the total number of compounds present in the chemical database as per the following equation. The active value yield can be used for comparing the performance of two or more pharmacophore screening algorithms when screening is performed on the same chemical database:

$$\text{Yield of actives} = \frac{\text{TP}}{n} \tag{14.4}$$

where TP is the detected true positives and n represents the total number of compounds, including their all conformations present in the chemical database.

14.3.2.3 Identification of the hits

This is the final step of the ligand-based virtual screening approach in which probable hits are identified based on the above screening approaches. Optimization of the hits is further performed to convert them further to possible lead molecules.

14.3.2.3.1 Advantages of the ligand-based virtual screening approach

- **Utilization of known ligands:** Ligand-based virtual screening leverages known ligands with desired pharmacological properties as reference molecules for screening large compound libraries, facilitating the identification of structurally similar compounds with potential biological activity.
- **Efficient filtering of compound libraries:** Ligand-based virtual screening efficiently filters large compound libraries to prioritize molecules with high similarity

to known active ligands, reducing the number of compounds requiring experimental testing and conserving resources.
- **Flexibility in target selection:** Ligand-based virtual screening can be applied to a wide range of drug targets, including proteins, enzymes, and receptors, making it adaptable to diverse therapeutic areas and enabling the exploration of multiple drug design possibilities.
- **Complementary to structure-based approaches:** Ligand-based virtual screening complements structure-based virtual screening approaches by focusing on ligand similarity rather than protein structure, allowing for the identification of bioactive compounds with diverse modes of action and binding sites.
- **Fast and cost-effective:** Ligand-based virtual screening offers a rapid and cost-effective screening method compared to experimental assays, accelerating the drug discovery process and reducing the time and resources required for lead identification and optimization.
- **Identification of novel chemical scaffolds:** Ligand-based virtual screening can identify novel chemical scaffolds and pharmacophore features that may not be readily apparent from protein structures alone, expanding the chemical diversity of potential drug candidates.
- **Prediction of SAR:** Ligand-based virtual screening enables the prediction of SARs by correlating ligand similarity with biological activity data, guiding the rational design of analogs with improved potency and selectivity.
- **Accessibility of computational tools:** Ligand-based virtual screening methods are supported by a wide range of computational tools and software packages, making them accessible to researchers with varying levels of expertise and resources.
- **Versatility in data types:** Ligand-based virtual screening can utilize various types of molecular descriptors, such as 2D fingerprints, 3D pharmacophore features, and molecular fingerprints, allowing for flexibility in data representation and analysis.
- **Integration with experimental data:** Ligand-based virtual screening can be integrated with experimental data, such as bioassay results and ADMET (absorption, distribution, metabolism, excretion, and toxicity) properties, to enhance the predictive power of virtual screening models and prioritize compounds with favorable drug-like properties.

14.3.2.3.2 Limitations of the ligand-based virtual screening approach

- **Dependence on available ligand data:** Ligand-based virtual screening relies heavily on the availability and quality of ligand data for constructing pharmacophore models or similarity searches. Limited or biased ligand datasets can lead to inaccurate predictions.

- **Lack of structural information:** Unlike structure-based methods, ligand-based approaches do not utilize structural information about the target protein, which can limit their ability to identify novel ligands that bind to different binding sites or exhibit allosteric effects.
- **Sensitivity to ligand conformation and flexibility:** Ligand-based methods may struggle to accurately predict ligand–receptor interactions when dealing with flexible ligands or binding site flexibility. Conformational changes in ligands and receptors can lead to false negatives or inaccuracies in binding affinity predictions.
- **Difficulty in handling diverse chemical spaces:** Ligand-based methods may struggle to identify structurally diverse ligands that do not resemble the compounds in the training set used to generate the pharmacophore models or similarity matrices.
- **Limited prediction accuracy:** The predictive accuracy of ligand-based virtual screening methods is influenced by factors such as the quality of the training data, choice of similarity metrics, and the reliability of the pharmacophore models. In some cases, these methods may yield false positives or false negatives.
- **Inability to capture protein flexibility:** Ligand-based approaches typically do not account for protein flexibility, which can influence ligand binding affinity and specificity. This limitation may lead to inaccuracies in predicting ligand–receptor interactions.
- **Challenge of handling multi-target ligands:** Ligand-based methods may struggle to identify ligands that interact with multiple targets or exhibit polypharmacology. Designing ligands with desired selectivity profiles can be challenging using ligand-based virtual screening alone.
- **Risk of overfitting:** Overfitting occurs when ligand-based models capture noise or spurious correlations in the training data, leading to poor generalization performance on unseen data. Careful validation and cross-validation procedures are necessary to mitigate this risk.
- **Computational resource intensiveness:** Ligand-based virtual screening methods can be computationally intensive, particularly when dealing with large ligand databases or complex similarity calculations. This may limit their practical utility for high-throughput screening applications.
- **Challenges in predicting binding modes:** Ligand-based methods may struggle to accurately predict the binding mode or orientation of ligands within the target binding site, which is critical for understanding structure-activity relationships and rational drug design.

References

[1] Walters WP, Stahl MT, Murcko MA. Virtual screening – an overview. Drug Discovery Today, 1998, 3(4), 160–78.
[2] Lengauer T, Lemmen C, Rarey M, Zimmermann M. Novel technologies for virtual screening. Drug Discovery Today, 2004, 9(1), 27–34.
[3] Lyne PD. Structure-based virtual screening: An overview. Drug Discovery Today, 2002, 7(20), 1047–55.
[4] Horvath D. A virtual screening approach applied to the search for trypanothione reductase inhibitors. Journal of Medicinal Chemistry, 1997, 40(15), 2412–23.
[5] Irwin JJ, Shoichet BK. ZINC– a free database of commercially available compounds for virtual screening. Journal of Chemical Information and Modeling, 2005, 45(1), 177–82.
[6] Chen J, Swamidass SJ, Dou Y, Bruand J, Baldi P. ChemDB: A public database of small molecules and related chemoinformatics resources. Bioinformatics, 2005, 21(22), 4133–9.
[7] Kim S, Thiessen PA, Bolton EE, et al. PubChem substance and compound databases. Nucleic Acids Research, 2016, 44(D1), D1202–13.
[8] Ayers M. ChemSpider: The free chemical database. Reference Reviews, 2012, 26(7), 45–6.
[9] Knox C, Wilson M, Klinger CM, et al. Drugbank 6.0: The drug bank knowledgebase for 2024. Nucleic Acids Research, 2024, 52(D1), D1265–75.
[10] Voigt JH, Bienfait B, Wang S, Nicklaus MC. Comparison of the NCI open database with seven large chemical structural databases. Journal of Chemical Information and Computer Sciences, 2001, 41(3), 702–12.
[11] Gaulton A, Hersey A, Nowotka M, et al. The ChEMBL database in 2017. Nucleic Acids Research, 2017, 45(D1), D945–54.
[12] https://www.emolecules.com/
[13] Korn M, Ehrt C, Ruggiu F, Gastreich M, Rarey M. Navigating large chemical spaces in early-phase drug discovery. Current Opinion in Structural Biology, 2023, 80, 102578.
[14] https://www.thermofisher.in/chemicals/en/brands/maybridge.html
[15] https://www.chemdiv.com/catalog/complete-list-of-compounds-libraries/
[16] https://www.asinex.com/
[17] Shoichet BK. Virtual screening of chemical libraries. Nature, 2004, 432(7019), 862–5.
[18] Lavecchia A, Di Giovanni C. Virtual screening strategies in drug discovery: A critical review. Current Medicinal Chemistry, 2013, 20(23), 2839–60.
[19] Reddy AS, Pati SP, Kumar PP, Pradeep HN, Sastry GN. Virtual screening in drug discovery-a computational perspective. Current Protein and Peptide Science, 2007, 8(4), 329–51.
[20] Lyne PD. Structure-based virtual screening: An overview. Drug Discovery Today, 2002, 7(20), 1047–55.
[21] Lionta E, Spyrou G, Vassil D, Cournia Z. Structure-based virtual screening for drug discovery: Principles, applications and recent advances. Current Topics in Medicinal Chemistry, 2014, 14(16), 1923–38.
[22] Maia EH, Assis LC, De Oliveira TA, Da Silva AM, Taranto AG. Structure-based virtual screening: From classical to artificial intelligence. Frontiers in Chemistry, 2020, 8, 343.
[23] Ghosh S, Nie A, An J, Huang Z. Structure-based virtual screening of chemical libraries for drug discovery. Current Opinion in Chemical Biology, 2006, 10(3), 194–202.
[24] Ma DL, Chan DS, Leung CH. Drug repositioning by structure-based virtual screening. Chemical Society Reviews, 2013, 42(5), 2130–41.
[25] Li Q, Shah S. Structure-based virtual screening. Protein Bioinformatics: From Protein Modifications and Networks to Proteomics, 2017, 111–24.

[26] Varela-Rial A, Majewski M, De Fabritiis G. Structure based virtual screening: Fast and slow. Wiley Interdisciplinary Reviews: Computational Molecular Science, 2022, 12(2), e1544.

[27] Ripphausen P, Nisius B, Bajorath J. State-of-the-art in ligand-based virtual screening. Drug Discovery Today, 2011, 16(9–10), 372–6.

[28] Hamza A, Wei NN, Zhan CG. Ligand-based virtual screening approach using a new scoring function. Journal of Chemical Information and Modeling, 2012, 52(4), 963–74.

[29] Rarey M, Hindle S, Maaß P, Metz G, Rummey C, Zimmermann M. Feature Trees: Theory and applications from large-scale virtual screening to data analysis. Pharmacophores and Pharmacophore Searches, 2006, 32, 81–116.

[30] Free SM, Wilson JW. A mathematical contribution to structure-activity studies. Journal of Medicinal Chemistry, 1964, 7(4), 395–9.

Index

https://doi.org/10.1515/9783111434858-015

www.ingramcontent.com/pod-product-compliance
Lightning Source LLC
Chambersburg PA
CBHW080911220326
41598CB00034B/5539

* 9 7 8 3 1 1 1 4 3 4 7 4 2 *